Reviews of Physiology, Biochemistry and Pharmacology

Springer-Verlag Berlin Heidelberg GmbH

Reviews of

127 Physiology Biochemistry and Pharmacology

Editors

M.P. Blaustein, Baltimore • H. Grunicke, Innsbruck
E. Habermann, Gießen • D. Pette, Konstanz
H. Reuter, Bern • B. Sakmann, Heidelberg
G. Schultz, Berlin • M. Schweiger, Berlin
E.R. Weibel, Bern

With 38 Figures and 19 Tables

Springer

ISBN 978-3-662-31190-5 ISBN 978-3-540-49453-9 (eBook)
DOI 10.1007/978-3-540-49453-9

Library of Congress-Catalog-Card Number 74-3674

Originally published by Springer-Verlag Berlin Heidelberg New York in 1996
Softcover reprint of the hardcover 1st edition 1996

SPIN: 10128282 27/3136/SPS – 5- 4 - 3 - 2 - 1 - 0 – Printed on acid free paper

Contents

Indexed in Current Contents

Rev. Physiol. Biochem. Pharmacol., Vol. 127

Protein Thiol Modification and Apoptotic Cell Death as cGMP-Independent Nitric Oxide (NO) Signaling Pathways

B. BRÜNE[1], S. MOHR, and U.K. MESSMER[1]

Contents

1 Introduction

Nitric oxide (NO) has gained wide recognition through a number of diverse activities in physiology and pathophysiology. This simple molecule, *Science* magazine's "molecule of the year 1992", is a key transducer of the vasodilator message from the endothelium to vascular cells, is a constitutent in central and peripheral neuronal transmission, and participates in the nonspecific immune defense. As a radical, NO displays high reactivity in a large number of processes attributed to both beneficial and harmful reactions. As

University of Konstanz, Faculty of Biology, P.O. Box 5560 M612, 78434 Konstanz, Germany
[1] *Present address:* University of Erlangen-Nürnberg, Faculty of Medicine, Department of Medicine IV – Experimental Division, Loschgestrasse 8, 91054 Erlangen, Germany

a steady stream of reports continue to expand its sphere of action, a demand for diverse signaling pathways arises in order to direct the pathophysiological actions of NO. The expression of a wide variety of effects is achieved through both coordinative interactions with metals and redox events. Reactions with oxygen (O_2), superoxide (O_2^-), and transition metals produce various NO_x species, peroxynitrite ($ONOO^-$), and metal-NO adducts, respectively. Additional nitrosative chemistry at nucleophilic centers broadens signaling pathways. However, for simplistic considerations, NO signaling can be categorized by cGMP-dependent versus cGMP-independent secondary pathways.

In this chapter, we only briefly describe the generation of NO and its main physiological effects, especially in relation to soluble guanylyl cyclase activation. We then focus on the NO-induced covalent modification of the glycolytic enzyme glyceraldehyde-3-phosphate dehydrogenase (GAPDH), paying attention to the interaction of NO-derived redox species with protein thiols, i.e., S-nitrosothiol formation. Finally, NO toxicity, identified as an apoptotic killing mechanism associated with accumulation of the tumor suppressor p53, is considered as a possible pathophysiological NO-mediated signaling mechanism.

2 Nitric Oxide-Releasing Compounds and Carbon Monoxide

Nitric oxide-releasing compounds, generally termed nitrovasodilators, have been used for more than 100 years in the treatment of congestive heart failure, pulmonary hypertension, angina pectoris, fibrinolytic therapy, and blood pressure control. It was in 1977 that the mechanism of action of such compounds as nitroglycerin became apparent. These agents, while releasing NO, activate soluble guanylyl cyclase (Schultz et al. 1977; Arnold et al. 1977), a heterodimeric heme protein with a molecular mass of 150 kDa. Activation of the enzyme is achieved by NO, or in some cases through S-nitrosothiol intermediates, via the formation of an NO-heme interaction eliciting a conformational change that enhances the enzyme's catalytic activity (Waldman and Murad 1987; Tremblay et al. 1988; Schmidt 1992).

Human platelets have been a preferred model for studying signal transduction mechanisms related to NO. After being produced in response to soluble guanylyl cyclase activation, cGMP is mainly under the control of various phosphodiesterases, degrading the second messenger to its inactive 5'-GMP derivative. The signaling of cGMP is mainly attributed to cGMP-dependent protein kinases, however, which become activated in response to increasing concentrations of the second messenger. Cyclic GMP kinases classified as types I and II are found in a number of different cells and are most abundant in smooth muscle, platelets, and cerebellum (Lincolin and Cornwell 1993). A distinctive pattern of phosphorylated proteins occurs in response to an activated cGMP-dependent protein kinase, causing vascular relaxation and inhibition of platelet aggregation in response to NO-releasing compounds. Recent findings suggest that phosphoproteins associated with actin filaments and focal contact areas serve as potential targets for regulation by cGMP-dependent protein kinases in human platelets (Reinhard et al. 1992). In addition to activating protein kinases, cGMP directly regulates cyclic nucleotide-gated ion channels and participates in visual as well as olfactory transduction. cGMP-responsive channels have also been identified in the pineal gland and the kidney (Walter 1989; Yau 1994).

Lipophilic cGMP analogs, inhibitors of soluble guanylyl cyclase or phosphodiesterase blockers, are valuable pharmacological tools mimicking NO actions. A possible interference with these agents substantiates the suggestion that activation of cGMP-dependent protein kinases is a cardinal mechanism of the signal transduction pathway of NO.

Activation of soluble guanylyl cyclase is achieved not only by NO, but also by another gaseous molecule, carbon monoxide (CO). The initial observation goes back to 1987, when we reported activation of soluble guanylyl cylase and concomitant cGMP formation as the underlying mechanism explaining inhibition of platelet activation by CO (Brüne and Ullrich 1987; Brüne et al. 1990). Originally stating that our finding may have neither physiological nor major toxicological implications, it seemed relevant to establish soluble guanylyl cyclase as being present as a ferrous hemoprotein. However, CO has recently been proposed as another neurotransmitter in the brain, taking part in long-term potentiation (Verma et al. 1993; Zhuo et al. 1993). Since its only physiological source is through the action of heme oxygenase types 1 and 2, the mRNA for these enzymes has been searched for, and for heme oxygenase type 2 it has been found selectively localized in distinctive brain regions. However, information about enzyme activity, formation, and degradation of the putative stable messenger CO is rather limited (see Dawson and Snyder 1994 for references).

Pharmacologically used NO donors are organic nitrates (i.e., nitroglycerin, NTG), sodium nitroprusside (SNP), 3-morpholinosydnonimine (SIN-1), S-nitrosothiols (i.e., S-nitrosoglutathione, GSNO or S-nitrosocysteine, CysNO), and the recently introduced NONOates (compounds containing the $N(O)NO^-$ functional group). Figure 1 shows some of the NO-releasing compounds.

Although all preserve NO in their molecular structure and all exhibit biological activity after NO release, these prodrugs vary considerably in their chemical nature. Variations in their biological activity and pharmacokinetic profile arise in part from differences in bioactivation, chemical

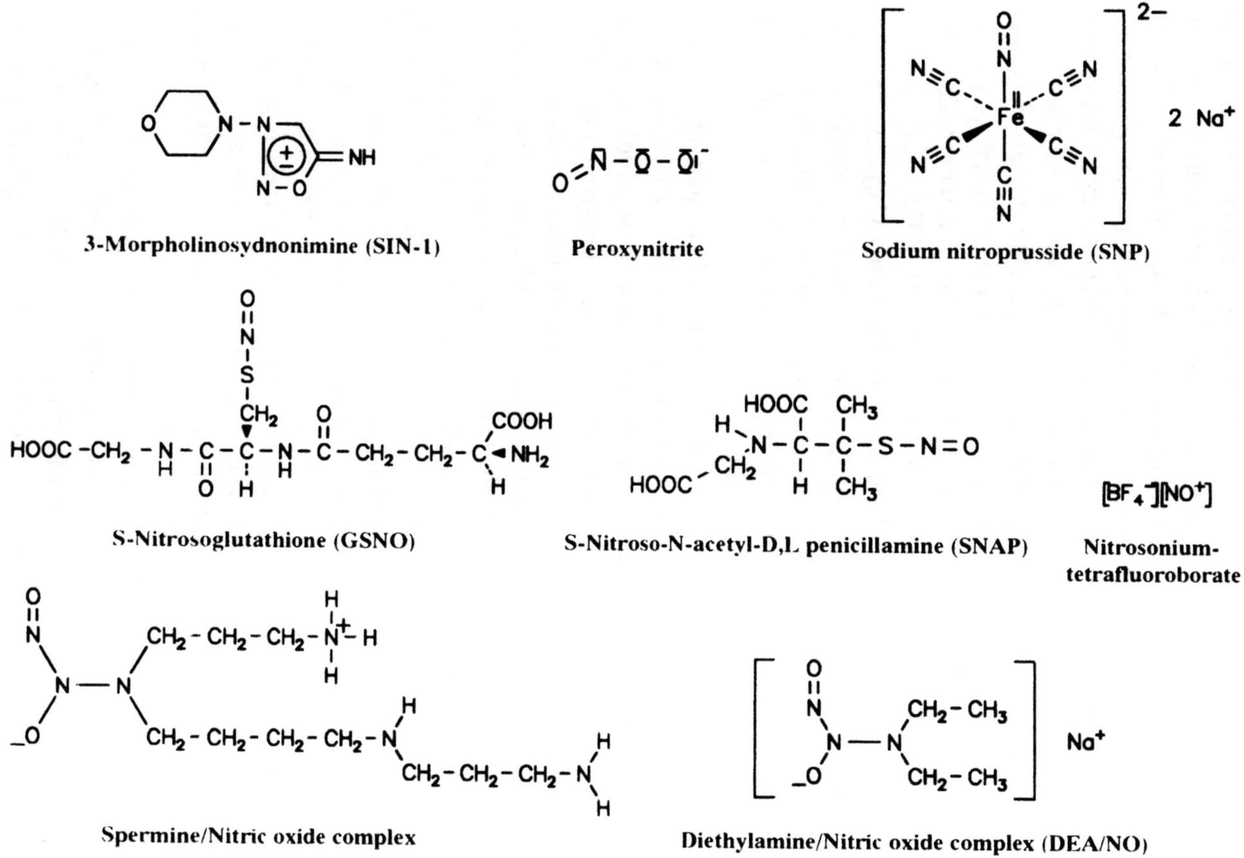

Fig. 1. Chemical structure of some NO releasing compounds

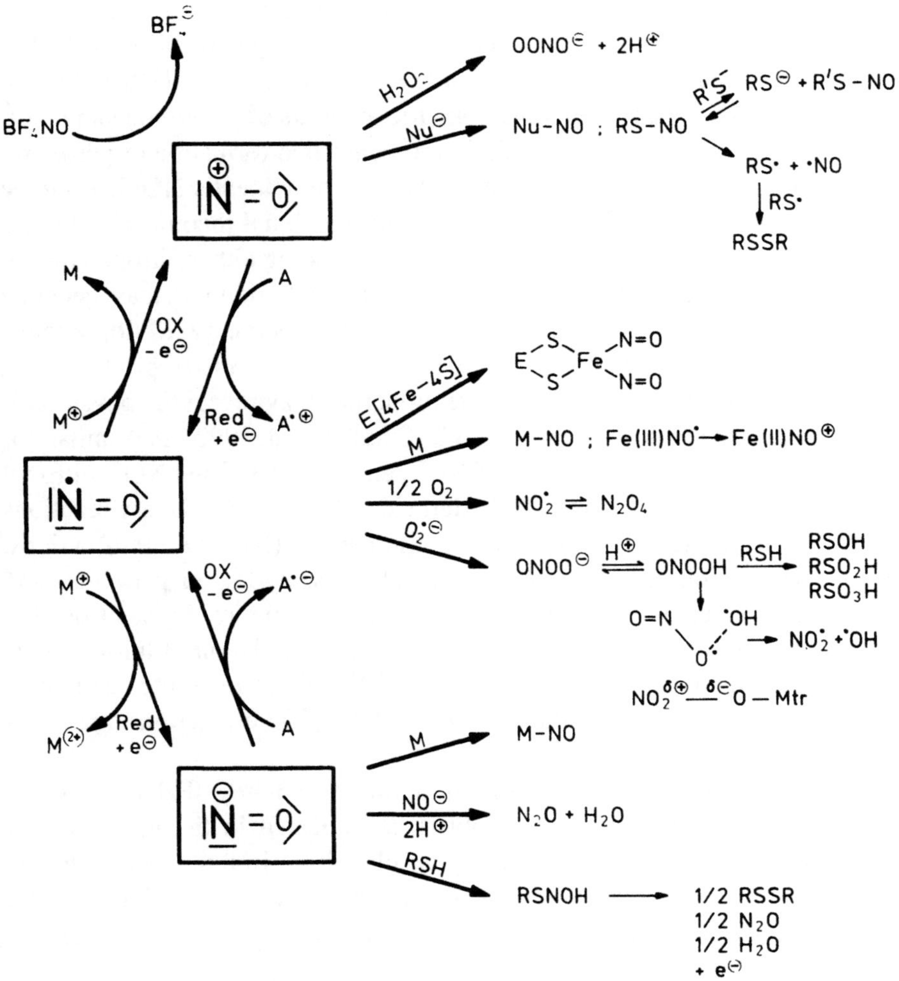

Fig. 2. Potential oxidation and reduction reactions of nitric oxide and possible target interactions as a framework of NO responses in biological systems

stability, and enzymatic clearance (Bennett et al. 1994). Biological activities associated with cytochrome P450 as well as the glutathione-S-transferase system seem to be involved in the enzymatic release of NO from several of these prodrugs. Some compounds like SNP do not necessarily require biotransformation prior to the release of NO. Additional variations which have to be considered in using nitrovasodilators are related to the NO redox species liberated from the individual compounds. For example, SNP may decompose spontaneously in the presence of light, or a one-electron reduction process may support decomposition, with thiol or even superoxide acting as reductants (see Noak and Murphy 1991 for references). As SNP contains $NO^{\bullet}$ (NO radical) or NO^+ (nitrosonium) bound to a ferrous ion coordinated with five cyanide anions, the release of cyanide prior to NO has to be considered, causing unwanted side effects (Bates et al. 1991). Decomposition of SIN-1 generates stoichiometric amounts of superoxide and $NO^{\bullet}$ at the same time, leading to the diffusion-controlled formation of peroxynitrite, which initiates reactions apart from those of authentic NO. During the breakdown of S-nitrosothiols the sulfur-nitrogen bond theoretically breaks up, either homolytically or heterolytically, generating either nitrogen monoxide ($NO^{\bullet}$) or nitrosonium (NO^+). Therefore, "NO donors" may generate different NO-related compounds with distinctive chemistry. Multiple interactions with targets via a rich redox and additive chemistry producing an array of chemical species with unique properties may reveal the physiological and pathophysiological faces of NO (Stamler et al. 1992). A summary of the one-electron redox chemistry of $NO^{\bullet}$ generating NO^+ and NO^- (nitroxyl anion), respectively, is given on the left side of Fig. 2. Secondary reactions of $NO^{\bullet}$, NO^+, and NO^- with nucleophilic centers (Nu^-; i.e., thiol anions, $RS^{-)}$, oxygen, metals (M), iron-sulfur enzymes, hydrogen peroxide, and superoxide are exemplified on the right side of the figure. Transnitrosation reactions, homolytic cleavage of the S-NO bond, oxidation of thiol groups, and the distinctive chemistry related to $ONOO^-$ or its protonated form (ONOOH), i.e., hydroxy radical ($^{\bullet}OH$) and $NO_2^{\bullet}$-like reactions, are envisioned.

The unique property of nitrosonium-tetrafluoroborate (BF_4NO), delivering exclusively NO^+ during hydrolysis, is indicated in the upper left of Fig. 2. The integration of this complex chemistry is fundamental to the understanding of current NO biochemistry, illuminating cGMP-dependent and especially cGMP-independent reactions related to thiol group modification and NO toxicity.

3 Enzymes Generating Nitric Oxide

Observations made at the beginning of the century suggested that human beings synthesize oxides of nitrogen, as detected in the urine (Mitchell et al. 1916). This observation was verified and extended in the early 1980s; it was shown that mammals produce NO_3^- and that the formation of this NO-oxidation product is enhanced by endotoxin treatment (Tannenbaum et al. 1978; Green et al. 1981). By 1985, evidence for induced nitrite and nitrate synthesis in murine macrophages was provided (Stuehr and Marletta 1985). Next, l-arginine was identified as the substrate for these metabolites, and the conversion of the substrate to various nitrogen oxides turned out to be closely associated with macrophage-induced cytotoxicity (Hibbs et al.

1987). In the meantime, the role of NO as a biologically active intermediate of acetylcholine-mediated smooth muscle relaxation accounting for endothelium-derived relaxing factor (EDRF) activity was advanced (Furchgott and Zawadzki 1980). However, it was not until 1987/1988 that the similarity between EDRF and NO was appreciated (Palmer et al. 1987, 1988). A nearly complete overlap between the biological properties of EDRF released from endothelial cells and the chemical properties of NO is now well established (Ignarro 1990; Feelisch et al. 1994).

Biological and pharmacological characterization of the NO release process and purification and intensive molecular studies of individual NOS (nitric oxide synthase) isoenzymes provided major insights into endogenous NO formation and its mechanism of action. A number of reviews published on this subject in recent years may serve as a general reference for more detailed information (Nussler and Billiar 1993; Knowles and Moncada 1994; Snyder 1992; Nathan and Xie 1994a; Schmidt and Walter 1994; Stamler 1994; Marletta 1994).

For NOS, eight cDNA sequences have been reported deriving from three known NOS genes in four species. Enzymes have been isolated and cloned from neurons, endothelial cells, and macrophages. However, these isoforms are not restricted to only one organ or one cell type, making a classification of isoenzymes according to cell-type designation exceedingly confusing. A nomenclature based on a combination of names and genes and/or numbers is now widely accepted (see Nathan and Xie 1994a for references). NOS isoenzymes are grouped as shown in Table 1.

For reasons of simplicity, NOS isoforms can be categorized by descriptive terms based on the dependence on intracellular calcium transients required for full activity: (a) a constitutive form regulated by calcium transients (Ca^{2+} requirement above the resting value, which is normally around 100 n*M*) and calmodulin, which leads to the release of NO over several minutes, and (b) a cytokine-inducible, calcium-independent isoform ex-

Table 1. Nomenclature for classification of isoenzymes

Numerical	*Descriptive*	*Definition*
I	ncNOS	Activity depends on Ca^{2+} above the "resting" level; first identified in neurons
II	iNOS	Activity is independent of Ca^{2+} above the "resting" level
III	ecNOS	Activity depends on Ca^{2+} above the "resting" level; first identified in endothelial cells

The "resting" level of calcium is normally around 100 n*M*.

pressed in many cell types after challenge with immunological or inflammatory stimuli, generating large amounts of NO for up to several days.

The reaction catalyzed by NOS is the oxidation of l-arginine to produce citrulline and stoichiometric amounts of NO. Besides the substrate, the reaction requires molecular oxygen and reducing equivalents in the form of NADPH. All NOS isoenzymes are P450-like hemoproteins, contain binding sites for FAD (flavin adenine dinucleotide) and FMN (flavin mononucleotide), and carry out a five-electron oxidation of the substrate in the presence of tetrahydrobiopterin. Multiple transcriptional as well as translational control mechanisms allow augmentation or suppression of NO production, providing for fine tuning of NO formation, accumulation, and action (Nathan and Xie 1994b).

Several NOS inhibitors are used to interfere pharmacologically with NO production and to trace back individual actions to the NO signaling system. Structurally, most of these inhibitors are analogs of the naturally occurring substrate l-arginine, with no preference for either the constitutive or the inducible form. N^G-monomethyl-l-arginine (NMMA) is just one classical example of an NOS inhibitor. NO-liberating drugs in combination with NOS inhibitors thus provide important biochemical tools, because they permit investigation of the role of NO in diverse pathophysiological situations.

4 Nitric Oxide in the Brain

Nitric oxide is a unique messenger that is not stored, but rather synthesized on demand with actions that are not confined to anatomical structures but rather restricted by transcellular diffusion. This makes NO an interesting messenger in the brain. The existence of NO in the brain was first demonstrated in cerebellar slices that, upon activation, released a factor with properties resembling those of NO; also, high amounts of NOS activity were recognized in the cerebellum (see Garthwaite et al. 1988 Garthwaite 1993; Dawson and Snyder 1994 for references). In the cerebellum NOS is present in granule cells as well as in basket cells, but is absent from Purkinje cells. Cyclic nucleotide levels increase severalfold in response to activation of the N-methyl-d-aspartate (NMDA) subtype of the glutamate receptor that is present on granule cells and basket cells and which conceivably triggers the formation of NO that diffuses to the adjacent Purkinje cells to activate soluble guanylyl cyclase. Attention has now shifted to identification of possible roles of NO production in response to stimulation of excitatory amino acid receptors. In the central nervous system a participation in synaptogenesis, sensory input processing, neurotransmitter release, synaptic plasticity and learning, has been postulated. A function of NO in the peripheral nervous system may be in the myenteric plexus and throughout the whole gastrointestinal pathway. In the latter, NO is released by stimulation of nonadrenergic-noncholinergic (NANC) neurons, mediating relaxation of the intestines as one component of peristalsis. Moreover, NO mediates vasodilation in the cerebral arteries released from both endothelial cells and the autonomic nerves within the adventitia.

In the nervous system, NO engenders several characteristics of an unusual neurotransmitter: It is synthesized on demand only; it diffuses to surrounding cells; and it binds to its intracellular receptor, soluble guanylyl cyclase, causing cGMP-mediated downstream effects.

5 Nitric Oxide and Neurotoxicity

Although the numerous physiological roles of NO, including those in the nervous system, are well substantiated, excessive production of NO in response to inflammatory cytokines or continuously elevated excitatory amino acids in the brain suggest pathophysiological roles in excito- and neurotoxicity, migraine, epilepsy, stroke, and other neurodegenerative diseases.

In the central nervous system, massive glutamate release and/or accumulation underlies neurodegenerative disorders, including ischemic hypoxic insults and Alzheimer's and Huntington's diseases (Lowenstein and Snyder 1992; Choi 1993; Dawson and Snyder 1994). Obviously, NO plays a central role in neuronal damage, although the precise mechanism by which NO overproduction kills neurons remains to be clarified. Continuous NMDA receptor stimulation, followed by pronounced NO formation, apparently constitutes the initiating process, as either NMDA receptor-blocking agents or NOS-inhibitors decrease the toxic insult. Mechanisms of toxicity seem to overlap with the cytotoxic actions of macrophage-derived NO. Free radical formation and energy depletion as a result of mitochondrial iron-sulfur enzyme inhibition, and NAD^+ depletion resulting from activation of poly-ADP-ribosyltransferase, are likely involved in the destructive process (Zhang et al. 1994). An as yet unresolved paradox is the fact that NOS-containing neurons are selectively spared from various neurotoxic insults, including NO.

6 Cytotoxic and Cytostatic Effects of Nitric Oxide

The release and action of EDRF is the main physiological regulatory system controlling vascular tone. In septic shock and chronic inflammation, there is apparently an overproduction of NO. As a result of the induction of the high-output system for NO, a severe life-threatening hypotension is manifested. Impaired or dysregulated cell functions related to massive NO generation are linked to important vascular diseases such as atherosclerosis, hypotension, and diabetes, as well as to reperfusion injury, bypass surgery, and transplantation. Moreover, it is plausible that NO production is part of the nonspecific defense system directed against intracellular parasites, microbes, tumor cells, and alloantigens (Reddy et al. 1983; Nussler and Billiar 1993). NO formation may also be induced in the target cell itself, i.e., during autodestruction of tumor cells, β-cell damage, epithelial autotoxicity, and macrophage elimination (Heiss et al. 1994; Ankarcrona et al. 1994; Albina et al. 1993). NO may also suppress lymphocyte and smooth muscle proliferation. The mechanisms by which NO generates a cytostatic or cytotoxic effect include inhibition of iron-sulfur proteins of mitochondrial respiration, inhibition of aconitase, inhibition of ribonucleotide reductase, GAPDH-modification, DNA deamination reactions, inhibition of protein synthesis, toxic radical formation when NO combines with superoxide, and alterations of the intracellular iron status (Henry et al. 1993; Schmidt and Walter 1994; Nathan 1992; for references). Examples of conditions linking NO to toxicity are type-I diabetes mellitus, when pancreatic β-cells become irreversibly damaged (Kröncke et al. 1993; Fehsel et al. 1993; Eizirik et al. 1993), or cardiac dysfunction (Schulz et al. 1992).

The unique chemical properties of NO determined by the ambient redox milieu that can convert NO• to NO^+ or NO^- cause a variety of biological responses, conferring on NO the ability to act as a physiological regulator molecule or as a toxic agent. Classically, the l-arginine:NO pathway in macrophages acts as a primary defense mechanism against tumor cells, intracellular micro-organisms, and pathogens. The beneficial roles of NO may turn into potentially toxic actions when macrophage NO becomes involved in cell killing. As NO turns out to be cytotoxic in some and cytostatic in other cells, the sensitivity to NO varies considerably from one cell to another. This might depend on cellular characteristics, including dependence on iron-sulfur enzymes, total energy stores, free radical defense systems, repair mechanisms, and the vulnerability of intracellular targets towards NO.

7 NAD^+: A Multifunctional Molecule

NAD^+ (nicotinamide-adenine-dinucleotide), a molecule composed of ADP (adenosine-diphosphate), ribose, and nicotinamide either acts as a redox cofactor or a substrate cofactor, or participates as a precursor of an intracellular messenger in the following reactions:

- As a hydride acceptor of dehydrogenase reactions
- As a precursor for cyclic ADP-ribose
- As a cofactor for NAD^+ glycohydrolases
- As a cofactor for mono-ADP-ribosylation reactions
- As a cofactor for poly-ADP-ribosylation reactions
- In the covalent NO-dependent modification of GAPDH

NAD^+ functions mainly as a redox cofactor for oxidation-reduction reactions utilized by a variety of dehydrogenases and reductases that occupy critical positions in metabolic pathways. Dehydrogenase reactions catalyze the transfer of hydrogen and electrons to NAD^+, producing NADH. Reversible formation of NADH is thereby coupled to the formation of ATP in the mitochondrial respiratory chain. However, under anaerobic glycolytic conditions, glyceraldehyde-3-phosphate dehydrogenase produces ATP, when NAD^+ acts as a redox cofactor linked to the conversion of glyceraldehyde-3-phosphate to 3-phosphoglycerate.

NAD^+ is decomposed by glycohydrolases, breaking up the β-glycosidic bond and producing free ADP-ribose, nicotinamide, and a proton. Free ADP-ribose is known to ADP-ribosylate proteins in a nonenzymatic fashion (Hilz et al. 1984). Studied examples are actin, for which nonenzymatic ADP-ribosylation at a cysteine residue inhibits actin polymerization (Just et al. 1994), or a mitochondrial protein that becomes labeled in response to oxidants (Richter and Frei 1988). NAD^+ is also consumed by enzymes that produce cyclic ADP-ribose (Lee 1994). Cyclic ADP-ribose is known to interact with the ryanodine receptor at the sarcoplasmic reticulum, promoting calcium release from intracellular stores in a manner analogous to inositol 1,4,5-trisphosphate.

Beyond these functions, the ADP-ribose portion of NAD^+ is used by mono- and poly-ADP-ribosyltransferases during the post-translational modification of proteins. ADP-ribosylations are ubiquitous, and the high group transfer potential of the β-glycosidic bond between nicotinamide and ribose drives these reactions. Nuclear poly-ADP-ribose synthase, also referred to as poly-ADP-ribose polymerase, transfers ADP-ribose to histones, chromatin, or the enzyme itself, and then elongates the ADP-ribose chain in a linear or branched polymer. The physiological function of the nuclear enzyme includes the regulation of DNA repair and probably regulates the conformation of the chromatin. In addition, a role during transcription and cell differentiation has been reported. The enzyme has a specific requirement for ends of DNA strands. Intact DNA does not activate the poly-ADP-ribose polymerase. Nicotinamide, benzamide, and its derivative, 3-aminobenzamide, are valuable pharmacological inhibitors (Althaus and Richter 1987; Jacobson and Jacobson 1989).

Mono-ADP-ribosylations occur in bacteria, viruses, plants, animals, and human beings. Reactions are visualized by the action of bacterial toxins and endogenous mammalian enzymes. All known transferase reactions seem to transfer ADP-ribose from NAD^+ to individual amino acid residues like arginine, histidine, asparagine, and cysteine. Heterotrimeric G proteins, low-molecular-weight G proteins, cytoplasmic actin, or other specific proteins like the eukaryotic elongation factor 2 serve as substrates for ADP-ribosylation reactions catalyzed by a large number of bacterial toxins. Examples are cholera toxin, pertussis toxin, pseudomonal toxin, and botulinus toxin. Alterations brought about by bacterial toxins are well documented, because their pathophysiological actions affect cell function and are often related to severe medical symptoms. Endogenous mono-ADP-ribosyltransferases have been characterized and isolated. Although some of them utilize the same target proteins as the bacterial toxins, detailed physiological functions remain obscure (see Jacobson and Jacobson 1989 for references).

8 GAPDH: A Target for Nitric Oxide Action

In 1989 we showed that NO induces an NAD^+-dependent modification of the glycolytic enzyme GAPDH (Brüne and Lapetina 1989). Initially, we incubated human platelet cytosol with $[^{32}P]NAD^+$ in the presence of NO-releasing compounds and observed the radioactive labeling of a 39-kDa protein. Increased incorporation of radioactivity in the presence of NO donors like SNP or SIN-1 was time and dose dependent, with maximal modification of the protein after 90–120 min. Protein modification withstood SDS gel electrophoresis (SDS-PAGE) performed under reducing conditions occurring not only in platelet cytosol but also in the 100l000x*g* supernatant of various rat tissues (brain, liver, heart, and lung). Consistently, the 39-kDa protein was the most prominent one modified, showing the greatest stimulatory response towards NO donors. Although a few other proteins were radioactively labeled as well, their activity was marginal compared with that of the 39-kDa protein. To determine whether NO generated from an active NOS would cause a comparable protein modification, we used rat cerebellum cytosol fortified with all cofactors needed to continuously activate the constitutive brain enzyme (Dimmeler and Brüne 1991). The active brain NOS stimulated incorporation of radioactivity from $[^{32}P]NAD^+$ into the 39-kDa protein, while NOS inhibitors blocked protein labeling. Thus, an active NOS and several chemically distinct NO-releasing compounds share the ability to promote radioactive labeling of a cytosolic 39-kDa protein in the presence of $[^{32}P]NAD^+$.

Unexpectedly, the reaction was independent of a cGMP response. This was rationalized, as neither was activation of soluble guanylyl cyclase involved, nor did cGMP or lipophilic cGMP analogs substitute for NO-liberating agents in stimulating this reaction. A direct cGMP-independent signaling pathway was envisioned. Since hemoglobin blocked the action of EDRF/NO, we used the hemeprotein for the purpose of capturing NO, thereby again linking the 39-kDa protein modification to NO actions. Increasing concentrations of hemoglobin dose dependently suppressed the NO donor-induced radioactive labeling of the 39-kDa protein. We considered this further proof that NO directly signals this NAD^+-dependent covalent protein modification. Other groups, working with brain tissue (Duman et al. 1991; Williams et al. 1992), confirmed our initial results. In homogenates and cytosol from rat brain SNP and an active brain, NOS caused the labeling of a 39-kDa protein in the presence of $[^{32}P]NAD^+$. Additionally, in rat cerebral cortex the NO donor SNP stimulated the incorporation of radioactivity from $[^{32}P]NAD^+$ into several proteins, a 36-kDa protein being the most prominent one. Moreover, a 41-kDa protein became modified in response to NO-releasing agents in *Dictyostelium discoideum* (Tao et al. 1992).

The identity of the 39-kDa protein with the glycolytic enzyme glyceraldehyde-3-phosphate dehydrogenase (GAPDH) was revealed when the purified protein was subjected to N-terminal sequence analysis. The first 29 amino acids of the protein purified from human platelet cytosol and the N-terminal portion of GAPDH matched completely (Dimmeler et al. 1992). Moreover, the homogeneous purified protein showed GAPDH enzyme activity, which is lost in the presence of NAD^+ and NO-releasing compounds. It was interesting to note that the modification of the purified protein with $[^{32}P]NAD^+$ and NO-donating agents occurred in a time- and dose-dependent manner, indicative of an automodification process. Two other groups established the identity of the 39-kDa protein as GAPDH as well, working with brain and erythrocyte cytosolic fractions, respectively (Zhang and Snyder 1992; Kots et al. 1992).

Initial characterization of the reaction leading to increased incorporation of $[^{32}P]NAD^+$ radioactivity into GAPDH seemed consistent with an ADP-ribosylation reaction (Brüne and Lapetina 1989; Kots et al. 1992; Zhang and Snyder 1992). This notion came from the following observations:

a) Free ADP-ribose does not serve as a cofactor for this reaction;
b) Snake venom phosphodiesterase cleavage of the $[^{32}P]NAD^+$-modified protein releases 5'-AMP;
c) GAPDH as a major protein on the surface of group A streptococci transfers the ADP-ribose moiety of NAD^+ to free cysteine. This reaction requires the cleavage of the β-glycosidic bond in NAD^+ in analogy to

bacterial toxin-catalyzed ADP-ribosylation reactions (Pancholi and Fischetti 1993);

d) There is a close correlation between GAPDH ADP-ribosylation and inhibition of the enzyme from *Dictyostelium discoideum* (Tao et al. 1994).

Contrary to the well-established notion that GAPDH is the predominant target for this NO-driven automodification in various cells and tissues, the exact mechanism behind this covalent modification using [^{32}P]NAD$^+$ is still being debated. In some experiments labeling of GAPDH is observed using [^{32}P]NAD$^+$ or [U-^{14}C]NAD$^+$, but not when [carbonyl-^{14}C]NAD$^+$ is employed as a substrate (Kots et al. 1992; Dimmeler and Brüne 1992). A recent study, however, proposed the covalent binding of the whole molecule NAD$^+$ to GAPDH, as NO-releasing agents like SNP promote some radiolabeling of GAPDH with [nicotinamide-^{14}C]NAD$^+$ (McDonald and Moss 1993). Subsequent reports indicate that GAPDH, depending on assay conditions such as pH value and cofactor composition of the buffer, generates free ADP-ribose (Tao et al. 1993). More stringent assay conditions may additionally direct the transfer of ADP-ribose from NAD$^+$ to various amino acids, eventually giving rise to incorporation of radioactivity from [^{32}P]NAD$^+$ into GAPDH in an NO-independent fashion (Tao et al. 1993; Zocchi et al. 1993).

Although NO-induced GAPDH modification resembles some features of ADP-ribosylation reactions, conditions for optimal protein modification are different from those employed in the toxin-labeling assays. This especially applies for compounds like ATP and thymidine and for high phosphate concentrations (Brüne and Lapetina 1990). In contrast, other agents like NADPH, normally not involved in ADP-ribosylation reactions, drastically enhance GAPDH labeling carried out using human platelet cytosol (Brüne et al. 1992).

9 GAPDH Labeling Systems

GAPDH modification can be studied easily using cytosolic fractions of various cells or employing the commercially available purified enzyme. Nitric oxide-releasing compounds may be used as the driving force to enhance protein modification in the presence of [^{32}P]NAD$^+$. Normally, reactions are run for 30–90 min, followed by protein precipitation, separation of the protein(s) by 11% SDS-PAGE, and detection of radioactivity using the phosphor imager system. Alternatively, a permanently active constitutive brain NOS may generate sufficient NO to enhance modification of GAPDH. This is exemplified using rat cerebellum cytosol (Dimmeler and Brüne 1991) and in rat brain with the addition of purified constitutive NOS

(Zhang and Snyder 1992). Besides proving that catalytically active NOS provides enough NO, the use of NOS inhibitors, while blocking GAPDH modification, substantiates the participation of the constitutive NO-forming enzyme. Taking into account that a permanently active constitutive NOS in vitro lacks the regulatory calcium component, we performed experiments focusing on GAPDH modification by activating the inducible NOS. The pancreatic β-cell line RINm5F responds to the cytokine interleukin-1 (IL-1) with an increased output of nitrite, stemming from the inducible NOS isoform (Meßmer and Brüne 1994). In RINm5F cells treated with IL-1, GAPDH enzyme activity decreases by roughly 50% compared with controls, whereas NOS inhibitors restore the catalytic activity completely (Dimmeler et al. 1993). Along with decreased activity, one also finds less incorporation of [^{32}P]NAD$^+$ into GAPDH in the cytosolic fraction of IL-1-treated cells. This may reflect endogenous modification of the enzyme, which is less responsive to the [^{32}P]NAD$^+$ modification in the cytosol afterwards. In recent experiments with homogenates from stimulated hippocampus, SNP-induced ADP-ribosylation was reduced at preconvulsive stage II and stage V (tonic-clonic seizures) of dentate gyrus kindling, compared with controls (Vezzani et al. 1994). As suggested, this probably reflects enhanced endogenous ADP-ribosylation, depending on the progressive activation of the NMDA receptors and the continued generation of NO. Besides probing for GAPDH modification using cytosolic fractions and cellular systems, we also studied advanced in vivo systems. Mice were injected with bacterial endotoxin (LPS) known to generate NO by an active cytokine-inducible NO synthase. GAPDH enzyme activity was significantly reduced in cytosolic fractions of heart and spleen, and incorporation of radioactivity from [^{32}P]NAD$^+$ into the protein was similarly decreased (Dimmeler et al. 1994). These experiments establish an association of GAPDH modification in vivo with enhanced NO production.

10 Nitric Oxide Redox Species

Nitrogen monoxide ($NO^•$) is susceptible to both oxidation and reduction, which produce nitrosonium (NO^+) or nitroxyl anion (NO^-), respectively (Stamler et al. 1992; for references). The efficacy of various NO-releasing compounds in inducing GAPDH modification compared with their $NO^•$-releasing potency revealed inconsistencies. Rates of $NO^•$ release and the extent of GAPDH modification seemed to be independent. For example, spermine-NO decomposes, releasing $NO^•$ with a half-life of roughly 40 min, while producing moderate GAPDH modification. In contrast, SNP causes efficient GAPDH modification but releases only a small amount of

$NO^{\bullet}$. Another exception is SIN-1; it releases $ONOO^-$ but gives rise to very efficient protein modification. With the notion that thiols (DTT) enhance GAPDH modification induced by most NO-donating agents, we proposed a mechanism involving NO^+. To verify the role of NO^+ and S-nitrosylation of active site thiol we probed the chemical mechanism of post-translational modification of GAPDH using nitrosonium tetrafluoroborate (BF_4NO) (Mohr et al. 1994). BF_4NO is unique, in that it hydrolyzes under controlled conditions in aqueous solution, producing only NO^+. BF_4NO efficiently causes GAPDH modification in the presence of [^{32}P]NAD^+. These experiments implicate NO^+ as the species transferred to the active-site thiol, causing an S-nitrosylated intermediate, which then leads to the covalent modification of the enzyme in the presence of radioactive NAD^+. For maximal labeling studies, BF_4NO, in contrast to to SNP or SIN-1, does not require another thiol compound to be present. Application of thiol together with BF_4NO attenuates GAPDH modification due to competing S-nitrosating reactions with DTT, present to a much greater degree than enzyme. Considering conflicting results regarding the mechanism of GAPDH modification using SIN-1, which is believed to produce peroxynitrite, we examined the effect of $ONOO^-$ in greater detail (Mohr et al. 1994). Interestingly, for $ONOO^-$ induced GAPDH modification, thiols (DTT) are ultimately required. With thiols (DTT) in relative excess to peroxynitrite, conditions are conducive to formation of RS-NO. Therefore, the likely scenario employing SIN-1 or peroxynitrite directly probably involves S-nitrosation of DTT, followed by NO^+ donation to the reactive thiol of GAPDH. In contrast, when peroxynitrite is present in relative excess to thiols (DTT), it oxidizes protein thiols to sulfenic, sulfinic, and sulfonic acids (Radi et al. 1991), thereby inhibiting subsequent covalent modification of GAPDH by [^{32}P]NAD^+. Our findings are in keeping with a role for nitrosonium (NO^+)- or nitronium (NO_2^+)-like species in the process that initiates covalent [^{32}P]NAD^+-dependent GAPDH modification (Molina y Vedia et al. 1992). The present concept proposes that S-nitrosylation precedes, and thereby inititates, the covalent protein modification and emphasizes the role of NO group transfer chemistry during NO signaling.

11 The Cofactor NAD^+ Versus NADH

Recent studies indicate that GAPDH modification is stimulated by RS-NO and related nitrosating agents, rather than by $NO^{\bullet}$ itself. The transfer of NO^+ to active-site thiol (S-nitrosylation) promotes subsequent modification by NAD^+. We attempted to reconcile nitrosative chemistry with GAPDH modification; the reaction mechanism seemed much more rational if NADH,

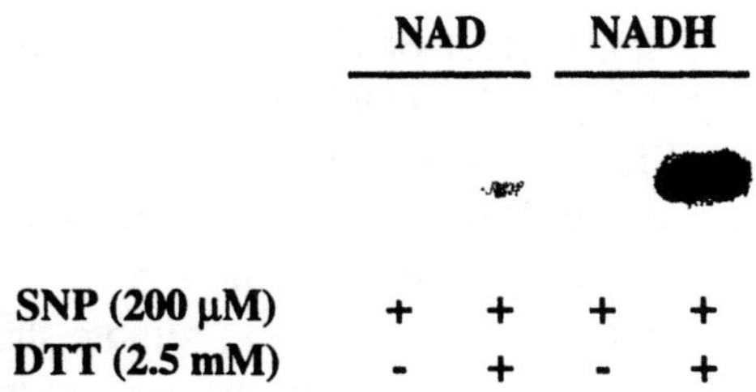

Fig. 3. Covalent modification of GAPDH by SNP in the presence of NAD^+ and NADH. GAPDH (10 μg/assay) was incubated at 37°C for 20 min in 100 m*M* Hepes buffer, pH 7.5, containing 10 μ*M* NAD^+/NADH and 2001000 cpm of $[^{32}P]NAD^+/[^{32}P]NADH$ in a total volume of 65 μl, with dithiothreitol (DTT) and sodium nitroprusside (SNP) as indicated. Preparation of samples and detection of radioactivity were performed essentially as described by Mohr et al. (1994)

rather than NAD^+, was the substrate. To test the hypothesis, radioactive NAD^+ was enzymatically reduced to produce $[^{32}P]$NADH, followed by HPLC purification. Using a standard GAPDH modification assay with thiol (DTT) and SNP present, it turned out that NADH was much more potent at causing radioactive enzyme modification (Fig. 3). The labeling is estimated on the basis of the same amount and specific radioactivity of the cofactors $[^{32}P]NAD^+$ and $[^{32}P]$NADH.

Reduction of NAD^+ to NADH makes the nicotinamide moiety susceptible to nitrosative (NO^+) attack. One would envision transnitrosation from active-site RS-NO to NADH (NO^+ attacks C5 of the nicotinamide moiety), thereby facilitating protein thiolate attack on the nucleotide (e.g., attack at C6 of the nicotinamide moiety or making nicotinamide a better leaving group, e.g., attack at ribose C1') (Mohr et al., unpublished data). The proposed reaction sequence allows for binding of the whole molecule NAD^+ (McDonald and Moss 1993), at the same time explaining the hydrolysis of the β-glycosidic bound of NAD^+, as reported by others (Pancholi and Fischetti 1993).

12 The Role of Thiols

Free, low-molecular-weight thiols and protein thiol groups are critical factors influencing NO-mediated GAPDH modification (Brüne and Lapetina 1990). Thiol addition suppresses basal, i.e., in the absence of NO donors, observed enzyme modification (Brüne and Lapetina 1990; McDonald and Moss 1993; Tao et al. 1993). In contrast, the NO-stimulated reaction is dramatically enhanced by thiols, DTT again being most effective (Brüne and Lapetina 1990; Vaidyanathan et al. 1993). However, DTT at concentrations of 5 m*M* or higher reversed the stimulatory effect of NO-releasing compounds in modifying the enzyme with [^{32}P]NAD$^+$ (McDonald and Moss 1993). Competing S-nitrosylation of DTT, rather than the attack of NO$^+$ at GAPDH, may explain this effect. Alternatively, one can speculate about the reversibility of the reaction in the presence of high thiol concentrations.

For some NO donors, such as SNP, thiols are required to promote the one-electron-driven release of NO from the compound, whereas in with others, e.g., BF_4NO, thiols blunt the reaction. This seems rational, assuming S-nitrosothiol formation of NO$^+$ with competing excessive thiols rather than GAPDH (Mohr et al. 1994). Reduced protein thiol groups, especially cysteine 149 in the active site of GAPDH, are a prerequisite for NO-stimulated enzyme modification. Any oxidation of critical protein thiols would automatically decrease incorporation of radioactivity associated with Cys-149. When investigating the effect of ONOO$^-$, this needs to be considered. In this case thiols are primarily required to suppress the oxidation of protein thiol groups. A similar explanation applies for SIN-1, when thiols consistently increase [^{32}P]NAD$^+$-dependent GAPDH labeling.

Protein thiol groups are key factors during covalent modification of GAPDH induced by NO$^+$ and NAD$^+$/NADH. Post-translational protein modifications resembling ADP-ribosylation reactions or ADP-ribosylation-like reactions are characterized by the stability of the newly formed ADP-ribose-protein or NAD$^+$-protein bond to $HgCl_2$, neutral hydroxylamine, and alkali. Covalent modification of GAPDH involves thiol groups, as indicated by the fact that radiolabel incorporated into the protein is sensitive to $HgCl_2$ cleavage (Dimmeler and Brüne 1992; McDonald and Moss 1993). A concentration of 2–10 m*M* $HgCl_2$, incubated at 37°C for up to 60 min with [^{32}P]NAD$^+$-labeled GAPDH, removed more than 90% of the radioactivity associated with GAPDH. The sensitivity towards hydroxylamine or alkali was negligible. However, pertussis toxin ADP-ribosylated cysteine residues of human platelet membranes seem much more labile toward mercury cleavage compared with the GAPDH situation. The use of thiol-blocking agents such as dithionitrobenzene or N-ethylmaleimide (Brüne and Lapetina 1990; Dimmeler et al. 1992) hindered basal as well as NO-stimulated GAPDH modification, further substantiating the role of reduced protein thiol groups during covalent, post-translational modification of the glycolytic enzyme GAPDH.

Compatible results came from experiments using the sesquiterpene antibiotic koninjic acid (McDonald et al. 1993). Koninjic acid binds with some preference to GAPDH at the Cys-149 of the active site and thereby blocks NO-stimulated enzyme modification. Although other dehydrogenases share with GAPDH the active-site thiol group and the NAD$^+$ binding site, they do not respond in a way comparable to GAPDH (Zhang and Snyder 1992; Dimmeler and Brüne 1993). Use the thiol alkylating agent N-ethylmaleimide (NEM) to inhibit enzyme activity of several dehydrogenases revealed that, compared with alcohol dehydrogenase (ADH) or lactate dehydrogenase (LDH), GAPDH was much more sensitive towards inhibition. This is paralleled by the degree of NO-stimulated dehydrogenase modification, GAPDH being much more susceptible than ADH, while LDH did not respond at all (Dimmeler and Brüne 1993). The focus on GAPDH is due at least partly to its intracellular prevalence; however, it also presents itself as an exceptional target due to its remarkably sensitive protein thiol group(s). Thus, the reactive Cys-149 at the active site of GAPDH is the most likely target for covalent NO-stimulated enzyme modification and is known for its extreme sensitivity towards all kinds of oxidants (Claiborne et al. 1993; Schuppe-Koistinen et al. 1994; for references). Oxidation of Cys-149 is easily achieved using various organic or inorganic peroxides, or is observed under "oxidative stress conditions". Oxidative modification of GAPDH is associated with the generation of intramolecular disulfides, sulfenic acids, or oxidant-specific S-thiolation. Oxidation of Cys-149 leads to decreased enzyme activity, although in a reversible manner only.

For NO-stimulated GAPDH modification the amount of radioactivity stemming from $[^{32}P]NAD^+$ which is bound to the protein-reactive cysteine residue varies considerably. The amount of activity incorporated into GAPDH ranges from below 1% up to 45%, calculated as the molar protein/NAD^+ ratio (Dimmeler and Brüne 1992; Brüne et al. 1994; McDonald and Moss 1993; Tao et al. 1994). For stoichiometric considerations, one has to take into account the portion of cold/unlabeled NAD^+ bound to the protein. The amount of $[^{32}P]NAD^+$ attached to the enzyme is underestimated by the amount of cold NAD^+ when radioactivity is measured.

NAD^+-dependent, NO-stimulated reactions do not seem to be restricted to GAPDH modification. In human neutrophils, actin becomes ADP-ribosylated when $[^{32}P]NAD^+$ is used in the presence of NO (Clancy et al. 1993), whereas in an NG108-15 (mouse neuroblastomarat glioma somatic hybrid cell) cell homogenate (Boyd et al. 1993) NO promotes the NAD^+ labeling of poly-ADP-ribose polymerase. In brain, NO-stimulated endogenous ADP-ribosylation has been linked to long-term potentiation (Duman et al. 1993), and changes in the ADP-ribosylation status of some hippocampal proteins are linked to kindling progression (Vezzani et al. 1994).

13 The Role of GAPDH as a Multifunctional Enzyme

NO-dependent inactivation of the glycolytic enzyme GAPDH by an automodification process in the presence of NAD^+ seems relevant not only in vitro, but also in intact cells and in whole animals. Besides its normal function in glycolysis, GAPDH may also be involved in one of the following functions unrelated to its glycolytic activity (see references in Brüne et al. 1994; Antonietta de Matteis et al. 1994; Ravichandran et al. 1994):

- Binding and transport of tRNA associated with nuclear localization of GAPDH
- DNA-repair activity, i.e., uracil DNA glycosylase
- Activation of transcription in neurons
- Interaction with tubulin and microtubules
- Transport of nitric oxide
- Serving as a substrate for brefeldin A-stimulated ADP-ribosylation
- Serving as a major constitutent of synaptic vesicles
- Representing a target for S-thiolation during the generation of hydrogen peroxide

Some of these alternative functions of GAPDH, just like NO-mediated modification of the enzyme, are related to the NAD^+-binding site of the protein. Therefore, it will be intriguing to search for the significance of these activities in relation to cGMP-independent NO actions.

NO-dependent modification of GAPDH, as well as S-thiolation of the enzyme, inhibits enzyme activity (Dimmeler et al. 1993; Ravichandran et al. 1994). However, assuming that GAPDH is a non-rate-limiting glycolytic step, it remains to be established whether these post-translational modification mechanisms alter the metabolism of glucose through glycolysis, thereby affecting the overall rate of energy production. Our preliminary data obtained from studying macrophage cell death after iNOS induction would imply that neither inhibition of iron sulfur enzymes like aconitase nor inhibition of the glycolytic activity of GAPDH directly signals for cell destruction (Meßmer and Brüne, unpublished). Therefore, alternative GAPDH functions in relation to NO actions need to be addressed. Alterations in the intracellular distribution of GAPDH after NO application have been observed (unpublished results), and variations in the GAPDH protein level seem achievable in response to hypoxia (Graven et al. 1994). The functional explanation of these observations remains to be clarified, however.

14 Toxicity of Nitric Oxide

Nitric oxide is a pathophysiological mediator, and, as a potentially toxic molecule, NO can cause cell death. Neuronal cells and pancreatic islets, i.e., β-cells, are highly susceptible to endogenously formed or chemically generated NO (Dawson and Snyder 1994; Lipton et al. 1993; Kröncke et al. 1993;

Delaney et al. 1993; Dawson et al. 1993). Mechanistically, NO may cause DNA-deaminating reactions (Wink et al. 1991), or it may cause toxicity, at least in part, by reaction with superoxide anion, apparently leading to the formation of $ONOO^-$ (Beckmann 1991). Generally, a pathophysiological action of NO can be defined by a relative excess of one or more of the following compounds:

- NO_x reaction products of NO and oxygen
- M-NO reaction products of NO and metal centers
- $ONOO^-$ reaction product of NO and superoxide

As these products are at the same time intermediates during the physiological NO signaling cascade, the term "relative excess" seems to describe the regulatory device between physiology and toxicity. The toxicity of NO depends on the chemistry it undergoes in a given biological/cellular milieu, which may vary considerably among cellular systems. Reactions are determined by the relative concentration of reaction partners as well as by the capacity of cellular defense systems.

Cell death is believed to occur by one of two mechanisms: (a) necrosis or (b) apoptosis. Necrotic cells typically exhibit distinctive morphological and biochemical characteristics. Early changes include swelling of the cytoplasma and organelles, especially the mitochondria, with only slight changes in the nucleus. Cellular changes ultimately lead to organelle and cell dissolution, allowing the intraorganelle and cytoplasmic contents to leak out into the extracellular space. Altered ion homeostasis due to a disappearance of membrane ion-pump activities may be the result of direct membrane damage, or it may be secondary to energy depletion. There is a rapid decrease in protein, RNA, and DNA levels. DNA is exposed to lysosomal nucleases, causing DNA degradation. Fragments display a continuous spectrum of sizes. Necrosis typically affects a group of cells, and an inflammatory response usually develops in adjacent viable tissue in response to the cellular debris released from lysed cells (Schartzman and Cidlowski 1993; Orrenius et al. 1988).

The term apoptosis was introduced by Kerr and co-workers in 1972 (Kerr et al. 1972) to describe a morphologically distinct, spontaneous form of cell death occurring under various conditions. The word "apoptosis" comes from the ancient Greek *apo* (away from, with the implication of separation) and the root *ptosis* (to fall). Apoptosis describes the process of "falling off, being separated," as leaves fall away from trees. Death occurs spontaneously in many cases and can be induced by physiological or noxious stimuli. The idea that cell killing is self-programmed leads to the expression "programmed cell death" as another name for this process. Apoptosis must be regarded as a basic physiological process that serves as the counterpart to cell division by mitosis. Discrete stages characterize the death pathway. First, nuclear and cytoplasmic condensation as characteristic morphological alterations are observed. The cell forms localized protrusions of the cell surface, which separate into multiple membrane-bound bodies, also known as "apoptotic bodies", containing nuclear remnants and intact organelles. In the second stage these cell fragments are phagocytosed and rapidly degraded by neighboring cells (Schwartzman and Cidlowski 1993; Corcoran et al. 1994; for references). In contrast to necrosis, apoptosis usually occurs in isolated, single cells, without giving rise to inflammatory responses in the surrounding tissue. Although morphological criteria remain an important standard for documentation of apoptosis, DNA cleavage patterns as biochemical markers of apoptosis have been established. DNA fragmentation begins early in the death process, appearing several hours before cell viability starts to decline. DNA from apoptotic cells is separated by agarose gel electrophoresis into a "ladder" of fragments, generated by an endonuclease cleaving the linker regions of DNA into 180- to 200-bp fragments and multiples of these fragments. However, biochemical markers, i.e., DNA ladder formation, are not an ultimate necessary sign of apoptosis (Collins et al. 1992).

For NO, both necrotic and apoptotic cell death pathways seem feasible. Although exact conditions are difficult to compare, apoptosis and necrosis may occur either concurrently or sequentially. A determinant may again be the cellular redox milieu.

For neuronal cells formation of NO is linked to toxicity (Dawson and Snyder 1994). This also applies for glutamate neurotoxicity, which is attributable, at least in part, to NOS activation evoked by elevated intracellular calcium. Experiments have been done with inhibitors of NOS, calmodulin antagonists, and reduced hemoglobin, all of which attenuate neurotoxicity, linking cell death and NO formation. Superoxide dismutase, which scavenges the superoxide anion, decreases glutamate and NO-mediated toxicity, which implies the reduced oxygen radical O_2^- and concomitant $ONOO^-$ formation in components of this process (Dawson and Snyder 1994; Lipton et al. 1993). Contradictory reports providing evidence that NO may protect neurons exist (Lipton et al. 1993). Different NO species are attributed to both beneficial and harmful reactions. NO^+ is considered to protect neurons by causing S-nitrosylation of the NMDA receptor, not allowing glutamate stimulation, whereas $ONOO^-$ seems involved in cell killing. Recently obtained evidence suggests that DNA damage is the key to NO neurotoxicity (Zhang et al. 1994). Damaged DNA activates the nuclear enzyme poly-ADP-ribose synthetase, utilizing NAD^+ in order to form ADP-ribose polymers attached to histones and the enzyme itself. As a consequence, cellular ATP would be depleted, leading to cell death (Dawson and Snyder 1994). Accordingly, cell lysis, measured mostly by the trypan blue exclusion assay as an indication of cell death, represents necrosis rather than apoptosis. In contrast, for at least some cells, such as RINm5F cells (cell line of pancreatic origin), mast cells, and macrophages, NO-mediated cell death occurs by apoptosis (Albina et al. 1993; Sarih et al. 1993; Kitajima et al. 1994; Ankarcrona et al. 1994). Islet DNA as a target for inflammatory attack by NO has been proposed, although apoptotic DNA fragmentation was not evidenced in experiments with endonuclease inhibitors (Fehsel et al. 1993). Our experiments show that IL-1β induces apoptosis in the pancreatic β-cell line RINm5F (Ankarcrona et al. 1994). DNA fragmentation into the typical DNA ladder, nuclear condensation, and apoptotic body formation are unquestionable indications of apoptosis. The production of NO precedes the appearance of these typical features, whereas NOS inhibitors blocked IL-1β-induced cell death. NO production activates the cell death program, which seems relevant for type-I, insulin-dependent diabetes mellitus (IDDM) when pancreatic β-cells are progressively destroyed as a consequence of an autoimmune process. The finding that both DNA fragmentation and apoptotic body formation were prevented by NOS inhibitors supports the role of NO generation by an inducible NOS for triggering apoptosis in this system. A role of NAD^+ depletion consequent to DNA damage as a prerequisite for islet cell lysis is still being debated (Radons et al. 1994).

For macrophage cell death, activation of endogenous NO generation or exogenously applied NO causes apoptosis. Morphological criteria as well as biochemical analysis revealed characteristic apoptotic features (Meßmer and Brüne, unpublished). Cytokine-elicited apoptosis was prevented by blocking the inducible NOS by NMMA, establishing a link between NO formation and cell destruction. A chemically heterogeneous group of NO-releasing compounds like SNP, SIN-1, SNAP, spermine-NO, and DEA-NO produce a time- and concentration-dependent effect. However, compound specific differences became apparent, related to the biological half-life of these compounds and the NO redox species being set free. Apoptosis induced by exogenous NO sources was effectively antagonized by activation of protein kinase C using phorbol esters (TPA), or by application of lipophilic cAMP analogues.

With respect to the mechanism proposed for NO toxicity, its interaction with protein thiol groups, iron sulfur proteins, and a direct DNA-damaging activity seems plausible. The latter, regardless of whether it is induced by radiation or by drugs such as etoposide, can result in apoptosis. Expression of wild-type p53, a tumor-suppressor gene, seems to be closely linked to apoptosis caused by most of the DNA-damaging agents. Recently, we demonstrated that NO, either endogenously generated by the inducible NOS in macrophages (RAW 264.7) and β-cells (RINm5F) or exogenously supplied by various NO-releasing compounds, caused p53 accumulation (Meßmer et al. 1994). Accumulation of p53 clearly preceded apoptotic cell death, with

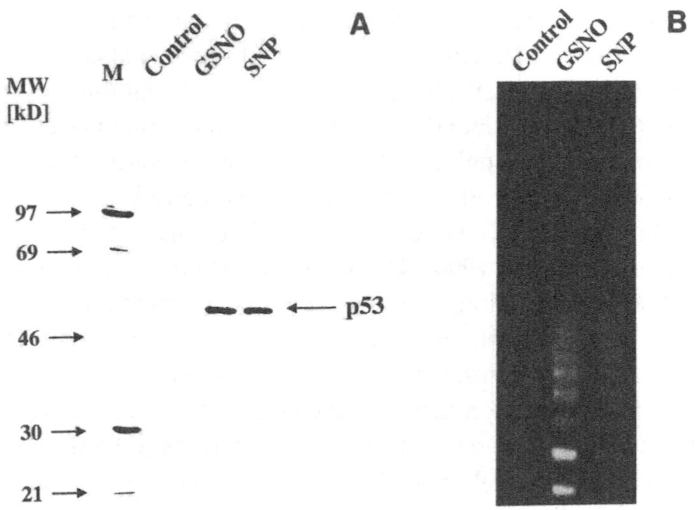

Fig. 4A,B. DNA fragmentation and p53 accumulation induced by NO-releasing compounds. RAW macrophages were incubated with S-nitrosoglutathione (1 m*M* GSNO) and sodium nitroprusside (1 m*M* SNP). **A** Detection of p53 (4 h incubation) and **B** visualization of DNA fragmentation (8 h incubation) were performed essentially as described by Meßmer et al. (1994)

NOS inhibitors preventing NO formation as well as p53 expression. As exemplified in Fig. 4, there is NO-induced DNA fragmentation, visualized by DNA-agarose gel electrophoresis, and p53 accumulation in response to the NO donor GSNO and SNP.

Accumulation of p53 may have the dual role of causing cell cycle arrest via downstream genes and stimulating DNA repair directly. However, intranuclear p53 accumulation following DNA damage can also be part of the signaling to apoptosis, either by directly acting on the DNA or, again, by causing cell cycle block (Meßmer et al. 1994; for references). Accumulation of p53 in response to NO – together with previous evidence that NO can cause growth arrest and cell necrosis and confers antitumor properties – supports the notion that NO may signal different forms of cell death, i.e., apoptosis versus necrosis. This may depend on the cell type and/or the steady-state concentration of the NO redox species involved.

15 Summary and Conclusions

Nitric oxide signaling is achieved through both cGMP-dependent and cGMP-independent mechanisms. The latter are exemplified by protein thiol modification followed by subsequent NAD^+-dependent automodification of the glycolytic enzyme GAPDH, or by mechanisms inducing accumulation of the tumor suppressor gene p53 and causing apoptotic cell death. Both cGMP-independent actions are initiated using NO-releasing compounds and an active LPS/cytokine-inducible NO synthase. NO-synthase inhibitors block the release of NO and hinder downstream signaling mechanisms; they are therefore valuable pharmacological tools linking a defined cellular response to various NO actions.

Signal transducing mechanisms elicited by NO can be studied using GAPDH as a representative example of NO-induced protein modification and are grouped as follows:

- S-Nitrosylation reactions initiated by NO^+
- NAD^+-dependent, post-translational covalent automodification of GAPDH
- Oxidative modification (thiol oxidation) and inhibition of GAPDH by NO-related agents, probably $ONOO^-$

GAPDH and several other protein targets may serve as molecular sensors of elevated NO concentrations and may transmit this message through post-translational modification and oxidation-induced conformational changes as cGMP-independent NO signaling pathways.

Toxicity of NO seems to be linked to both apoptosis and necrosis, depending on the chemistry of NO it undergoes in a given biological milieu. Toxicity manifests as a relative excess of NO_x, metal-NO interactions, and $ONOO^-$ formation in relation to cellular defense systems. Although accumulation of the tumor-suppressor gene product p53 in response to NO opens a regulatory mechanism known to be involved in apoptotic cell death, cGMP-independent signaling pathways remain to be elucidated. As NO-dependent modification of GAPDH would imply down-regulation of glycolysis and concomitant energy production followed by cell death, our data so far do not support this assumption.

In recent years, NO has proved to be a beneficial messenger with a potentially toxic activity. It will be challenging to investigate NO biochemistry in closer detail and to elucidate how NO targets biological systems, especially in relation to its pathophysiological role.

Acknowledgments. Continuous support for our studies during the past few years has come from the *Deutsche Forschungsgemeinschaft* and partly from the European Community. These institutions are gratefully acknowledged. We thank Professor Ullrich for reviewing the manuscript and for his interest in our work.

References

Albina JE, Cui S, Mateo RB, Reichner JS (1993) Nitric oxide-mediated apoptosis in murine peritoneal macrophages. J Immunol 150: 5080–5085

Althaus FR, Richter C (eds) (1987) ADP-ribosylation of proteins: enzymology and biological significance. Springer, Berlin Heidelberg New York

Ankarcrona M, Dypbukt JM, Brüne B, Nicotera P (1994) Interleukin-1b-induced nitric oxide production activates apoptosis in pancreatic RINm5F cells. Exp Cell Res 213: 172–177

Antonietta de Matteis M, di Girolamo M, Colanzi A, Pallas M, di Tullio G, McDonald LJ, Moss J, Santini G, Bannykh S, Corda D, Luini A (1994) Stimulation of endogenous ADP-ribosylation by brefeldin A. Proc Natl Acad Sci USA 91: 1114–1118

Arnold WP, Mittal CK, Katsuki S, Murad F (1977) Nitric oxide activates guanylate cyclase and increases guanosine 3',5'-cyclic monophosphate levels in various tissue preparations. Proc Natl Acad Sci USA 74: 3203–3207

Bates JN, Baker MT, Guerra R Jr, Harrison DG (1991) Nitric oxide generation from nitroprusside by vascular tissue. Biochem Pharmacol 42: 157–165

Beckmann J (1991) The double-edged role of nitric oxide in brain function and superoxide-mediated injury. J Dev Physiol 15: 53–59

Bennett BM, McDonald BJ, Nigam R, Simon WC (1994) Biotransformation of organic nitrates and vascular smooth muscle cell function. TIPS 15: 245–249

Boyd RS, Donnelly LE, Allport JR, MacDermot J (1993) Sodium nitroprusside promotes NAD^+ labelling of a 116-kDa protein in NG108-15 cell homogenates. Biochem Biophys Res Commun 197: 1277–1282

Brüne B, Lapetina EG (1989) Activation of a cytosolic ADP-ribosyltransferase by nitric oxide-generating agents. J Biol Chem 264: 8455–8458

Brüne B, Lapetina EG (1990) Properties of a novel nitric oxide-stimulated ADP-ribosyltransferase. Arch Biochem Biophys 279: 286–290

Brüne B, Ullrich V (1987) Inhibition of platelet aggregation by carbon monoxide is mediated by activation of guanylate cyclase. Mol Pharmacol 32: 497–504

Brüne B, Schmidt K-U, Ullrich V (1990) Activation of soluble guanylate cyclase by carbon monoxide and inhibition by superoxide anion. Eur J Biochem 192: 683–688

Brüne B, Dimmeler S, Lapetina EG (1992) NADPH: a stimulatory cofactor for nitric oxide-induced ADP-ribosylation reaction. Biochem Biophys Res Commun 182: 1166–1171

Brüne B, Dimmeler S, Molina y Vedia L, Lapetina EG (1994) Nitric oxide:a signal for ADP-ribosylation of proteins. Life Sci 54: 61–70

Choi DW (1993) Nitric oxide: foe or friend to the injured brain? Proc Natl Acad Sci USA 90: 9741–9743

Claiborne AL, Miller H, Parsonage D, Ross RP (1993) Protein-sulfenic acid stabilization and function in enzyme catalysis and gene regulation. FASEB J 7: 1483–1490

Clancy RM, Leszczynska-Piziak J, Abramson SB (1993) Nitric oxide stimulates the ADP-ribosylation of actin in human neutrophils. Biochem Biophys Res Commun 191: 847–852

Collins RJ, Harmon BV, Gobe GC, Kerr JFR (1992) Internucleosomal DNA cleavage should not be the sole criterion for identifying apoptosis. Int J Radiat Biol 61: 451–453

Corcoran GB, Fix L, Jones DP, Moslen MT, Nicotera P, Oberhammer F, Buttyan R (1994) Apoptosis: molecular control point in toxicity. Toxicol Appl Pharmacol 128: 169–181

Dawson TM, Snyder SH (1994) Gases as biological messengers: nitric oxide and carbon monoxide in the brain. J Neurosci 14: 5147–5159

Dawson VL, Dawson TM, Bartley DA, Uhl GR, Snyder SH (1993) Mechanisms of nitric oxide-mediated neurotoxicity in primary brain cultures. J Neurosci 13: 2651–2661

Delaney CA, Green MHL, Lowe JE, Green IC (1993) Endogenous nitric oxide induced by interleukin-1b in rat islets of Langerhans and HIT-T15 cells causes significant DNA damage as measured by the 'comet' assay. FEBS Lett 333: 291–295

Dimmeler S, Brüne B (1991) l-arginine stimulates an endogenous ADP-ribosyltransferase. Biochem Biophys Res Commun 178: 848–855

Dimmeler S, Brüne B (1992) Characterization of a nitric-oxide-catalysed ADP-ribosylation of glyceraldehyde-3-phosphate dehydrogenase. Eur J Biochem 210: 305–310

Dimmeler S, Brüne B (1993) Nitric oxide preferentially stimulates auto-ADP-ribosylation of glyceraldehyde-3-phosphate dehydrogenase compared to alcohol or lactate dehydrogenase. FEBS Lett 315: 21–24

Dimmeler S, Lottspeich F, Brüne B (1992) Nitric oxide causes ADP-ribosylation and inhibition of glyceraldehyde-3-phosphate dehydrogenase. J Biol Chem 267: 16771–16774

Dimmeler S, Ankarcrona M, Nicotera P, Brüne B (1993) Exogenous nitric oxide (NO)-generation or IL-1b induced intracellular NO production stimulates inhibitory auto-ADP-ribosylation of glyceraldehyde-3-phosphate dehydrogenase in RINm5F cells. J Immunol 150: 2964–2971

Dimmeler S, Meßmer UK, Tiegs G, Brüne B (1994) Modulation of glyceraldehyde-3-phosphate dehydrogenase in *Salmonella abortus equi* lipopolysaccharide-treated mice. Eur J Pharmacol 267: 105–112

Duman RS, Terwilliger RZ, Nestler EJ (1993) Alterations is nitric oxide stimulated endogenous ADP-ribosylation associated with long-term potentiation in rat hippocampus. J Neurochem 61: 1542–1545

Eizirik DL, Sandler S, Palmer JP (1993) Repair of pancreatic b-cells. Diabetes 42: 1383–1391

Feelisch M, Te Poel M, Zamora R, Deussen A, Moncada S (1994) Understanding the controversy over the identity of EDRF. Nature 368: 62–65

Fehsel K, Jalowy A, Qi S, Burkart V, Hartmann B, Kolb H (1993) Islet cell DNA is a target of inflammatory attack by nitric oxide. Diabetes 42: 496–500

Furchgott RF, Zawadzki JV (1980) The obligatory role of endothelial cells in the relaxation of arterial smooth muscle by acetylcholine. Nature 288: 373–376

Garthwaite J (1993) Nitric oxide signalling in the nervous system. The Neurosciences 5: 171–180

Garthwaite J, Charles SL, Chess-Williams R (1988) Endothelium-derived relaxing factor release on activation of NMDA receptors suggests role as intracellular messenger in the brain. Nature 336: 385–388

Graven KK, Troxler RF, Kornfeld H, Panchenko MV, Farber HW (1994) Regulation of endothelial cell glyceraldehyde-3-phosphate dehydrogenase expression by hypoxia. J Biol Chem 269: 24446–24453

Green LC, Tannenbaum SR, Goldman P (1981) Nitrate synthesis in the germ-free and conventional rat. Science 212: 56–58

Heiss LN, Lancaster JR Jr, Corbett JA, Goldman WE (1994) Epithelial autotoxicity of nitric oxide: role in the respiratory cytopathology of pertussis. Proc Natl Acad Sci USA 91: 267–270

Henry Y, Lepoivre M, Drapier J-C, Ducrocq C, Boucher JL, Guissani A (1993) EPR characterization of molecular targets for NO in mammalian cells and organelles. FASEB J 7: 1124–1134

Hibbs JB Jr, Taintor RR, Vavrin Z (1987) Macrophage cytotoxicity: role for l-arginine deiminase and imino nitrogen oxidation to nitrite. Science 235: 473–476

Hilz H, Koch R, Fanick W, Klapproth K, Adamietz P (1984) Nonenzymatic ADP-ribosylation of specific mitochondrial polypeptides. Proc Natl Acad Sci USA 81: 3929–3933

Ignarro LJ (1990) Biosynthesis and metabolism of endothelium-derived nitric oxide. Annu Rev Pharmacol Toxicol 30: 535–560

Jacobson MK, Jacobson EL (eds) (1989) ADP-ribose transfer reactions: mechanisms and biological significance. Springer, Berlin Heidelberg New York

Just I, Wollenberg P, Moss J, Aktories K (1994) Cystein-specific ADP-ribosylation of actin. Eur J Biochem 221: 1047–1054

Kerr JFR, Wyllie AH, Currie AR (1972) Apoptosis: a basic biological phenomenon with wide-ranging implications in tissue kinetics. Br J Cancer 26: 239–257

Kitajima I, Kawahara K, Nakajima T, Soejima Y, Matsuyama T, Maruyama I (1994) Nitric oxide-mediated apoptosis in murine mastocytoma. Biochem Biophys Res Commun 204: 244–251

Knowles RG, Moncada S (1994) Nitric oxide synthases in mammals. Biochem J 298: 249–258

Kots AY, Skurat AV, Sergienko EA, Bulargina TV, Severin ES (1992) Nitroprusside stimulates the cysteine-specific mono(ADP-ribosylation) of glyceraldehyde-3-phosphate dehydrogenase from human erythrocytes. FEBS Lett 300: 9–12

Kröncke K-D, Brenner H-H, Rodriguez M-L, Etzkorn K, Noack EA, Kolb H, Kolb-Bachofen V (1993) Pancreatic islet cells are highly susceptible towards the cytotoxic effects of chemically generated nitric oxide. Biochim Biophys Acta 1182: 221–229

Lee HC (1994) Cyclic ADP-ribose: a new member of super family of signalling cyclic nucleotides. Cell Signal 6: 591–600

Lincolin TM, Cornwell TL (1993) Intracellular cyclic GMP receptor proteins. FASEB J 7: 328–338

Lipton SA, Choi Y-B, Pan Z-H, Lei SZ, Chen H-SV, Sucher NJ, Loscalzo J, Singel DJ, Stamler JS (1993) A redox-based mechanism for the neuroprotective and neurodestructive effects of nitric oxide and related nitroso-compounds. Nature 364: 626–632

Lowenstein CJ, Snyder SH (1992) Nitric oxide, a novel biologic messenger. Cell 70: 705–707

Marletta MA (1994) Nitric oxide synthase: aspects concerning structure and catalysis. Cell 78: 927–930

McDonald B, Reep B, Lapetina EG, Molina y Vedia L (1993) Glyceraldehyde-3-phosphate dehydrogenase is required for the transport of nitric oxide in platelets. Proc Natl Acad Sci USA 90: 11122–11126

McDonald LJ, Moss J (1993) Stimulation by nitric oxide of an NAD linkage to glyceraldehyde-3-phosphate dehydrogenase. Proc Natl Acad Sci USA 90: 6238–6241

Meßmer UK, Brüne B (1994) Modulation of inducible nitric oxide synthase in RINm5F cells. Cell Signal 6: 17–24

Meßmer UK, Ankarcrona M, Nicotera P, Brüne B (1994) p53 expression in nitric oxide-induced apoptosis. FEBS Lett 355: 23–26

Mitchell HH, Shonle HA, Grindley HS (1916) The origin of the nitrates in the urine. J Biol Chem 24: 461–490

Mohr S, Stamler JS, Brüne B (1994) Mechanism of covalent modification of glyceraldehyde-3-phosphate dehydrogenase at its active site thiol by nitric oxide, peroxynitrite and related nitrosating agents. FEBS Lett 348: 223–227

Molina y Vedia L, McDonald B, Reep B, Brüne B, DiSilvio M, Billiar TR, Lapetina EG (1992) Nitric oxide-induced S-nitrosylation of glyceraldehyde-3-phosphate dehydrogenase inhibits enzymatic activity and increases endogenous ADP-ribosylation. J Biol Chem 267: 24929–24932

Nathan C (1992) Nitric oxide as a secretory product of mammalian cells. FASEB J 6: 3051–3064

Nathan C, Xie Q-W (1994a) Nitric oxide synthases: roles, tolls, and controls. Cell 78: 915–918

Nathan C, Xie Q-W (1994b) Regulation of biosynthesis of nitric oxide. J Biol Chem 269: 13725–13728

Noack E, Murphy M (1991) Vasodilation and oxygen radical scavenging by nitric oxide/EDRF and organic nitrovasodilators. In: Sies H (ed) Oxidative stress; oxidants and antioxidants. Academic, San Diego, pp 445–489

Nussler AK, Billiar TR (1993) Inflammation, immunoregulation, and inducible nitric oxide synthase. J Leukoc Biol 54: 171–178

Orrenius S, McConkey D, Jones DP, Nicotera P (1988) Ca^{2+}-activated mechanisms in toxicity and programmed cell death. ISI atlas of science: pharmacology, pp 318–324

Palmer RMJ, Ashton DS, Moncada S (1988) Vascular endothelial cells synthesize nitric oxide from l-arginine. Nature 333: 664–666

Palmer RMJ, Ferrige AG, Moncada S (1987) Nitric oxide release accounts for the biological activity of endothelium-derived relaxing factor. Nature 327: 524–526

Pancholi V, Fischetti VA (1993) Glyceraldehyde-3-phosphate dehydrogenase on the surface of group A streptococci is also an ADP-ribosylating enzyme. Proc Natl Acad Sci USA 90: 8154–8158

Radi R, Beckman JS, Bush KM, Freeman BA (1991) Peroxynitrite oxidation of sulfhydryls. J Biol Chem 266: 4244–4250

Radons J, Heller B, Bürkle A, Hartmann B, Rodriguez M-L, Kröncke K-D, Burkart V, Kolb H (1994) Nitric oxide toxicity in islet cells involves poly(ADP-ribose) polymerase activation and concomitant NAD^+ depletion. Biochem Biophys Res Commun 199: 1270–1277

Ravichandran V, Seres T, Moriguchi T, Thomas JA, Johnston RB Jr (1994) S-Thiolation of glyceraldehyde-3-phosphate dehydrogenase induced by the phagocytosis-associated respiratory burst in blood monocytes. J Biol Chem 269: 25010–25015

Reddy D, Lancaster JR Jr, Cornforth DP (1983) Nitrite inhibition of *Clostridium botulinum*: electron spin resonance detection of iron-nitric oxide complexes. Science 221: 769–770

Reinhard M, Halbrügge M, Scheer U, Wiegand C, Jockusch BM, Walter U (1992) The 46/50-kDa phosphoprotein VASP purified from human platelets is a novel protein associated with actin filaments and focal contacts. EMBO J 11: 2063–2070

Richter C, Frei B (1988) Calcium release from mitochondria induced by prooxidants. Free Radic Biol Med 4: 365–375

Sarih M, Souvannavong V, Adam A (1993) Nitric oxide induces macrophage death by apoptosis. Biochem Biophys Res Commun 191: 503–508

Schmidt HHHW (1992) NO, CO and OH endogenous soluble guanylyl cyclase-activating factors. FEBS Lett 307: 102–107

Schmidt HHHW, Walter U (1994) NO at work. Cell 78: 919–925

Schultz K-D, Schultz K, Schultz G (1977) Sodium nitroprusside and other smooth muscle relaxants increase cyclic GMP levels in rat ductus deferens. Nature 265: 750–751

Schulz R, Nava E, Moncada S (1992) Induction and potential biological relevance of a calcium-independent nitric oxide synthase in the myocardium. Br J Pharmacol 105: 575–580

Schuppe-Koistinen I, Moldeus P, Bergman T, Cotgrave IA (1994) S-thiolation of human endothelial cell glyceraldehyde-3-phosphate dehydrogenase after hydrogen peroxide treatment. Eur J Biochem 221: 1033–1037

Schwartzman RA, Cidlowski JA (1993) Apoptosis: the biochemistry and molecular biology of programmed cell death. Endocr Rev 14: 133–151

Snyder SH (1992) Nitric oxide: first in a new class of neurotransmitters? Science 257: 494–496

Stamler JS (1994) Redox signaling: nitrosylation and related target interactions of nitric oxide. Cell 78: 931–936

Stamler JS, Singel DJ, Loscalzo J (1992) Biochemistry of nitric oxide and its redox-activated forms. Science 258: 1898–1902

Stuehr DJ, Marletta MA (1985) Mammalian nitrate biosynthesis: mouse macrophages produce nitrite and nitrate in response to *Escherichia coli* lipopolysaccharide. Proc Natl Acad Sci USA 82: 7738–7742

Tannenbaum SR, Fett D, Young VR, Land PD, Bruce WR (1978) Nitrite and nitrate are formed by endogenous synthesis in the human intestine. Science 200: 1487–1489

Tao Y, Howlett A, Klein C (1992) Nitric oxide stimulates the ADP-ribosylation of a 41-kDa cytosolic protein in *Dictyostelium discoideum*. Proc Natl Acad Sci USA 89: 5902–5906

Tao Y, Howlett AC, Klein C (1993) Endogenous ADP-ribosylation of glyceraldehyde-3-phosphate dehydrogenase that is not regulated by nitric oxide in *Dictyostelium discoideum*. Cell Signal 5: 763–775

Tao Y, Howlett A, Klein C (1994) Nitric oxide regulation of glyceraldehyde-3-phosphate dehydrogenase activity in *Dictyostelium discoideum* cells and lysates. Eur J Biochem 224: 447–454

Tremblay J, Gerzer R, Hamet P (1988) Cyclic GMP in cell function. Adv Second Messenger Phosphoprotein Res 22: 319–383

Vaidyanathan VV, Sastry PS, Ramasarma T (1993) Inverse relationship of the dehydrogenase and ADP-ribosylation activities in sodium-nitroprusside-treated glyceraldehyde-3-phosphate dehydrogenase is coincidental. Biochim Biophys Acta 1203: 36–44

Verma A, Hirsch DJ, Glatt CE, Ronnett GV, Snyder SH (1993) Carbon monoxide, a putative neural messenger. Science 259: 381–384

Vezzani A, Sparvoli S, Rizzi M, Zinetti M, Fratelli M (1994) Changes in the ADP-ribosylation status of some hippocampal proteins are linked to kindling progression. Neuroreport 5: 1217–1220

Waldman SA, Murad F (1987) Cyclic GMP synthesis and function. Pharmacol Rev 39: 163–196

Walter U (1989) Physiological role of cGMP and cGMP-dependent protein kinase in the cardiovascular system. Rev Physiol Biochem Pharmacol 113: 41–88

Williams MB, Li X, Gu X, Jope RS (1992) Modulation of endogenous ADP-ribosylation in rat brain. Brain Res 592: 49–52

Wink DA, Kasprzak KS, Maragos CM, Elespuru RK, Misra M, Dunams TM, Cebula TA, Koch WH, Andrews AW, Allen JS, Keefer LK (1991) DNA-deaminating ability and genotoxicity of nitric oxide and its progenitors. Science 254: 1001–1003

Yau K-W (1994) Cyclic nucleotide-gated channels: an expanding new familiy of ion channels. Proc Natl Acad Sci USA 91: 3481–3483

Zhang J, Snyder SH (1992) Nitric oxide stimulates auto-ADP-ribosylation of glyceraldehyde-3-phophate dehydrogenase. Proc Natl Acad Sci USA 89: 9382–9385

Zhang J, Dawson VL, Dawson TM, Snyder SH (1994) Nitric oxide activation of poly(ADP-ribose) synthetase in neurotoxicity. Science 263: 687–689

Zhuo M, Small SA, Kandel ER, Hawkins RD (1993) Nitric oxide and carbon monoxide produce activity-dependent long-term synaptic enhancement in hippocampus. Science 260: 1946–1950

Zocchi E, Guida L, Franco L, Silvestro L, Guerrini M, Benatti U, de Flora A (1993) Free ADP-ribose in human erythrocytes: pathways of intra-erythrocytic conversion and non-enzymatic binding to membrane proteins. Biochem J 295: 121–130

Editor-in-Charge: Professor H. Grunicke and Professor M. Schweiger

Rev. Physiol. Biochem. Pharmacol., Vol. 127

Chemistry and Pathophysiology of Oxidation of LDL

H. ESTERBAUER and P. RAMOS

Contents

1 Introduction

The past decade has seen a series of remarkable studies suggesting that oxidation of low-density lipoprotein (LDL) might be a risk factor in atherosclerosis. Many reviews published recently in journals (Juergens et al. 1987; Steinberg et al. 1989; Steinbrecher et al. 1990; Bruckdorfer 1990; Lyons 1991; Carpenter et al. 1991; Esterbauer et al. 1992; Leake 1993; Esterbauer and Juergens 1993; Aviram 1993; Halliwell 1993; Chait and Heinecke 1994) and in books (Gebicki et al. 1991; Haberland and Steinbrecher 1992; Chisolm 1992) reflect the considerable interest in this new concept. Early atherosclerotic lesions are characterized by massive accumulation of lipid-laden foam cells in the subendothelial space of arteries. Most of the foam cells are derived from monocyte-macrophages, and much of the interest in oxidized LDL (oLDL) stems from the discovery that it exhibits in vitro properties which could explain the immigration of monocytes into the arterial wall, their differentiation into resident macrophages, and their conversion to foam cells. Most significant in this respect is the fact that oLDL bypasses the normal tight control exercised by the classical LDL receptor but is avidly endocytosed via the scavenger receptor pathway of macro-

Institute of Biochemistry, University of Graz, Schubertstr. 1, 8010 Graz, Austria

phages. It may also be very significant that oLDL contains highly cytotoxic lipid peroxidation products; the release of such diffusible toxins from oLDL deposited in the arterial wall would be a constant irritant for the endothelial cell layer and would provoke a number of other deleterious effects, such as endothelial cell death, platelet aggregation, release of growth factors, disturbances of eicosanoid homeostasis, accumulation of inflammatory cells, and increased infiltration of LDL. Inhibition by oLDL of endothelial-derived relaxing factor (EDRF)-mediated relaxation of smooth muscle cells, the immunogenicity of oLDL, and the stimulation (preferentially by minimally oxidized LDL) of endothelial cells to express and/or release a number of

Table 1. Biological effects of minimally (MM-LDL) and extensively oxidized LDL

1. Unregulated uptake by macrophages, induction of foam cell formation (reviewed in Steinberg et al. 1989).
2. Cytotoxic to fibroblasts, EC and SMC (reviewed in Chisolm 1992).
3. Chemotactic for monocyte (Quinn et al. 1987) and SMC (Autio et al. 1990).
4. MM-LDL stimulates release of monocyte-chemotactic protein-1 (MCP-1) from EC (Cushing et al. 1990).
5. MM-LDL increases adhesion of monocytes to EC (Berliner et al. 1990; Kim et al. 1994).
6. Systemic administration into hamster initiates immmediate leukocyte adhesion to capillary endothelium (Lehr et al. 1991).
7. MM-LDL stimulates expression of colony stimulating factors for monocytes (MCSF) and granulocytes (GCSF) by EC (Rajavashisth et al. 1990).
8. MM-LDL injected into mice increases serum and tissue levels of MCP-1 and CSF (Liao et al. 1991).
9. Increases tissue factor (TF) expression by cultured EC and suppresses protein C (Drake et al. 1991; Weis et al. 1991).
10. Increases EC expression of plasminogen activator inhibitor-1 (Kugiyama et al. 1993).
11. Inhibits production of platelet-derived growth factor (PDGF) by monocyte-macrophages (Malden et al. 1991) and endothelial cells (Fox et al. 1987).
12. Increases PDGF expression by smooth muscle cells (Zwijsen et al. 1992a; Stiko-Rahm et al. 1992).
13. Inhibits tumor necrosis factor (TNF) expression by monocyte-macrophages (Hamilton et al. 1990).
14. Stimulates release of interleukin-1 beta from monocyte-macrophages (Thomas et al. 1994a).
15. Induces in monocytic cells expression of interleukin-8 and activates T-lymphocytes (Frostegard et al. 1992; Terkeltaub et al. 1994).
16. Stimulates (low concentration) or inhibits prostacyclin production by smooth muscle cells (Zhang et al. 1990; Daret et al. 1993).
17. Induces increased expression of stress proteins by macrophages (Yamaguchi et al. 1993).
18. Inhibits endothelial cell dependent arterial relaxation (Ohgushi et al. 1993) and inhibits NO activation of guanylate cyclase (Schmidt et al. 1990).
19. Is immunogenic and induces formation of antibodies (Palinski et al. 1989).
20. Reduces motility of macrophages and may thus inhibit egression of macrophages and foam cells from the arterial lesion (Quinn et al. 1987).

EC, Endothelial cells; *SMC*, smooth muscle cells.

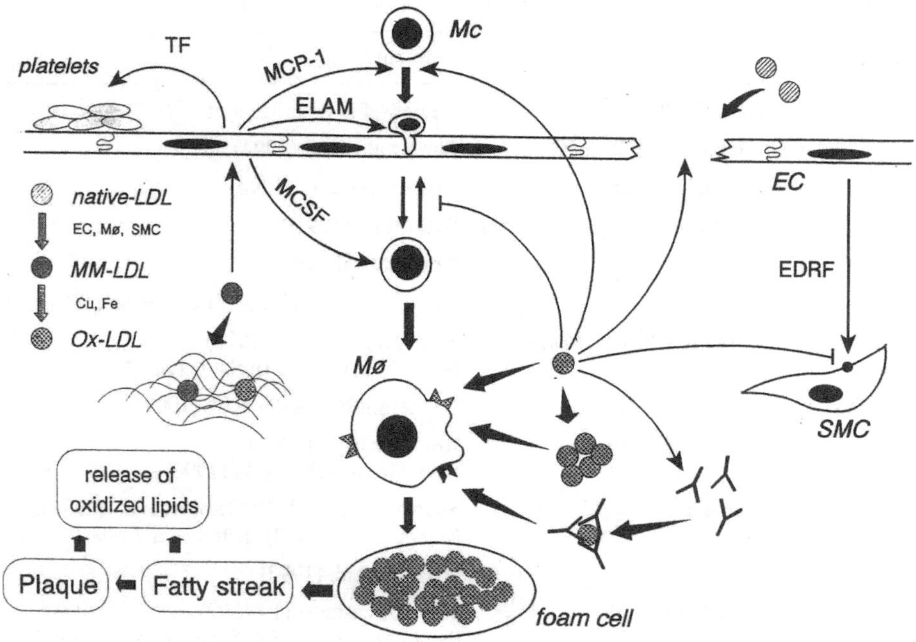

Fig. 1. Hypothetical scenario of events triggered by mildly and extensively oxidized LDL in the arterial wall. Mildly oxidized LDL (*MM-LDL*) mediates recruitment of monocyte-macrophages (*Mc*) in the subendothelial space by stimulation of endothelial cells (*EC*) to release monocyte chemotactic protein-1 (*MCP-1*), endothelial leukocyte adhesion molecule (*ELAM*), and monocyte colony stimulating factor (*MCSF*). Increased release of tissue factor (*TF*) can induce platelet aggregation. Uncontrolled uptake of extensively oxidized LDL (*Ox-LDL*) by macrophages (M_q) leads to foam cells and plaques. Aggregated oxidized LDL and oxidized LDL immunecomplexes are endocytosed via the scavenger receptor or by phagocytosis. Cytotoxic products of oxidized LDL cause endothelial cell injury, vasoconstriction, and immune response (*EDRF*, endothelial derived relaxing factor; *SMC*, smooth muscle cells)

biologically active factors such as monocyte chemotactic proteins, endothelial leukocyte adhesion molecules (ELAMS), and growth factors for monocytes also support the hypothesis that the atherogenicity of LDL increases when it becomes oxidized.

Furthermore, it has been shown that oLDL can activate T lymphocytes in atherosclerotic lesions (Frostegard et al. 1992) and stimulate proliferation of smooth muscle cells by inducing the expression of the gene coding for the A-chain of platelet-derived growth factor (Zwijsen et al. 1992a). Table 1 lists biological effects of minimally (or mildly) and extensively oxidized LDL that have so far been reported. Taken altogether, these data on functional and biological properties of oxidized LDL strongly support the hypothesis of its atherogenic role. Fig. 1 shows a hypothetical scenario of events initiated by oLDL in the arterial wall.

Table 2. Prooxidants mediating oxidation of LDL in absence of cells

System	Reference
Cu^{++} ions ($CuSO_4$, $CuCl_2$)	Reviewed in Esterbauer et al. (1992)
Ceruloplasmin (native, with 7 Cu^{++})	Ehrenwald et al. (1994)
Iron in transferrin (acidic pH)	Lamb and Leake (1994)
Hem in presence of H_2O_2 or lipid peroxides	Balla et al. (1991)
Peroxynitrite	Graham et al. (1993)
Hypochlorite	Panasenko et al. (1994); Hazell et al. (1994)
Peroxyradicals from AAPH or AMVN	Bowry and Stocker (1993); Noguchi et al. (1993); Sato et al. (1990)
Defined oxygen radicals (OH, HO_2)	Bonnefont-Rousselot et al. (1993)
Thiols in presence of Cu^{2+} or Fe^{3+}	Heinecke et al. (1993); Sparrow and Olszewski (1993)
Lipoxygenases (sojabean Lox, 15-Lox)	Sparrow et al. (1988); Cathcart et al. (1991); Belkner et al. (1993); Kühn et al. (1994)
Horseradish peroxidase + H_2O_2	Wieland et al. (1993)
Myeloperoxidase + H_2O_2 * Cl^-	Stelmaszynska et al. (1992); Panasenko et al. 1994; Savenkova et al. (1994)
Cholesterol oxidase	Aviram (1992)
UV-B-light	Salmon et al. (1991); Nègre-Salvayre and Salvayre (1992)

LDL oxidation can be initiated in vitro by its incubation with macrophages, endothelial cells, smooth muscle cells, or lymphocytes (for review see Gebicki et al. 1991; Chait and Heinecke 1994) or in a cell-free system utilizing a variety of prooxidants such as lipoxygenase, myeloperoxidase, defined oxygen radicals, UV light, gamma irradiation, heme, copper ions, or hypochlorous acid (Table 2). Traces of transition metals in free form or in redox-active complexes are generally agreed to be essential for producing oLDL with the properties described above. The mechanism of initiation and progression of LDL oxidation in vivo is largely a matter of speculation. It is believed to occur not in the circulation but within the arterial wall itself, where LDL is sequestered by proteoglycans and other extracellular matrix constituents (Steinberg et al. 1989; Haberland and Steinbrecher 1992). LDL isolated from the arterial wall shares some functional and biological properties with LDL oxidized in vitro (reviewed in Esterbauer et al. 1992).

It is one thing to demonstrate altered biological properties of oLDL, but it is quite another thing to study the complex chemistry of LDL oxidation, to analyze the structure of the large number of oxidation products, and to identify the substances in oLDL responsible for the various biological effects. This topic, although of central importance for the oxidation hypothesis, has so far not received the broad attention it would deserve, and thus the nature of the "atherogenic" compounds in oLDL is largely unknown.

Oxidation of LDL is a lipid peroxidation chain reaction driven by free radicals. As such, LDL oxidation possesses the general characteristics of lipid peroxidation reactions and free radical reactions. What makes the process and its dynamics so complex is the fact that all components of LDL, i.e., antioxidants, phospholipids, cholesteryl ester, triglycerides, and apolipoprotein B, participate at certain stages, leading to multiple secondary and tertiary reactions.

This contribution focuses mainly on in vitro oxidation of LDL by Cu^{++} ions. It is generally believed that LDL oxidized by copper ions exhibits biological properties very similar, if not identical, to those of cell-oxidized LDL.

2 Composition of Native LDL

Human LDL is defined as the population of lipoproteins which can be isolated from plasma by ultracentrifugation within a density gradient of 1.019–1.063 g/ml. LDL molecules are large spherical particles with a diameter of 19–25 nm and molecular weights between 1.8 and 2.8 million, with an average of 2.5 million. The mean chemical composition (weight%) deduced from various reports (reviewed in Esterbauer et al. 1992) is 22.3% phospholipids, 5.9% triglycerides, 9.6% free cholesterol, 42.2% cholesteryl ester, and 22.0% protein. The lipid and fatty acid composition is shown in Table 3. The mean total cholesterol content is 34.7%. Based on a molecular weight of 2.5 million, each LDL particle would contain about 1600 molecules of cholesteryl ester and 170 molecules of triglycerides, which form a central lipophilic core. The core is surrounded by a monolayer of about 700 phospholipid molecules and 600 molecules of free cholesterol. The main phospholipids are phosphatidylcholine (63%) and sphingomyelin (26%). Of importance for the oxidation resistance of LDL might be the plasmalogen content. The major one is ethanolamine plasmalogen, which was reported to act as an antioxidant (Vance 1990; Engelmann et al. 1994). The total amount of fatty acids in an LDL molecule is roughly 2600, and about one half of them are polyunsaturated fatty acids (86% linoleic acid 18:2, 12% arachidonic acid 20:4, 2% docosahexaenoic acid 22:6). The standard deviations in Table 3 indicate a rather strong variation in fatty acid distribution, which might be significant for the variation observed in the oxidation resistance of LDL.

Table 3. Lipid composition and individual fatty acids in native LDL. [Compiled from data in Esterbauer et al. (1992) and references given in the footnotes]

	nmol/mg LDL protein mean	±SD	mol/mol LDL mean
Total phospholipids	1300	±227	700
Phosphatidylcholine	818		450
Phosphatidylethanolamine	19		10
Lysophosphatidylcholine [a]	30		16
Sphingomyelin	336		185
Ethanolamine plasmalogen [b]	43		24
Choline plasmalogen [b]	4		2
Triglycerides	304	±140	170
Free cholesterol	1130	± 82	600
Cholesteryl ester	2960	±220	1600
Total cholesterol	4090		2200
Free fatty acids	48		26
Palmitic acid	1260	±375	693
Palmitoleic acid	80	± 4	44
Stearic acid	260	±118	143
Oleic acid	825	±298	454
Linoleic acid	2000	±541	1100
Arachidonic acid	278	±100	153
Docosahexaenoic acid	53	± 31	29
Total fatty acids	4756		2616
Total PUFAS	2330		1280

[a] Jougasaki et al. (1992)
[b] Sommer et al. (1992)

The antioxidants contained in LDL are listed in Table 4. On a molar base, the amount of α-tocopherol, by far the major one, is 11.58 nmol/mg protein, equal to about 6 molecules per LDL particle. Ethanolamine plasmalogen amounts to 24 molecules/LDL, but it remains to be determined whether it indeed shows chain-breaking antioxidant activity in LDL. All other compounds with potential antioxidant activity such as carotenoids and ubiquinol are present in much smaller amounts than α-tocopherol. The reported ubiquinol-10 content varies widely from 0.1 to 1.0 mol/mol LDL. A number of studies by Stocker's group (Stocker et al. 1991; Mohr et al. 1992; Bowry and Stocker 1993) revealed that ubiquinol protects LDL more efficiently than α-tocopherol against oxidation. Tribble et al. (1994) showed that dense LDL, which is more susceptible to oxidation than buoyant LDL, contains

Table 4. Antioxidants in native LDL. [Data in upper part of the table are from Esterbauer et al. (1992); the ubiquinol data are from the references given in the footnotes]

Antioxidant	nmol/mg LDL protein mean	±SD	mol/mol LDL mean
α-Tocopherol	11.58	±3.34	6.37
γ-Tocopherol	0.93	±0.36	0.51
β-Carotene	0.53	±0.47	0.29
α-Carotene	0.22	±0.25	0.12
Lycopene	0.29	±0.20	0.16
Cryptoxanthine	0.25	±0.23	0.14
Cantaxanthine	0.04	±0.07	0.02
Lutein + zeaxanthine	0.07	±0.05	0.04
Ubiquinol-10	0.18	±0.18	0.10
ubiquinol-10 (buoyant LDL) [a]	0.24	±0.14	0.12
ubiquinol-10 (dense LDL) [a]	0.14	±0.12	0.07
ubiquinol-10 [b]	0.64	±0.28	0.32
ubiquinol-10 + ubiquinone-10 [c]	1.08	±0.30	0.54
ubiquinol-10 [d]	1.0 – 2.0		0.5 – 1.0

[a] Tribble et al. (1994) (*n*= 8)
[b] Frei and Gaziano (1993) (*n*= 62)
[c] Kontush et al. (1994) (*n*= 20)
[d] Stocker et al. (1991); Bowry and Stocker (1993); Mohr et al. (1992)

significantly less ubiquinol. About 85% of ubiquinol-10 is in the reduced form; the remaining part is ubiquinone-10.

An excellent recent article by Yang and Pownall (1993) reviewed the structure and function of the apolipoprotein B of LDL. Apo B is one of the largest monomeric proteins known; the single polypeptide chain contains 4536 amino acids, with a calculated molecular weight of 512.937 daltons. The number of amino acid residues per apo B are Ala 266, Asp plus Asn 478, Arg 148, Cys 25, Glu plus Gln 529, Gly 207, His 115, Ile 288, Leu 523, Lys 356, Met 78, Phe 223, Pro 169, Ser 393, Thr 298, Trp 37, Tyr 152, and Val 251. The apo B is glycosylated, and the carbohydrate content can amount to 9–10 weight% of apo B, with galactose, mannose, N-acetylglucosamine, and sialic acid residues. Camejo et al. (1985) reported that the sialic acid content is important for the sequestering of LDL by proteoglycans in the arterial wall. Of the 25¯ cysteine residues, it is likely that seven have the free sulfhydryl group, whereas the rest form disulfide bonds. Two of the sulfhydryl groups are exposed to the LDL surface and they could play a role in reductive activation of transition metal ions, e.g., Cu^{++}+ RSH→Cu^{+}+ 1/2 RSSR.

A final important fact is that LDL with the buoyant density 1.019–1.063 g/ml consists of subfractions differing in size, molecular weight, density, and composition (Kraus 1991; Esterbauer et al. 1992; Campos et al. 1992). Most individuals (about 75%) belong to phenotype A, characterized by an LDL profile with predominance of larger and less dense LDL (1.025–1.038 g/ml). Individuals with phenotype B (about 25% of a population) have an LDL profile with predominance of smaller and denser LDL (d 1.038 g/ml). Epidemiological studies have shown that phenotype B is associated with an increased risk of myocardial infarction and coronary artery disease. A possible explanation for the higher atherogenicity of the dense LDL predominant in phenotype B might be that it is more susceptible to oxidation than the less dense LDL (De Graaf et al. 1991; Tribble et al. 1992).

3 Principles of Lipid Peroxidation

Oxidation of LDL is a free radical-driven lipid peroxidation process. It starts with the removal by an initiating radical $X^{\bullet}$ of a hydrogen atom from one of the polyunsaturated fatty acids (PUFAS) contained in the LDL lipids, according to reaction 1 and Fig. 2b (note that LH in reactions is a PUFA bound to one of the LDL lipids). The rate of hydrogen removal (Ri) determines the rate of initiation. This initiation is one of the key steps. However, despite intensive work, the nature of the short-lived primary radical $X^{\bullet}$ is still a mystery in most in vitro systems (e.g., copper-mediated oxidation) and even more in vivo. Once formed, the carbon-centered lipid radical $L^{\bullet}$ reacts very quickly with molecular oxygen, yielding a lipid peroxyl radical $LOO^{\bullet}$ (reaction 2). The $LOO^{\bullet}$ radical, in turn, abstracts a hydrogen atom from an adjacent lipid LH, yielding a lipid hydroperoxide LOOH and a new lipid radical $L^{\bullet}$ (reaction 3). This reaction is termed chain propagation and proceeds with a given rate constant k_p.

An interesting feature of such chain reactions is that a single initiating event (reaction 1) could convert a large number of lipids to lipid hydroperoxides. The number of lipid molecules oxidized per one initiated radical $X^{\bullet}$ depends on several factors, especially on the presence of antioxidants (reaction 4) and the rate of chain termination, when two $LOO^{\bullet}$ radicals combine to form nonradical products (reaction 5). The nonradical products (NRP) produced in reaction 5 by the Russel mechanism are ketones and secondary alcohols with conjugated double bonds, singlet oxygen, and light (Fig. 2b). If the system contains antioxidants, for example vitamin E (TOH), the $LOO^{\bullet}$ radical can be scavenged according to reaction 4, with the consequence that the chain is terminated,

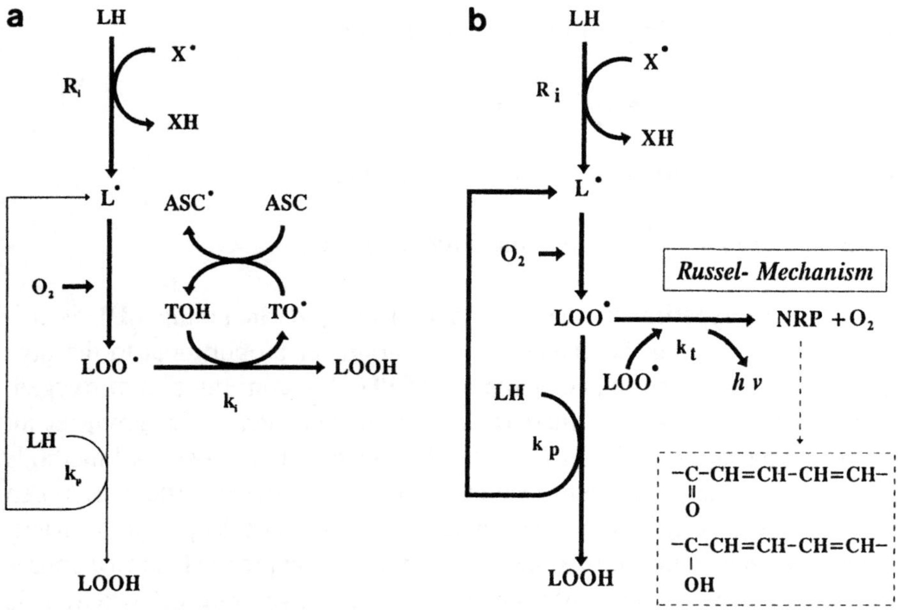

Fig. 2a,b. Major reaction steps during the inhibited (**a**) and uninhibited (**b**) phase of LDL oxidation. *LH,* a lipid (e.g., cholesteryl ester or phospholipid) containing an oxidizable polyunsaturated fatty acid; *LOO*$^•$, a lipid peroxyl radical; *LOOH,* a lipid hydroperoxide (e.g., cholesteryl-13-hydroperoxy-octadecadienoic acid); *TOH, TO*$^•$, tocopherol and tocopheroxyl radical; *ASC, ASC*$^•$, ascorbate and ascorbyl radical.
a The antioxidants TOH and ASC inhibit chain propagation by scavenging LOO radicals. LH is converted only slowly to LOOH.
b Uninhibited lipid peroxidation at a fast rate commences when the LDL is depleted of antioxidants. Note that this is a chain reaction. The structure of the nonradical products (*NRP*) produced by recombination of two LOO radicals is also shown

because the tocopheroxyl radical $TO^•$ has very low reactivity and, under most conditions, does not propagate the lipid peroxidation chain. If ascorbate is present, it recycles vitamin E by reducing the $TO^•$ radical to TOH (Fig. 2a).

reaction 1: initiation

$$LH + X^• \xrightarrow{\text{rate= Ri}} L^• + XH$$

reaction 2: oxygen addition

$$L^• + O2 \xrightarrow{\text{rate very fast}} OO^•$$

reaction 3: chain propagation

$$LOO^• + LH \xrightarrow{kp} LOOH + L^•$$

reaction 4: scavenging of $LOO^{\bullet}$ by antioxidants

$$LOO^{\bullet} + TOH \xrightarrow{ki} LOOH + TO^{\bullet}$$

reaction 5: termination via the Russel mechanism

$$LOO^{\bullet} + LOO^{\bullet} \xrightarrow{kt} \text{nonradical products} + O_2$$

In typical experiments the autoxidation of lipids including LDL is followed by measuring the change of one parameter proportional to the progress of the reaction, i.e., consumption of PUFAS, consumption of oxygen, increase of peroxides, or increase of conjugated dienes (Cosgrove et al. 1987; Niki 1987; Esterbauer et al. 1992; Noguchi et al. 1993). All methods give more or less equivalent results, but chemists prefer methods (oxygen uptake, dienes) which allow a continuous monitoring of the progress curves. If the system contains antioxidants, the autoxidation proceeds in two consecutive phases with quite different rates. Initially, the rate of oxidation is slow, because the antioxidants scavenge $LOO^{\bullet}$ radicals (reaction 4) and consequently compete with the propagation. This initial phase is termed lag time (t_{inh}) or inhibition period. The reaction sequence in the presence of antioxidants is shown in Fig. 2a. As the reaction proceeds, the antioxidants are consumed with time and the rate of the competing reaction slows down. As a consequence, the rate of propagation accelerates until a maximum rate of the uninhibited autoxidation is reached. The reaction sequence in the absence of antioxidant is shown in Fig. 2b. The rate of autoxidation during the lag phase is determined by Eq. 1; the length of the lag phase is directly proportional to the concentration of the antioxidants TOH (Eq. 2), and the maximal rate under uninhibited conditions (i.e., when no antioxidants are present) is described by Eq. 3.

$$V_{inh} = \frac{d\,[LOOH]}{dt} = \frac{k_p\,[LH]\,R_i}{n\,k_{inh}\,[TOH]} \quad (1)$$

$$t_{inh} = \text{length of lag-phase} = \frac{n\,[TOH]}{R_i} \quad (2)$$

$$V_{max} = \frac{d\,[LOOH]}{dt} = \frac{k_p\,[LH]\,R_i^{1/2}}{(2k_t)^{1/2}} \quad (3)$$

In Eq. 2 the factor n is defined as the number of peroxyl radicals $LOO^{\bullet}$ trapped by each molecule of antioxidants. For vitamin E the value n is 2, since both vitamin E and vitamin E radical (tocopheroxyl radical) trap

$LOO^{\bullet}$. Equation 2 also shows that the length of the lag phase is inversely proportional to the rate R_i, by which the initiating radicals are formed.

An intrinsic problem in determining rate constants in lipid peroxidation is the uncertainty about the rate of initiation R_i, and it is clear that without knowing R_i the absolute rate constants cannot be obtained. One possibility of overcoming this problem is to introduce into the reaction mixture a compound which decomposes at a constant rate to free radicals ($X^{\bullet}$) capable of abstracting a hydrogen atom from the PUFAS according to reaction 1 and consequently initiating the autoxidation process. The compounds most frequently used for this are so-called azo-initiators (X-N=N-X), which thermally decompose to highly reactive carbon-centered radicals ($R^{\bullet}$); these, in turn, react rapidly with oxygen to give peroxylradicals ($ROO^{\bullet}$) which abstract hydrogen atoms from lipids (reaction 6).

reaction 6: $R-N=N-R \xrightarrow{k_d} 2R^{\bullet} + N_2$

$$R^{\bullet} + O_2 \xrightarrow{\text{fast}} ROO^{\bullet}$$

$$LH + ROO^{\bullet} \xrightarrow{\text{rate } R_i} L^{\bullet} + ROOH$$

The water-soluble azo-initiator AAPH, 2,2'-azobis-(2-amidinopropane) dihydrochlorid, can be used for producing radicals in the aqueous phase, whereas the lipid-soluble AMVN, 2,2'-azobis-(2,4-dimethylvaleronitrile), can be used to produce radicals in the lipid phase. AAPH decomposes with a first-order rate constant of $k_d = 6.6 \times 10^{-5}$ min^{-1} at 37°C, and the flux of free radicals is directly proportional to the AAPH concentration.

A crucial point is that the rate of initiation (R_i) by AAPH additionally depends on the initiator efficacy (e), that is, the number of primordial radicals ($X^{\bullet}$) that initiate according to the following formula: $R_i = (AAPH) \, . \, 2 \, k_d \, . \, e$.

The absolute measurement of the initiator efficacy by independent methods is extremely difficult; therefore, R_i is generally indirectly measured by the so-called induction period method, based on Eq. 2. Briefly, a defined amount of α-tocopherol or the water-soluble analog trolox is added to the sample to be oxidized. The oxidation is initiated by AAPH, and the length of the lag phase (t_{inh}) is measured. The R_i value is then assumed to be: $R_i = 2(\alpha\text{-tocopherol})/t_{inh}$. In the case of samples containing endogenous vitamin E (e.g., LDL) its initial concentration is used as an internal calibration standard for R_i measurements (Mino et al. 1989; Sato et al. 1990; Noguchi et al. 1993).

As stated above, AAPH has the advantage that any desired rate of initiation can be adjusted by variation of the AAPH concentration. For example (Noguchi et al. 1993), 1 m*M* AAPH solution gives a rate of initiation (R_i) of about 1.8×10^{-8} M min^{-1}. If the solution contains LDL in a concentration of 0.5×10^{-6} M, a concentration frequently used in AAPH oxidation studies (Noguchi et al. 1993), the rate of initiation per LDL particle is only 0.036 strikes/min, or, in other words, an LDL particle is hit by a free radical only every 27 min. In a thorough kinetic study, Noguchi et al. (1993) recently investigated oxidation of LDL (0.5 μ*M*) with 1 m*M* AAPH. The reaction was followed by measuring vitamin E consumption, oxygen uptake, increase of dienes, and increase of cholesteryl ester hydroperoxides and phospholipid hydroperoxides. The reaction proceeded clearly in two distinct consecutive phases. An inhibited lag phase, during which vitamin E was consumed, and an uninhibited propagation phase. The kinetic chain length for formation of cholesteryl ester hydroperoxides was 2.9 during the lag phase and 7.5 during the propagation phase. For formation of phospholipid hydroperoxides and oxygen consumption the kinetic chain lengths during the lag phase and the propagation phase were 0.6/1.3 and 6.3/50, respectively. This clearly indicates that vitamin E acted as a chain-breaking antioxidant. Similar results were found for oxidation of LDL (0.5 μ*M*) with 2 μ*M* copper. On the other hand, Bowry and Stocker (1993) reported that the kinetic chain length in LDL oxidation by AAPH is higher in the presence of vitamin E than in its absence. They also reported that vitamin E acts as a prooxidant by chain transfer through the reaction of the tocopheroxyl radical ($TO^\bullet$) with a lipid molecule as follows: $TO^\bullet + LH \rightarrow TOH + L^\bullet$.

4 Kinetics of Copper-Induced LDL Oxidation

A large number of kinetic experiments (reviewed in Esterbauer et al. 1992) have shown that the chronology of LDL oxidation by Cu^{++} ions can be divided into three consecutive time phases: lag time (or lag phase), propagation phase, and decomposition phase (Fig. 3). During the lag phase, the LDL becomes progressively depleted of its antioxidants, with α-tocopherol as the first one and β-carotene as the last. During this period, only minimal lipid peroxidation occurs in LDL, as evidenced by the measurement of PUFAS, thiobarbituric acid-reactive substances (TBARS), lipid hydroperoxides, fluorescence, and conjugated dienes. Macrophage-mediated LDL oxidation shows the same time sequence. When the LDL is depleted of its antioxidants the lipid peroxidation rapidly accelerates to a maximum rate of the uninhibited process given by Eq. 3. A lipid peroxide peak is reached when about 70%–80% of the LDL PUFAS are oxidized; thereafter, the peroxide content of LDL starts to decrease again, because decomposition reactions (e.g., formation of aldehydes) become predominant. The kinetics of the formation of lipid peroxides, TBARS, and fluorescence at 430 nm (excitation 360 nm)

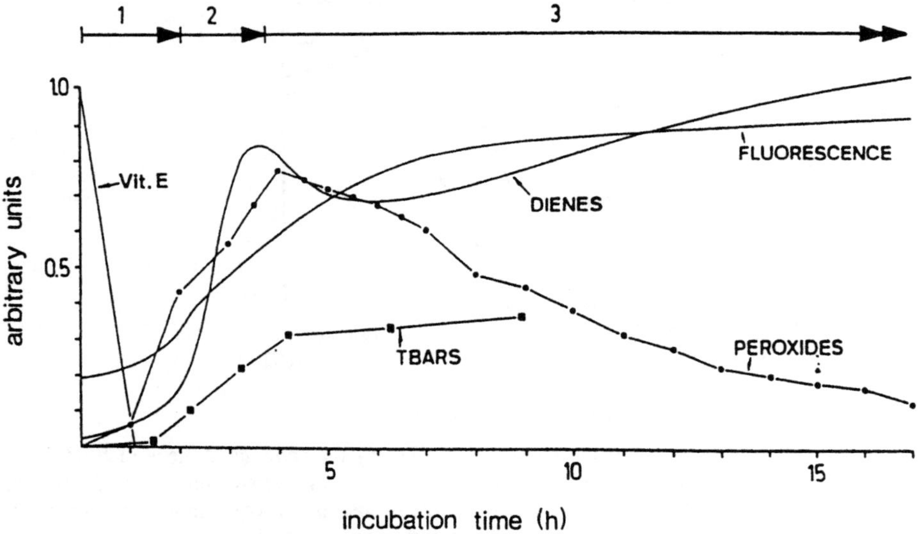

Fig. 3. Kinetics of copper-stimulated oxidation of LDL, measured by consumption of vitamin E, change of 430-nm fluorescence, lipid hydroperoxides, and TBARS. *1, 2, 3,* lengths of the lag, propagation and decomposition phases, respectively. (From Esterbauer et al. 1992)

during the lag and propagation phases closely follows the diene vs. time profile, and only after the diene maximum do the different indices separate and follow different kinetics. The second increase of the 234-nm absorption, seen shortly after the peroxide maximum, is due not to newly formed dienes, but to accumulation of decomposition products absorbing at this wavelength.

When PUFAS become oxidized to lipid hydroperoxides their isolated CC double bonds are converted to conjugated double bonds showing strong UV absorption at 234 nm. A convenient and very frequently used method for continuously monitoring the process of copper-induced LDL oxidation is to measure in a spectrophotometer the change of the diene absorption as a function of time, as first proposed by us (Esterbauer et al. 1989; Puhl et al. 1994). A spectrometer with an autocell holder allows the measurement of six or more LDL samples simultaneously, which has advantages for routine analyses. A typical example of such assays is shown in Fig. 4.

The indices which can easily be derived from the diene vs. time profiles are the lag phase (t_{inh}, Eq. 2), the maximum rate of oxidation (V_{max}, Eq. 3), and the maximum amount of dienes formed. If the diene vs. time profile is precisely measured, it should also be possible to determine the oxidation rates during the lag phase (Eq. 1) and the rate of peroxide decomposition. Many pitfalls may be encountered if conditions during copper oxidation are not strictly controlled. The most important factors are concentration of LDL

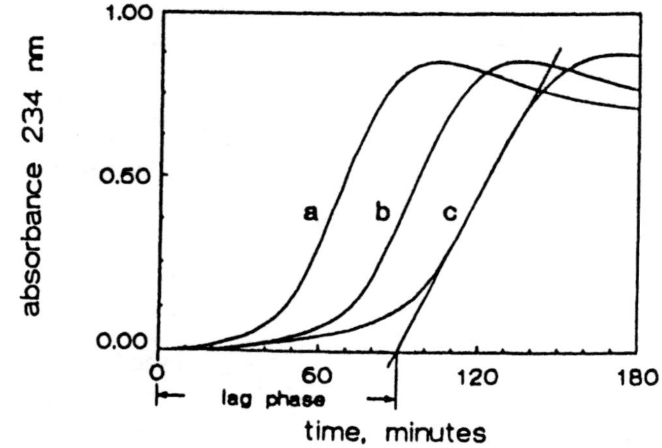

Fig. 4. Measurement of oxidation resistance of LDL by the conjugated diene method. LDL (0.1 μ*M*) isolated from three donors (*a, b, c*) in phosphate-buffered saline was mixed with 1.66 μ*M* $CuCl_2$ and the change of the 234-nm diene absorption was continuously recorded. Oxidation resistance, as measured by the lag phase, increases from donor *a* to donor *c*

and Cu^{++}, ratio of Cu^{++} to LDL, medium, and temperature (Kleinveld et al. 1992; Puhl et al. 1994). Most laboratories using the diene method agree that the method shows an excellent reproducibility, for example, between runs with the same batch of LDL the coefficient of variation for t_{inh} and V_{max} is less than 5% (Kleinveld et al. 1992; Esterbauer and Juergens 1993; Puhl et al. 1994). Nevertheless, the mean lag time for LDL from healthy controls, reported by various laboratories in recent papers, ranges from 68 to 124 min (Esterbauer and Juergens 1993). This wide variation probably reflects minor, but nevertheless important, methodological differences.

The mechanism by which Cu^{++} induces lipid peroxidation in LDL is still poorly understood. It seems very likely that Cu^{++} binds to discrete sites of the apo B and forms their centers for repeated free radical production, similar to the mechanism proposed for other biological systems (Chevion 1988). The exact number of binding sites in apo B is not known. Values ranging from 3 to about 10 were reported (Esterbauer et al. 1992; Kuzuya et al. 1992). Any compound which displaces copper from the apo B binding site (e.g., EDTA, histidine, certain proteins, and many other substances) by forming redox-inactive complexes should then inhibit or fully prevent LDL oxidation. Such inhibitory effects of copper chelators were frequently reported. Interaction of medium components with Cu^{++} explains why lag phases are much longer in Ham's F10 medium than in phosphate buffer (Thomas 1992).

Once bound, Cu^{++} must be reductively activated by a net transfer of one electron (reaction 7), and it is likely that the rate of this reaction is rate limiting and equal to the rate of initiation (R_i) of lipid peroxidation (reaction

1). The rate of initiation would then also be essential for the length of the lag phase (Eq. 2).

With LDL in phosphate buffer, the required reducing equivalents must be provided either from apo B (e.g., cysteine residues, reaction 7b) or from the lipids. In the latter case, the source of initiating radicals could be preformed peroxides. The presence of small "seed" peroxides has received considerable attention. Using highly sensitive HPLC methods, Bowry et al. (1992) reported 0.0004 mol cholesteryl ester hydroperoxides (CEOOH) per mol LDL; phospholipid hydroperoxides were below the detection limit. Thomas et al. (1994c) found in fresh LDL 0.003–0.006 mol CEOOH per mol LDL. Significantly higher peroxide levels were reported by Frei and Gaziano (1993). They obtained as an average value 0.24 ± 0.21 nmol CEOOH/mg protein (mean ± SD, n= 61) with a range of 0.013–0.886 nmol/mg protein. This is equal to 0.007–0.44 mol CEOOH/mol LDL. Frei and Gaziano (1993) also found that elevated levels of CEOOH are associated with increased susceptibility (lag phase) of LDL to Cu^{++} oxidation. It is assumed that Cu^{++} oxidation begins with preformed peroxides, yielding lipid peroxylradicals ($LOO^{\bullet}$), as shown in equation 7c. Another possibility proposed by Yoshida et al. (1994) is Cu^{++}-mediated formation of a tocopheroxyl radical (Eq. 7d). Whether Cu^{++} indeed initiates lipid peroxidation by one of the proposed mechanisms (reaction 7b, 7c, 7d) or by other elementary reactions remains to be established. In case of oxidation by cells, additional reducing components (e.g., thiols, superoxide anions) released by the cells probably enhance the rate of site-specific reduction of Cu^{++} and therefore accelerate initiation. An involvement of thiols in transition metal ion-dependent cell-mediated oxidation was first proposed by Heinecke et al. (1987). Sparrow and Olszewski (1993) recently presented evidence that oxidation of LDL by endothelial cells and macrophages in media containing transition metal ions is caused by the cell-dependent appearance of thiol in the medium. This topic has recently been reviewed by Chait and Heinecke (1994).

reaction 7: reductive activation of Cu^{++}

a. $Cu^{++} + e \xrightarrow{\text{rate Ri}} Cu^{+}$

b. $Cu^{++} + RSH \longrightarrow Cu^{+} + 1/2\ RSSR$

c. $Cu^{++} + LOOH \longrightarrow Cu^{+} + LOO\bullet + H^{+}$

d. $Cu^{++} + TOH \longrightarrow Cu^{+} + TO\bullet + H^{+}$

Cuprous ions (Cu^{+}) are strong prooxidants, which probably rapidly form the ultimate initiating radicals by a Fenton-type reaction (reaction 8a, 8b) or by a transition complex with molecular oxygen (reaction 8c).

reaction 8: initiation of lipid peroxidation (LPO) by Cu^+.

a. $Cu^+ + LOOH \rightarrow Cu^{++} + OH^- + LO^\bullet \rightarrow$ initiates LPO

b. $Cu^+ + HOOH \rightarrow Cu^{++} + OH^- + OH^\bullet \rightarrow$ initiates LPO

c. $Cu^+ + O_2 \rightarrow (Cu^{++} - O_2^{-\bullet}) \rightarrow$ initiates LPO or releases $O_2^{-\bullet}$

The principle difference between AAPH- and Cu^{++}-induced oxidation is that the former system produces a more or less random attack of free radicals in LDL, whereas the latter most likely involves a site-specific mechanism, which probably has more relevance for the in vivo situation (Gieseg and Esterbauer 1994). Prooxidant copper and iron are present in human atherosclerotic lesions (Smith et al. 1992), and the copper protein ceruloplasmin is a strong prooxidant for LDL (Ehrenwald et al. 1994). The lag phase in AAPH oxidation is determined in the first place by vitamin E, whereas in copper oxidation the overall lag phase most likely reflects the contribution of all antioxidants, the rate of initiation R_i, and perhaps structural parameters of LDL (Esterbauer and Juergens 1993; Thomas et al. 1994b).

5 Aldehydes

The decomposition of lipid hydroperoxides to aldehydes is a general phenomenon in fat autoxidation and lipid peroxidation in biological systems (Esterbauer et al. 1990). These secondary reactions are strongly accelerated by transition metal ions decomposing lipid hydroperoxides to lipid alkoxyl radicals in a Fenton-type reaction, e.g., $LOOH + Fe^{++} \rightarrow LO^\bullet + OH^- + Fe^{+++}$ (Fig. 5). The lipid alkoxyl radicals undergo β-cleavage reactions (homolytic scission) of the two C-C bonds on either side of the alkoxy group, yielding aldehydes and carbon-centered lipid radicals. If this mechanism is applied to the phospholipid and cholesteryl ester hydroperoxides contained in oLDL, the cleavage of the carbon bonds results in two classes of aldehydes: (a) aliphatic aldehydes derived from the methyl terminus of the fatty acid chain and (b) aldehyde still bound to the parent lipid molecule, i.e., so-called core aldehydes. The cholesteryl ester core aldehydes are discussed together with the oxysterols. Phospholipid core aldehydes are almost certainly present in oLDL. However, they have so far not been investigated. The aliphatic aldehydes demonstrated in oLDL are listed in Table 5. The time course of aldehyde accumulation in LDL oxidized by copper shows a lag phase of about 1–2 h with minimal formation of aldehydes, followed by a rapid increase lasting about 6 h; thereafter, the aldehyde content of LDL remains more or less constant (e.g., 4-hydroxyhexenal) or continues to increase slightly (e.g., malonaldehyde, hexanal, 4-hydroxynonenal). All aldehydes

Table 5. Aldehydes in LDL oxidized with copper 4–5 and 20–24 h, respectively. [Compiled from Esterbauer et al. (1992)]

	nmol / mg protein 4–5 h	20–24 h
Hexanal	52	229
Malonaldehyde	86	114
4-Hydroxynonenal	25	114
Nonanal	10	27
4-Hydroxyhexenal	8	49
4-Hydroxyoctenal	7	n.d.
Propanal	6	n.d.
Pentanal	5	n.d.
2,4-Heptadienal	5	n.d.
Butanal	4	n.d.
Octanal	1	5
Total aldehydes	209	538

n.d., Not determined.

listed in Table 5 except malonaldehyde are lipophilic compounds and remain largely associated with the LDL particle (Esterbauer et al. 1987). On the other hand, malonaldehyde exists at pH 7.4 as hydrophilic enolate anion ($^{-}$O-CH=CH-CHO) and about 80%–90% diffuses out of the LDL particle in the aqueous phase. It should be noted that what is measured with the TBA assay (i.e., TBARS) reflects very well the level of free malonaldehyde measured by HPLC, suggesting that other oxidized lipids present in oLDL interfere not at all or only to a low extent with the TBA assay.

The total amount of free aldehydes including malonaldehyde present in LDL after 24 h oxidation is in the range of 540 nmol/mg protein. This is the largest individual class of products identified so far in oLDL. The mol% distribution is: 42% hexanal, 21% malonaldehyde, 21% 4-hydroxynonenal, 9% 4-hydroxyhexenal, 6% other aldehydes. The concentrations of hexanal and 4-hydroxynonenal in the LDL lipid phase are extremely high, approximately 300 and 150 m*M*, respectively. Hexanal and 4-hydroxynonenal arise from oxidation of the omega-6 PUFAS, linoleic acid (18:2) and arachidonic acid (20:4). 4-Hydroxyhexenal and propanal arise from oxidation of the omega-3 PUFA docosahexaenoic acid (22:6). The major, if not the only, source of malonaldehydes are the fatty acids with more than 3 double bonds, i.e., 20:4 and 22:6. Frankel et al. (1992) recently proposed determination of hexanal by the rapid headspace gas chromatographic method for assessment of resistance of LDL to copper oxidation. The amount of hexanal found in oLDL samples (1 mg protein/ml, 8 μM Cu^{++}, 37°C, 4 h) from different donors correlated ($p < 0.05$, $r = 0.59$–0.61) with the PUFA and linoleic acid content of LDL, but no significant correlation was seen with vitamin E. The headspace chromatograms showed, besides hexanal, also peaks of pentane, propanal, and pentanal. Using the hexanal assay, Frankel et al. (1993) showed that inclusion of certain phenolic antioxidants into the LDL-copper system can inhibit formation of hexanal up to 100%. Red wine phenolics appear to be particularly powerful: 500- and 1000-fold dilutions inhibited hexanal formation by 96% and 100%, respectively (Frankel et al. 1993). H-Proton-NMR spectroscopy of lipid extracts from oLDL showed various signals indicative for aldehydes (Lodge et al. 1993).

Various lines of research suggest (Steinbrecher 1987; Steinbrecher et al. 1987; Juergens et al. 1987; Esterbauer et al. 1992; Hoff and O'Neil 1993) that important changes occurring in apo B during oxidation result from reaction of aldehydes with amino acid residues. The strong increase of the 430-nm fluorescence of apo B and the concomitant loss of free amino groups is, for example, likely caused by reaction of aldehydes with epsilon amino groups of lysine residues. Similarly, the strong increase of the net negative surface charge of the LDL particle is attributed to loss of positively charged amino groups by Schiff's base formation (RCHO + protein $NH_3^+ \rightarrow$ R-CH=N-protein + H_2O + H^+) or formation of Michael adducts with α,β-unsaturated aldehydes (R-CH=CH-CHO + protein $NH_3^+ \rightarrow$ R-CH(NH protein)-CH_2-CHO + H^+). Uchida and Stadtman (1992, 1993) recently showed that 4-hydroxynonenal can react with histidine, lysine, and cysteine residues in proteins. The products formed are Michael-type adducts in which the carbonyl function either is preserved or has undergone secondary reactions with amino groups to yield inter- and intramolecular cross-links. The complex pattern of reactions given by 4-hydroxynonenal is also evident from a study by Sayre et al. (1993), suggesting that primary amines condense with the aldehyde, at physiological conditions, to pyrrole derivatives.

Antibodies prepared against malonaldehyde and 4-hydroxynonenal-treated native LDL also react with copper- or cell-oxidized LDL, indicating that these aldehyde conjugates are indeed epitopes of apo B newly formed during oxidation (for review, see Esterbauer et al. 1992). Even more important, immunohistochemical methods clearly showed that malonaldehyde- and 4-hydroxynonenal-modified apo B occurs in atherosclerotic lesions of rabbits (Palinski et al. 1990) and human beings (Juergens et al. 1993). In human aorta the thickened intima of initial, transitional, and advanced lesions and atheromata showed predominantly extracellular staining with antibodies specific for 4-hydroxynonenal (HNE) epitopes (Juergens et al. 1993). Moreover, autoantibodies directed against malonaldehyde (MDA)- or HNE-modified proteins are present in the serum of rabbits and human beings (Esterbauer et al. 1992). The titer of autoantibodies to MDA-modified LDL was highly significantly correlated with progression of human carotid atherosclerosis (Salonen et al. 1992).

Many of the chemical (e.g., loss of NH_2 groups), physicochemical (e.g., increase of fluorescence and electrophoretic mobility), and biological (e.g., macrophage uptake, cytotoxicity) properties of cell- or copper-oxidized LDL can be reproduced in full or in part by treatment of previously nonoxidized LDL with aldehydes or aldehyde mixtures (Jessup et al. 1986; Juergens et al. 1987; Hoff et al. 1989; Hoff and Cole 1991; Esterbauer et al. 1992; Hoff and O'Neil 1993). Thomas et al. (1994a) recently showed that micromolar concentrations of aldehydes (e.g., decadienal, hydroxynonenal) induce interleukin-1 expression in human blood mononuclear cells. Aminoguanidine, a compound presently being used in clinical trials for inhibiting formation of advanced glycosylation products in chronic hyperglycemia, inhibits oxidative modification of LDL in vitro, presumably by binding reactive aldehydes and preventing their conjugation to apo B (Picard et al. 1992). The precise chemical structure of the aldehyde apo B conjugates and their location on the apo B polypeptide chain remain to be determined. Fruebis et al. (1992) examined the interaction of phospholipid hydroperoxides with peptides and proposed that in the first step adducts between lipid radicals and protein amino groups are formed which, in a second oxidation step, form protein-linked aldehydes.

6 Hydroperoxy and Hydroxy Fatty Acids

The change in lipid and fatty acid composition of LDL caused by copper oxidation is shown in Table 6. The amount of total hydroperoxides formed in LDL during copper oxidation in PBS was first measured iodometrically by a simple commercial assay (Esterbauer et al. 1989, 1992; El-Saadani et al. 1989). After a certain lag period of 1–2 h, total peroxides rapidly increased and reached a maximum of about 700–1000 nmol peroxides/mg LDL protein after 4–5 h. Thereafter, total peroxides decreased again and were hardly detectable 24 h after initiation of oxidation (Table 6). Using a similar peroxide assay, Jessup et al. (1990) found the same time course for total peroxides in oxidation of LDL by mouse peritoneal macrophages in Ham's F10 supplemented with 3 μM $FeSO_4$, and in cell free Ham's F10 medium supplemented with 100 μM Cu^{++}. Stocker et al. (1991) were among the first to employ a newly developed HPLC method, which allows separation and detection in a highly sensitive and selective manner of cholesteryl ester hydroperoxides (CEOOH), phospholipid hydroperoxides (PLOOH), and

Table 6. Change of lipid, fatty acid and antioxidant composition of LDL caused by copper oxidation. The data are from various sources as indicated in the footnotes

	Reference	Change observed
Phosphatidylcholine	1,2,9	decrease to 65–55%
Lysophosphatidylcholine	1,2,9	increase to about 200–400 nmol/mg protein
Sphingomyelin	9	no change
Triglycerides	1,2,3	decrease to 76%–52%
Free cholesterol	1,2,8	decrease to 90%–50%
Cholesteryl ester	3,8	decrease to 48%–25%
Total cholesterol	1,2,7	decrease to 78%–30%
Palmodic acid	4	no change
Stearic acid	4,5	decrease to 96%–79%
Oleic acid	4,5	decrease to 80%–46%
Linoleic acid	4,5,6	decrease to 15%–0%
Arachidonic acid	4,5,6	complete consumption
Docosahexaenoic acid	4	complete consumption
Vitamin E	4	complete consumption
Carotenoids	4	complete consumption

1. Steinbrecher et al. (1987b): 0.2–0.4 µ*M* LDL, 5 µ*M* Cu^{++} in PBS, 37°C, 24 h.
2. Barenghi et al. (1990): 0.4 µ*M* LDL, 5 µ*M* Cu^{++} in PBS, 37°C, 29 h.
3. Van Hinsbergh et al. (1986): 0.4 µ*M* LDL, 25 µ*M* Cu^{++} in PBS, 4°C, 48 h.
4. Esterbauer et al. (1992): 0.2–0.4 µ*M* LDL, 3–6 µ*M*Cu^{++} in PBS, ambient temp., 24 h.
5. Wang et al. (1992): 0.4 µ*M* LDL, 20 µ*M* Cu^{++} in Ham's F10, 37°C, 1.5 and 20 h.
6. Lenz et al. (1990): 30 µ*M* LDL, 5 µ*M* Cu^{++}, ambient temp., 24 h.
7. Malavasi et al. (1992): 0.4 µ*M* LDL, 20 µ*M* Cu^{++} in PBS, 24 h.
8. Zhang et al. (1990): 0.4 µ*M* LDL, 5 µ*M* Cu^{++} in PBS, 37°C, 20 h.
9. Liu et al. (1994): 0.6 µ*M* LDL, 5 µ*M* Cu^{++}, PBS, 20°C and 37°C, 20 h.

triglyceride hydroperoxides (TGOOH). LDL oxidized with 1 m*M* AAPH for 3 h contained substantial amounts of all three hydroperoxide classes with a molar ratio of about 5:2:1. This group did not investigate copper-oxidized LDL, but they reported that LDL exposed to activated PMN for 2 h also contained significantly increased levels of cholesteryl ester, phospholipid, and triglyceride hydroperoxides.

A recent study (Noguchi et al. 1993) measured the time course of the formation of total phospholipid hydroperoxides and total cholesteryl ester hydroperoxides with HPLC and detection at 234 nm (this wavelength is actually not specific for hydroperoxides, but rather for conjugated dienes, which could include both hydroperoxides and the corresponding reduced monohydroxy compound). It was found that in LDL (0.25 mg protein/ml)

Table 7. Lipid oxidation products demonstrated in copper-oxidized LDL. Note that the data are approximate values from various investigations as indicated in the footnotes

	Ref.	nmol/mg protein 4–5 h	nmol/mg protein 20–24 h
Total peroxides	1	1000	227
Phospholipid hydroperoxides	2	60	-
Cholesteryl ester hydroperoxides	2	180	-
Conjugated dienes	2	240	-
TBARS	1	85	114
Total aldehydes + TBARS	1	210	540
Hydroxy-octadecenoic acid[a]	3	7	50
Hydroxyoctadecadienoic acid[b]	3	110	30
Hydroxyeicosatetraenoic acid[c]	3	17	0
7-Hydroxy- and 7 Hydroperoxycholesterol	4,5,8	60	120–760
7-Ketocholesterol	4,5,6	n.q.	n.q.
Cholesteroloxoalkanoylester	7	n.q.	30
7-Ketocholesterolalkanoylester	7	n.q.	30
5,6-Epoxycholesterol	5,6	n.q.	traces
25-Hydroxycholesterol	6	n.q.	traces
Cholest-3,5-dien-7-one	4,6	n.q.	n.q.

n.q., Not quantified
1. Esterbauer et al. (1992): 0.1 μ*M* LDL, 1.66 μ*M* Cu^{++} in PBS, ambient temp.
2. Noguchi et al. (1993): 0.5 μ*M* LDL, 2 μ*M* Cu^{++} in PBS, 37°C.
3. Wang et al. (1992): 0.4 μ*M* LDL, *20* μ*M* Cu^{++} in Ham's F10, 37°C.
4. Malavasi et al. (1992): 0.4 μ*M* LDL, *20* μ*M* Cu^{++} in PBS, 37°C.
5. Zhang et al. (1990): 0.4 μ*M* LDL, *5* μ*M* *Cu*$^{++}$ in PBS, 37°C.
6. Bhadra et al. (1991): 0.2–0.4 μ*M* LDL, *?* μ*M* *Cu*$^{++}$ in M199.
7. Kamido et al. (1992): 30 μ*M* LDL, *5* μ*M* *Cu*$^{++}$, ambient temp.
8. Carpenter et al. (1994): 0.2 μ*M* LDL, *5–25* μ*M* *Cu*$^{++}$ (or mouse macrophages), F-10, 24 h.

[a] Sum of 8-, 10- and 11-hydroxy-derivatives of oleic acid.
[b] Sum of 9-HODE and 13-HODE.
[c] Sum of 5-HETE, 8-HETE, 9-HETE, 11-HETE, 12-HETE and 15-HETE.

exposed to copper (2 μ*M*) cholesteryl ester hydroperoxides and phospholipid hydroperoxides rise slowly during the first 2 h; then the increase occurs more rapidly to a level of about 45 μ*M* (=180 nmol CEOOH/mg protein) and 15 μ*M* (=60 mg PLOOH/mg protein) (Table 7). After 5 h the cholesteryl ester hydroperoxides tended to decrease, whereas phospholipid hydroperoxides still increased. In LDL oxidized for 5 h with AAPH (1 m*M*), cholesteryl ester and phospholipid hydroperoxide content were approximately 120 nmol/mg protein and 33 nmol/mg protein, respectively. AMVN gave mainly cholesteryl ester hydroperoxides (about 60 nmol/mg protein) with only trace amounts of phospholipid hydroperoxides.

Both phospholipid hydroperoxides and cholesteryl ester hydroperoxides are still a complex mixture of hydroperoxides differing in type of fatty acids (18:2, 20:4, 22:6) and position of the hydroperoxy group (e.g., 9-HODE or 13-HODE). New methods have been developed which would allow further separation and quantification of the individual species present in the pool of phospholipid or cholesteryl ester peroxides (Kritharides et al. 1993; Sattler et al. 1994). Investigators interested in more detailed analyses frequently reduce the hydroperoxides present in the bulk LDL lipid extract to the corresponding alcohols, saponify the lipids, and separate the monohydroxy fatty acids by gas chromatography. Of course, this does not allow conclusions regarding the parent lipid molecule to which the hydroperoxy fatty acid was bound. The first study in this direction was reported by Lenz et al. (1990). In LDL (1.5 mg protein/ml) oxidized with 5 μM $CuSO_4$ for 24 h they found *trans, cis* and *trans, trans* 13-hydroxy-9,11-octadecadienoic acid, 13-HODE (92 nmol/mg protein) and *trans, cis* and *trans, trans* 9-hydroxy-10,12-octadecadienoic acid, 9-HODE (86 nmol/mg protein) and the hydroxy fatty acids derived from arachidonic acid, i.e., 5-HETE (4.3 nmol), 8-HETE + 9-HETE (5.9 nmol), 12-HETE (4.5 nmol) and 15-HETE (4.1 nmol). The time course showed no increase of HODE and HETE during the first 4 h of incubation. Thereafter, the increase was more or less linear with time up to 24 h. The hydroxy fatty acids accounted for approximately 70% of the linoleate consumed during LDL oxidation and represented 45-fold more products than were measured with the TBARS analyses. It should be noted that in this study (Lenz et al. 1990) strong variation among the four investigated LDL samples was observed. For example, in one LDL sample most of the linoleic acid and arachidonic acid was still present after 24 h. This is not in agreement with other studies (Esterbauer et al. 1992; Wang et al. 1992), which found that linoleic acid and arachidonic acid are more or less completely consumed after 24 h copper oxidation.

Wang et al. (1992) studied by GC/MS the time course of the formation of monohydroxy derivatives of arachidonic acid, linoleic acid, and oleic acid during oxidation of LDL (0.2 mg protein/ml) by 20 μM copper and by endothelial cells. Oxidation of LDL by copper (Table 7) resulted after a lag phase of about 1 h in a strong increase of the linoleic acid oxidation products (9-HODE, 13-HODE) and the arachidonic acid oxidation products (5-, 8-, 9-, 11-, 12-, and 15-HETE). Interestingly, considerable amounts of oxidation products of oleic acid (8-, 10-, and 11-hydroxyoctadecenoic acid) were also found. The HETES and HODES reached a maximum at 5 h and decreased nearly to zero after 20 h, whereas the hydroxy derivatives of oleic acid increased for up to 24 h. The quantities measured at 5 h oxidation were 110 nmol HETES/mg protein, 17 nmol HODES/mg protein, and, after 20 h, 50 nmol hydroxy derivatives of oleic acid. TBARS values after 20 h reached 60 nmol/mg protein. In this study it was also demonstrated that 99% of the hydroxy fatty acids remain associated with the LDL particle, as evidenced by re-isolation of the oLDL by ultracentrifugation. Oxidation of LDL by endothelial cells gave a product pattern similar to that of oxidation by copper ions and there was little positional specificity, suggesting that also in endothelial cell oxidation the majority of the fatty acid oxidation products are formed via nonenzymatic lipid peroxidation processes.

7 Oxysterols

Oxidation of LDL lipids is not restricted to the PUFAS; a number of more recent studies have demonstrated that the cholesterol moiety can also be oxygenated. Zhang et al. (1990) reported that Cu^{++}-oxidized LDL contains 7-ketocholesterol, 7-hydroxycholesterol, and 5,6-epoxycholesterol, with 7-ketocholesterol comprising the abundant sterol oxidation products. Unlike Zhang et al. (1990), Bhadra et al. (1991) found in copper-oxidized LDL cholest-3,5-dien-7-one as the major product, with smaller amounts of 5,6-epoxycholesterol, 7-hydroxycholesterol, 7-ketocholesterol, and 25-hydroxy cholesterol. Endothelial cell-oxidized LDL contained only the 5,6-epoxycholesterol (Bhadra et al. 1991). Malavasi et al. (1992) followed the time course of cholesterol oxidation (0.2 mg protein/ml) with 20 μM $CuSO_4$ in PBS at 37°C and found that 7-hydroperoxy cholesterol (7αOOH, 7βOOH) is largely prevalent at early times of oxidation. The concentration of the hydroperoxides decreases with oxidation time with concomitant formation of 7-hydroxycholesterol (7αOH, 7βOH), cholest-3,5-dien-7-one. The total amount of oxysterols increased during the first 12 h from zero to about 50 μg/mg protein and then started to decrease again. After 24 h about 54% of the total LDL cholesterol had been consumed. Incubation of plasma with copper ions (0.5–2 μM, 24 h, 37°C) gives rise to the formation of free and esterified oxysterols, mainly 7-ketocholesterol, 7-hydroxycholesterol, and 5,6-epoxycholesterol (Tamasawa and Takebe 1992). Our group (Esterbauer et al. 1990, 1992; Gebicki et al. 1991) has repeatedly suggested that oxidation of phospholipids or cholesteryl ester yields, in addition to the aldehyde fragments derived from the methyl terminus of fatty acids, the counterpart aldehydes, where the fragmented fatty acid chains are still bound to the parent lipid molecules (Fig. 5). Kamido et al. (1992) first demonstrated that lipid peroxidation of cholesteryl esters and phospholipids leads to formation of cholesteryl-oxoalkanoates and phospholipid-oxoalkanoates. This new

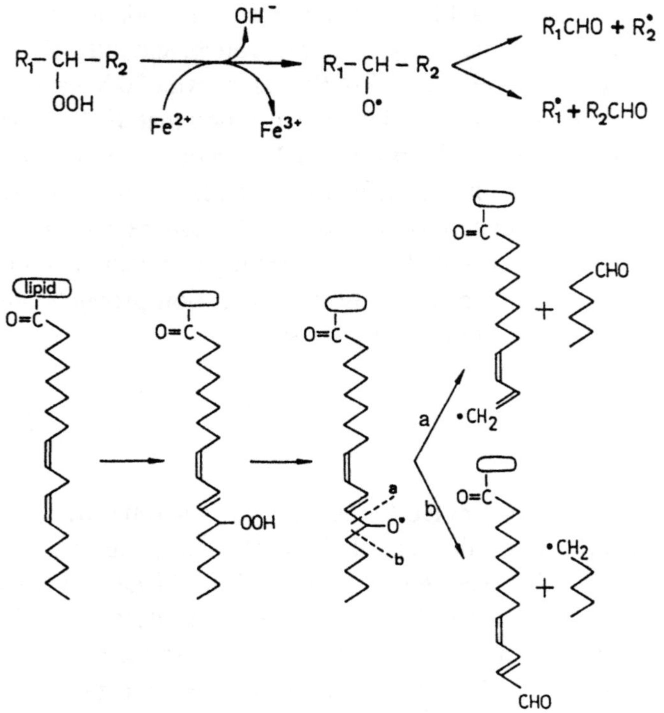

Fig. 5. Decomposition of lipid hydroperoxides by -cleavage yield aldehydes. *Bottom:* Two classes of aldehydes are formed, *(a)* aldehydes derived from the methylterminus of the fatty acids (shown is hexanal) and *(b)* core aldehydes bound to the parent lipid molecule

class of lipid-derived aldehydes was given the name "core aldehyde". About 1%–2% of the cholesteryl linoleate and cholesteryl arachidonate consumed upon copper oxidation of LDL are such core aldehydes. Identified were 7-cholesteryl ester core aldehydes and 7-ketocholesteryl ester core aldehydes. In both series the aldehydes had chain lengths of 4, 5, 6, 7, 8, 9, and 10 carbon atoms. The 9-carbon atom core aldehydes (cholesteryl-9-oxo-nonanoate, 7-ketocholesteryl-9-oxo-nonanoate) were with about 60% the main products, followed by the 8-carbon atom core aldehydes (~20%) and 5-carbon atom core aldehydes (8%–10%). The parent lipids are most likely cholesteryl linoleate for the C_9 aldehydes and cholesteryl arachidonate for the C_5 aldehydes. The C_8 aldehydes might arise from traces of cholesteryl eicosatrienoic acid (n-6) or double-bond migration during peroxidation.

The identification and quantification of oxysterols and cholesteryl ester core aldehydes is still at an early stage. Since some of these products are biologically very active, this subject deserves further attention. For example, Carpenter et al. (1990) have shown that cultured human monocyte macrophages oxidize a cholesteryl linoleate albumin complex to 7-hydroxycholesterol and 9- and 13-HODE. In lipids extracted from atheroma 26-hydroxycholesterol, 7-hydroxycholesterol and isomeric HODES were detected (Carpenter et al. 1993). Hodis et al. (1992) reported that the antioxidant probucol reduces plasma and aortic wall oxysterol levels (7-hydroxycholesterol, 5,6-epoxycholesterol and 3,5,6-cholestanetriol) in cholesterol-fed rabbits. A similar observation was made in probucol-treated WHHL rabbits (Stalenhoef et al. 1993), where a significant reduction of plasma levels of 7α–hydroxycholesterol, 3,5,6-cholestanetriol, 7-ketocholesterol, and 25-hydroxycholesterol was found. Rabbits treated with the antioxidant BHT showed decreased plasma levels of 7-ketocholesterol and 5,6-epoxycholesterol (Björkhem et al. 1991). It appears that some of the oxysterols found in human plasma are also found in the dietary cholesterol, for example, cholesterol-5β,6β-epoxide, while others are not. This strongly suggests that some of the cholesterol oxidation products are formed by in vivo oxidation (Sevanian et al. 1994). Cholestanetriols (e.g., 3β, 5α, 6β) and 25-hydroxycholesterol have been shown to cause injury to endothelial cells and smooth muscle cells and to alter LDL receptor functions (Peng et al. 1991).

Some oxysterols (7-hydroxycholesterol, 3,7,22-cholestanetriol, 3,5,6-cholestanetriol, 5,6-epoxycholesterol) are highly toxic for endothelial cells at concentrations in the range of 15–50 nmol/ml (Sevanian et al. 1991; Petterson et al. 1991). Zwijsen et al. (1992b) have shown that oxysterols inhibit gap-junctional intercellular communications of smooth muscle cells by more than 40% at concentrations of 1–10 nmol/ml. The inhibitory activity increased in the order: 5,6-epoxycholesterol < 7-ketocholesterol < 3,5,6-cholestanetriol < 25-hydroxycholesterol. A disturbance of intercellular communication may result in disturbances of growth and induction of proliferation of smooth muscle cells.

8 Lipoxygenase-Mediated Oxidation of LDL

In the first study of lipoxygenase-mediated LDL oxidation it was reported (Sparrow et al. 1988) that soybean lipoxygenase, a plant-derived 15-lipoxygenase, in combination with phospholipase A_2 is capable of converting LDL into a form with increased TBARS, relative electrophoretic mobility (REM), and macrophage uptake. Cathcart et al. (1991) then showed that soybean lipoxygenase can oxidize LDL also in the absence of phospholipase A_2, as evidenced by conjugated dienes, TBARS, and cytotoxicity. Several attempts were later made to discover whether lipoxygenases are involved in cell-mediated oxidation of LDL. Lipoxygenase inhibitors indeed blocked modification of LDL by rabbit endothelial cells (Parthasarathy et al. 1989), human monocytes (McNally et al. 1990), and mouse peritoneal macrophages

(Rankin et al. 1991). However, the high concentration of inhibitors required and their rather nonspecific character made the involvement of lipoxygenases in cell-mediated oxidation questionable. In several studies it was clearly demonstrated that 5-lipoxygenases are not responsible for oxidation of LDL by mouse peritoneal macrophages (Jessup et al. 1991; Sparrow and Olszewski 1992) and human monocytes (Folcik and Cathcart 1993). On the other hand, it was shown that 15-lipoxygenase is present in human and rabbit atherosclerotic lesions and co-localized with deposits of oLDL (Ylä-Herttuala et al. 1990, 1991). Belkner et al. (1991) studied the oxygenation of LDL by purified rabbit reticulocyte 15-lipoxygenase and recombinant human 15-lipoxygenase. This enzyme converts free arachidonic acid and linoleic acid into the 15-hydroperoxy eicosatetraeinoic acid (15-HPETE) and 13-hydroperoxy octadecadienoic acid (13-HPODE). The enzyme is also capable of converting arachidonic acid and linoleic acid bound to phospholipids and cholesteryl ester to the corresponding 15- and 13-hydroperoxy lipids. Incubation of LDL with 15-lipoxygenase in a molar ratio of 17:1 (molecular mass of lipoxygenase 75 kDa) led within 30 min to the oxidation of about 0.5% of the linoleic acid residues in LDL. This corresponds to a turnover of roughly 3 linoleate molecules/min produced by one lipoxygenase molecule. Analyses of the oxygenated polyenic fatty acids (after reduction and alkaline hydrolysis) revealed 13-hydroxy-9*cis*, 11-*trans*-octadecadienoic acid (13-HODE, Z,E) as the main product (71%), with lower amounts (10%) of 15-hydroxy-5,8,11,13-eicosatetraenoic acid (15-HETE), 5% 13-hydroxy 9-*trans*, 11-*trans*-octadecadienoic acid (13-HODE), and 17% 9-hydroxy-10, 12-octadecadienoic acid (9-HODE). At low LDL concentration predominantly the S-isomer of 13-HODE and 15-HETE were formed. More than 90% of the hydroxy fatty acids were contained in the esterified lipid fraction, particularly in the cholesteryl esters. Long-term treatment (20 h) or high concentrations of lipoxygenase also altered the apo B, as demonstrated by the increased electrophoretic mobility and the increased content of Schiff's bases in apo B. This suggests that a fraction of the hydroperoxides decompose, most likely catalyzed by traces of transition metal ions present in the incubation mixture. It is important to mention that also in the lipoxygenase-mediated oxidation, oxygen uptake is about twofold higher than formation of hydroxy fatty acids. A similar observation was made in AAPH- or copper-induced LDL oxidation (Noguchi et al. 1993). The fate of this additionally consumed oxygen and where it is bound are not clear.

In human plasma incubated with reticulocyte 15-lipoxygenase 13-HODE (main product), 9-HODE and 15-HETE esterified with cholesterol were formed (Belkner et al. 1991). Moreover, lipids extracted from pieces of thoracic aortas of subjects who died of acute heart failure contain considerable amounts of cholesterol esterified with keto- and hydroxy-octadecadienoic acid (Kühn et al. 1992). The nonspecific product pattern of the arterial wall material as regards positional and stereoisomers suggests that in vivo, nonenzymatic peroxidation processes are responsible for formation of the majority of these oxygenated fatty acids. Of course, cellular 15-lipoxygenase of endothelial cells or monocyte macrophages could play an important role in the initiation of the nonenzymatic lipid peroxidation process by providing seed-hydroperoxides in LDL. Even small amounts of peroxides would render LDL susceptible to a subsequent nonenzymatic oxidation mediated by free or complexed transition metal ions.

Oxidation of LDL by human umbilical vein endothelial cells (Ham's F10 medium, 20 h) also yielded an oxygenated fatty acid pattern characteristic for a nonenzymatic peroxidation (Wang et al. 1992). The main products identified were the monohydroxy derivatives of linoleic acid (9-HODE, 13-HODE) and arachidonic acid (5-HETE, 8-HETE, 9-HETE, 11-HETE, 12-HETE, 15-HETE). Small amounts of hydroxylated derivatives of oleic acid (8-hydroxy-, 10-hydroxy- and 11-hydroxy-octadecenoic acid) were also formed. The total amount of monohydroxy fatty acids was about 22 nmol/mg LDL protein (=12 mol/mol LDL) with 54% HODES and 45% HETES. The isomer distribution in LDL oxidized 5 h with Cu^{++} was identical to that in the endothelial cell-oxidized LDL. The total amount of monohydroxy fatty acids in copper-oxidized LDL, however, was around 150 nmol/mg LDL protein.

Acknowledgements. The authors' work has been supported by the Austrian Science Foundation (FWF), project SO7102, by the *Jubiläumsfonds der Österreichischen Nationalbank*, project no. 4484, and by a European Community grant PL931790. P. Ramos received a Lise-Meitner Fellowship from the FWF.
This article is a modified and updated version of an article published previously by H. Esterbauer in "Oxidative stress, lipoproteins and cardiovascular dysfunction", Portland Press 1995. Permission granted by Portland Press London is acknowledged.

References

Autio I, Jaakkola O, Solakivi T, Nikkari T (1990) Oxidized low-density lipoprotein is chemotactic for arterial smooth muscle cells in culture. FEBS Lett 277: 247–249

Aviram M (1992) Low density lipoprotein modification by cholesterol oxidase induces enhanced uptake and cholesterol accumulation in cells. J Biol Chem 267: 218–225

Aviram M (1993) Modified forms of low density lipoprotein and atherosclerosis. Atherosclerosis 98: 1–9

Balla G, Jacob HS, Eaton JW, Belcher JD, Vercellotti GM (1991) Hemin: a possible physiological mediator of low density lipoprotein oxidation and endothelial injury. Arterioscler Thromb 11: 1700–1711

Barenghi L, Bradamante S, Giudici GA, Vergani C (1990) NMR analysis of low-density lipoprotein oxidatively modified in vitro. Free Radic Res Commun 8: 175–183

Belkner J, Wiesner R, Kühn H, Lankin VZ (1991) The oxygenation of cholesterol esters by the reticulocyte lipoxygenase. FEBS Lett 279: 110–114

Belkner J, Wiesner R, Rathman J, Barnett J, Sigal E, Kühn H (1993) Oxygenation of lipoproteins by mammalian lipoxygenase. Eur J Biochem 213: 251–261

Berliner JA, Territo MC, Sevanian A, Ramin S, Kim JA, Ramshad B, Esterson M, Fogelman AM (1990) Minimally modified low density lipoprotein stimulates monocyte endothelial interactions. J Clin Invest 85: 1260–1266

Bhadra S, Arshad MAQ, Rymaszewski Z, Norman E, Wherley R, Subbiah MTR (1991) Oxidation of cholesterol moiety of low density lipoprotein in the presence of human endothelial cells or Cu^{+2} ions: identification of major products and their effects. Biochem Biophys Res Commun 176: 431–440

Björkhem I, Henriksson-Freyschuss A, Breuer O, Diczfalusy U, Berglund L, Henriksson P (1991) The antioxidant butylated hydroxytoluene protects against atherosclerosis. Arterioscler Thromb 11: 15–22

Bonnefont-Rousselot D, Gardès-Albert M, Delattre J, Ferradini C (1993) Oxidation of low-density lipoproteins by $OH^{\bullet}$ and $OH/O_2^{-\bullet}$ free radicals produced by gamma radiolysis. Radiat Res 134: 271–282

Bowry VW, Stocker R (1993) Tocopherol-mediated peroxidation – the prooxidant effect of vitamin E on the radical initiated oxidation of human low-density lipoprotein. J Am Chem Soc 115: 6029–6044

Bowry VW, Stanley KK, Stocker R (1992) High density lipoprotein is the major carrier of lipid hydroperoxides in human blood plasma from fasting donors. Proc Natl Acad Sci USA 89: 10316–10320

Bruckdorfer KR (1990) Free radicals, lipid peroxidation and atherosclerosis. Curr Opin Lipidol 1: 529–535

Camejo G, Lopez A, Lopez F, Quinones J (1985) Interaction of low density lipoproteins with arterial proteoglycans. The role of charge and sialic acid content. Atherosclerosis 55: 93

Campos H, Blijlevens E, McNamara JR, Ordovas JM, Posner BM, Wilson PWF, Castelli WP, Schaffer EJ (1992) LDL particle size distribution. Arterioscler Thromb 12: 1410–1419

Carpenter KLH, Ballantine JA, Fussell B, Enright JH, Mitchinson MJ (1990) Oxidation of cholesteryl linoleate by human monocyte-macrophages in vitro. Atherosclerosis 83: 217–229

Carpenter KLH, Brabbs CE, Mitchinson MJ (1991) Oxygen radicals and atherosclerosis. Klin Wochenschr 69: 1039–1045

Carpenter KLH, Taylor SE, Ballantine JA, Fussell B, Halliwell B, Mitchinson MJ (1993) Lipids and oxidised lipids in human atheroma and normal aorta. Biochim Biophys Acta 1167: 121–130

Carpenter KLH, Wilkins GM, Fussel B, Ballantine JA, Taylor SE, Mitchinson MJ, Leake DS (1994) Production of oxidized lipids during modification of low-density lipoprotein by macrophages or copper. Biochem J 304: 625–633

Cathcart MK, McNally AK, Chisolm GM (1991) Lipoxygenase-mediated transformation of human low density lipoprotein to an oxidized and cytotoxic complex. J Lipid Res 32: 63–70

Chait A, Heinecke JW (1994) Lipoprotein modification: cellular mechanisms. Curr Opin Lipidol 5: 365–370

Chevion M (1988) A site-specific mechanism for free radical induced biological damage: the essential role of redox-active transition metals. Free Radic Biol Med 5: 27–37

Chisolm GM (1992) The oxidation of lipoproteins: implications for atherosclerosis. In: Spatz L, Bloom AD (eds) Biological consequences of oxidative stress. Implications for cardiovascular disease and carcinogenesis. Oxford University Press, New York, pp 78–106

Cosgrove JP, Church DF, Pryor WA (1987) The kinetics of the autoxidation of polyunsaturated fatty acids. Lipids 22: 299–304

Cushing SD, Berliner JA, Valente AJ, Territo MC, Navab M, Parhami F, Gerrity R, Schartz CJ, Fogelman AM (1990) Minimally modified low density lipoprotein induces monocyte chemotactic protein 1 in human endothelial cells and smooth muscle cells. Proc Natl Acad Sci USA 87: 5134–5138

Daret D, Blin P, Dorian B, Rigaud M, Larrue J (1993) Synthesis of monohydroxylated fatty acids from linoleic acid by rat aortic smooth muscle cells and tissues: influence on prostacyclin production. J Lipid Res 34: 1473–1482

De Graaf J, Hak-Lemmers HLM, Hectors MPC, Demacker PNM, Hendricks JCM, Stalenhoef AFH (1991) Enhanced susceptibility to in vitro oxidation of the dense low density lipoprotein subfraction in healthy subjects. Arterioscler Thromb 11: 298–306

Drake TA, Hannani K, Fei H, Lavi S, Berliner JA (1991) Minimally oxidized low-density lipoprotein induces tissue factor expression in cultured human endothelial cells. Am J Pathol 138: 601–607

Ehrenwald E, Chisolm GM, Fox PL (1994) Intact human ceruloplasmin oxidatively modifies low density lipoprotein. J Clin Invest 93: 1493–1501

El-Saadani M, Esterbauer H, El-Sayed M, Goher M, Nasser AY, Juergens G (1989) A spectrophotometric assay for lipid peroxides in serum lipoproteins using a commercially available reagent. J Lipid Res 30: 627–630

Engelmann B, Bräutigam C, Thiery J (1994) Plasmalogen phospholipids as potential protectors against lipid peroxidation of low density lipoproteins. Biochem Biophys Res Commun 204: 1235–1242

Esterbauer H, Juergens G, Quehenberger O, Koller E (1987) Autoxidation of human low density lipoprotein: loss of polyunsaturated fatty acids and vitamin E and generation of aldehydes. J Lipid Res 28: 495–509

Esterbauer H, Juergens G (1993) Mechanistic and genetic aspects of susceptibility of LDL to oxidation. Curr Opin Lipidol 4: 114–124

Esterbauer H, Striegl G, Puhl H, Rotheneder M (1989) Continuous monitoring of in vitro oxidation of human low density lipoprotein. Free Radic Res Commun 6: 67–75

Esterbauer H, Zollner H, Schaur R (1990) Aldehydes formed by lipid peroxidation: mechanism of formation, occurrence and determination. In: Pelfrey C (ed) Membrane lipid oxidation. CRC Press, Boca Raton, pp 239–268

Esterbauer H, Gebicki J, Puhl H, Juergens G (1992) The role of lipid peroxidation and antioxidants in oxidative modification of LDL. Free Radic Biol Med 13: 341–390

Folcik VA, Cathcart MK (1993) Assessment of 5-lipoxygenase involvement in human monocyte-mediated LDL oxidation. J Lipid Res 34: 69–79

Fox PL, Chisolm GM, DiCorleto PE (1987) Lipoprotein-mediated inhibition of endothelial cell production of platelet-derived growth factor-like protein depends on free radical lipid peroxidation. J Biol Chem 262: 6046–6054

Frankel EN, German JB, Davis PA (1992) Headspace gas chromatography to determine human low density lipoprotein oxidation. Lipids 27: 1047–1051

Frankel EN, Kanner J, German JB, Parks E, Kinsella JE (1993) Inhibition of oxidation of human low-density lipoprotein by phenolic substances in red wine. Lancet 341: 454–457

Frei B, Gaziano JM (1993) Content of antioxidants, preformed lipid hydroperoxides, and cholesterol as predictors of the susceptibility of human LDL to metal ion-dependent and -independent oxidation. J Lipid Res 34: 2135–2145

Frostegard J, Wu R, Giscombe R, Holm G, Lefvert AK, Nilsson J (1992) Induction of T-cell activation by oxidized low density lipoproteins. Arterioscler Thromb 12: 461–467

Fruebis J, Parthasarathy S, Steinberg D (1992) Evidence for a concerted reaction between lipid hydroperoxides and polypeptides. Proc Natl Acad Sci USA 89: 10588–10592

Gebicki J, Juergens G, Esterbauer H (1991) Oxidation of low density lipoprotein in vitro. In: Sies H (ed) Oxidative stress. Academic, London, pp 371–397

Gieseg SP, Esterbauer H (1994) Low density lipoprotein is saturable by pro-oxidant copper. FEBS Lett 343: 188–194

Graham A, Hogg N, Kalyanaraman B, O'Leary VJ, Darley-Usmar VM, Moncada S (1993) Peroxynitrite modification of low density lipoprotein leads to recognition by the macrophage scavenger receptor. FEBS Lett 330: 181–185

Haberland ME, Steinbrecher UP (1992) Modified low density lipoproteins: diversity and biological relevance in atherogenesis. In: Lusis AJ, Rotter JI, Sparks RS (eds) Molecular genetics of coronary artery disease. Candidate genes and processes in atherosclerosis. Monogr Hum Genet 14: 35–61

Halliwell B (1993) The role of oxygen radicals in human disease, with particular reference to the vascular system. Haemostasis 23 [Suppl 1]: 118–126

Hamilton TA, Ma GP, Chisolm GM (1990) Oxidized low density lipoprotein suppresses the expression of tumor necrosis factor-alpha mRNA in stimulated murine peritoneal macrophages. J Immunol 144: 2343–2350

Hazell LJ, Van den Berg JJM, Stocker R (1994) Oxidation of low-density lipoprotein by hypochlorite causes aggregation that is mediated by modification of lysine residues rather than lipid oxidation. Biochem J 302: 297–304

Heinecke JW, Rosen H, Suzuki LA, Chait A (1987) The role of sulfur-containing amino acids in superoxide production and modification of low density lipoprotein by arterial smooth muscle cells. J Biol Chem 262: 10098–10103

Heinecke JW, Kawamura M, Suzuki L, Chait A (1993) Oxidation of low density lipoprotein by thiols: superoxide-dependent and -independent mechanisms. J Lipid Res 34: 2051–2061

Hodis HN, Chauhan A, Hashimoto S, Crawford DW, Sevanian A (1992) Probucol reduces plasma and aortic wall oxysterol levels in cholesterol fed rabbits independently of its plasma cholesterol lowering effect. Atherosclerosis 96: 125–134

Hoff HF, Cole TB (1991) Macrophage uptake of low-density lipoprotein modified by 4-hydroxynonenal. Lab Invest 64: 254–264

Hoff HF, O'Neil J (1993) Structural and functional changes in LDL after modification with both 4-hydroxynonenal and malondialdehyde. J Lipid Res 34: 1209–1217

Hoff HF, O'Neill J, Chisolm GM, Cole TB, Quehenberger O, Esterbauer H, Juergens G (1989) Modification of LDL with 4-hydroxynonenal, a propagation product of lipid peroxidation, induces uptake of LDL by macrophages. Arteriosclerosis 9: 538–549

Jessup W, Juergens G, Lang J, Esterbauer H, Dean RT (1986) The interaction of 4-hydroxynonenal-modified low density lipoproteins with the fibroblast apo B/E receptor. Biochem J 234: 245–248

Jessup W, Rankin SM, De Whalley CV, Hoult JRS, Scott J, Leake DS (1990) α-Tocopherol consumption during low-density lipoprotein oxidation. Biochem J 265: 399–405

Jessup W, Darley-Usmar V, O'Leary V, Bedwell S (1991) 5-Lipoxygenase is not essential in macrophage-mediated oxidation of low-density lipoprotein. Biochem J 278: 163–169

Jougasaki M, Kugiyama K, Saito Y, Nakao K, Imura H, Yasue H (1992) Suppression of endothelin-1 secretion by lysophosphatidyl choline in oxidized low density lipoprotein in cultured vascular endothelial cells. Circ Res 71: 614–619

Juergens G, Hoff HF, Chisolm GM, Esterbauer H (1987) Modification of human serum low density lipoprotein by oxidation-characterization and pathophysiological implications. Chem Phys Lipids 45: 315–336

Juergens G, Chen Q, Esterbauer H, Mair S, Ledinski G, Dinges HP (1993) Atherogenic lipoproteins in man. Immunostaining of human autopsy aorta with antibodies to modified apolipoprotein B and apolipoprotein (a). Arterioscler Thromb 13: 1689–1699

Kamido H, Kuksis A, Marai L, Myher JJ (1992) Identification of cholesterol-bound aldehydes in copper-oxidized low density lipoprotein. FEBS Lett 304: 269–272

Kim JA, Territo MC, Wayner E, Carlos TM, Parhami F, Smith CW, Haberland ME, Fogelman AM, Berliner JA (1994) Partial characterization of leukocyte binding molecules on endothelial cells induced by minimally oxidized LDL. Arterioscler Thromb 14: 427–433

Kleinveld HA, Hak-Lemmers HLM, Stalenhoef AFH, Demacker PNM (1992) Improved measurement of low-density lipoprotein susceptibility to copper-induced oxidation: application of a short procedure for isolating low-density lipoprotein. Clin Chem 38: 2066–2072

Kontush A, Hübner C, Finckh B, Kohlschütter A, Beisiegel U (1994) Low density lipoprotein oxidizability by copper correlates to its initial ubiquinol-10 and polyunsaturated fatty acid content. FEBS Lett 341: 69–73

Krauss RM (1991) Low-density lipoprotein subclasses and risk of coronary artery disease. Curr Opin Lipidol 2: 248–252

Kritharides L, Jessup W, Gifford J, Dean RT (1993) A method for defining the stages of low-density lipoprotein oxidation by the separation of cholesterol- and cholesteryl ester-oxidation products using HPLC. Anal Biochem 213: 79–89

Kugiyama K, Sakamoto T, Misumi I, Sugiyama S, Ohgushi M, Ogawa H, Horiguchi M, Yasue H (1993) Transferable lipids in oxidized low-density lipoprotein stimulate plasminogen activator inhibitor-1 and inhibit tissue-type plasminogen activator release from endothelial cells. Circ Res 73: 335–343

Kühn H, Belkner J, Wiesner R, Schewe T, Lankin VZ, Tikhaze AK (1992) Structure elucidation of oxygenated lipids in human atherosclerotic lesions. Eicosanoids 5: 17–22
Kühn H, Belkner J, Zaiss S, Fährenklemper T, Wohlfeil S (1994) Involvement of 15-lipoxygenase in early stages of atherogenesis. J Exp Med 179: 1903–1911
Kuzuya M, Yamada K, Hayashi T, Funaki C, Naito M, Asai K, Cuzuya F (1992) Role of lipoprotein-copper complex in copper-catalyzed peroxidation of low-density lipoprotein. Biochim Biophys Acta 1123: 334–341
Lamb DJ, Leake DS (1994) Iron released from transferrin at acidic pH can catalyse the oxidation of low density lipoprotein. FEBS Lett 352: 15–18
Leake DS (1993) Oxidised low density lipoproteins and atherogenesis. Br Heart J 69: 476–478
Lehr HA, Hübner C, Finckh B, Angermüller S, Nolte D, Beisiegel U, Kohlschütter A, Messmer K (1991) Role of leukotrienes in leukocyte adhesion following systemic administration of oxidatively modified human low density lipoprotein in hamsters. J Clin Invest 88: 9–14
Lenz ML, Hughes H, Mitchell JR, Via DP, Guyton JR, Taylor AA, Gotto AM, Smith C (1990) Lipid hydroperoxy and hydroxy derivatives in copper-catalyzed oxidation of low density lipoprotein. J Lipid Res 31: 1043–1050
Liao F, Berliner JA, Mehrabian M, Navab M, Demer LL, Lusis AJ, Fogelman AM (1991) Minimally modified low density lipoprotein is biologically active in vivo in mice. J Clin Invest 87: 2253–2257
Liu SY, Lu X, Choy S, Dembinski TC, Hatch GM, Mymin D, Shen X, Angel A, Choy PC, Man RYK (1994) Alteration of lysophosphatidylcholine content in low density lipoprotein after oxidative modification: relationship to endothelium dependent relaxation. Cardiovasc Res 28: 1476–1481
Lodge JK, Patel SU, Sadler PJ (1993) Aldehydes from metal ion-and lipoxygenase-induced lipid peroxidation: detection by ^{1}H-NMR spectroscopy. Biochem J 289: 149–153
Lyons TJ (1991) Oxidized low density lipoproteins: a role in the pathogenesis of atherosclerosis in diabetes? Diabetic Med 8: 411–419
Malavasi B, Rasetti MF, Roma P, Fogliatto R, Allevi P, Catapano AL, Galli G (1992) Evidence for the presence of 7-hydroperoxy cholest-5-en-3β-ol in oxidized human LDL. Chem Phys Lipids 62: 209–214
Malden LT, Chait A, Raines EW, Ross R (1991) The influence of oxidatively modified low density lipoproteins on expression of platelet-derived growth factor by human monocyte-derived macrophages. J Biol Chem 266: 13901–13907
McNally AM, Chisolm GM, Morel DW, Cathcart MK (1990) Activated human monocytes oxidize low-density lipoprotein by a lipoxygenase-dependent pathway. J Immunol 145: 254–259
Mino M, Miki M, Miyake M, Ogiahara T (1989) Nutritional assessment of vitamin E in oxidative stress. Ann NY Acad Sci 570: 296–310
Mohr D, Bowry VW, Stocker R (1992) Dietary supplementation with coenzyme Q_{10} results in increased levels of ubiquinol-10 within circulating lipoproteins and increased resistance of human low-density lipoprotein to the initiation of lipid peroxidation. Biochim Biophys Acta 1126: 247–254
Nègre-Salvayre A, Salvayre R (1992) UV-treated lipoproteins as a model system for the study of biological effects of lipid peroxides on cultured cells. 4. Calcium is involved in the cytotoxicity of UV-treated LDL on lymphoid cell lines. Biochim Biophys Acta 1123: 207–215
Niki E (1987) Lipid antioxidants: how they may act in biological systems. Br J Cancer 55: 153–157
Noguchi N, Gotoh N, Niki E (1993) Dynamics of the oxidation of low density lipoprotein induced by free radicals. Biochim Biophys Acta 1168: 348–357
Ohgushi M, Kugiyama K, Fukunaga K, Murohara T, Sugiyama S, Miyamoto E, Yasue H (1993) Protein kinase C inhibitors prevent impairment of endothelium-dependent relaxation by oxidatively modified LDL. Arterioscler Thromb 13: 1525–1532
Palinski W, Rosenfeld ME, Ylä-Herttuala S, Gurtner GC, Socher SS, Butler SW, Parthasarathy S, Carew TE, Steinberg D, Witztum JL (1989) Low density lipoprotein undergoes oxidative modification in vivo. Proc Natl Acad Sci USA 86: 1372–1376

Palinski W, Ylä-Herttuala S, Rosenfeld ME, Butler SW, Socher SA, Parthasarathy S, Curtiss LK, Witztum JL (1990) Antisera and monoclonal antibodies specific for epitopes generated during oxidative modification of low density lipoprotein. Arteriosclerosis 10: 325–335

Panasenko OM, Evgina SA, Aidyraliev RK, Sergienko VI, Vladimirov YA (1994) Peroxidation of human blood lipoproteins induced by exogenous hypochlorite or hypochlorite generated in the system of "myeloperoxidase + H_2O_2 + C^-". Free Radic Biol Med 16: 143–148

Parthasarathy S, Wieland E, Steinberg D (1989) A role for endothelial cell lipoxygenase in the oxidative modification of low density lipoprotein. Proc Natl Acad Sci USA 86: 1046–1050

Peng S, Hu B, Morin RJ (1991) Angiotoxicity and atherogenicity of cholesterol oxides. J Clin Lab Anal 5: 144–152

Petterson KS, Boberg KM, Stabursvik A, Prydz H (1991) Toxicity of oxygenated cholesterol derivatives toward cultured human umbilical vein endothelial cells. Arteriocler Thromb 11: 423–428

Picard S, Parthasarathy S, Fruebis J, Witztum JL (1992) Aminoguanidine inhibits oxidative modification of low density lipoprotein protein and the subsequent increase in uptake by macrophage scavenger receptors. Proc Natl Acad Sci USA 89: 6876–6880

Puhl H, Waeg G, Esterbauer H (1994) Methods to determine oxidation of low-density lipoproteins. Methods Enzymol 233: 425–441

Quinn MT, Parthasarathy S, Fong LG, Steinberg D (1987) Oxidatively modified low density lipoproteins: a potential role in recruitment and retention of monocyte/macrophages during atherogenesis. Proc Natl Acad Sci USA 84: 2995–2998

Rankin SM, Parthasarathy S, Steinberg D (1991) Evidence for a dominant role of lipoxygenase(s) in the oxidation of LDL by mouse peritoneal macrophages. J Lipid Res 32: 449–456

Rajavashisth TB, Andalibi A, Territo MC, Berliner JA, Navab M, Fogelman AM, Lusis AJ (1990) Induction of endothelial cell expression of granulocyte and macrophage colony-stimulating factors by modified low-density lipoproteins. Nature 344: 254–257

Salmon S, Mazière JC, Santus R, Morliere P (1991) A mechanistic study of the interaction of UVB radiations with human serum lipoproteins. Biochim Biophys Acta 1086: 1–6

Salonen J, Ylä-Herttuala S, Yamamoto R, Butler S, Korpela H, Salonen R, Nyyssönen K, Palinski W, Witztum JL (1992) Autoantibody against oxidised LDL and progression of carotid atherosclerosis. Lancet 339: 883–887

Sato K, Niki E, Shimasaki H (1990) Free radical-mediated chain oxidation of low density lipoprotein and its synergistic inhibition by vitamin E and vitamin C. Arch Biochem Biophys 279: 402–405

Sattler W, Mohr D, Stocker R (1994) Rapid isolation of lipoproteins and assessment of their peroxidation by high-performance liquid chromatography postcolumn chemiluminescence. Methods Enzymol 233: 469–489

Savenkova MI, Mueller DM, Heinecke JW (1994) Tyrosyl radical generated by myeloperoxidase: a physiological catalyst for the initiation of lipid peroxidation in low density lipoprotein. J Biol Chem 269: 20394–20400

Sayre LM, Arora PK, Iyer RS, Salomon RG (1993) Pyrrole formation from 4-hydroxynonenal and primary amines. Chem Res Toxicol 6: 19–22

Schmidt K, Graier WF, Kostner GM, Mayer B, Kukovetz WR (1990) Activation of soluble guanylate cyclase by nitrovasodilators is inhibited by oxidized low-density lipoprotein. Biochem Biophys Res Commun 172: 614–619

Sevanian A, Berliner J, Petterson H (1991) Uptake, metabolism, and cytotoxicity of isomeric cholesterol-5,6-epoxides in rabbit aortic endothelial cells. J Lipid Res 32: 147–155

Sevanian A, Seraglia R, Traldi P, Rossato P, Ursini F, Hodis H (1994) Analysis of plasma cholesterol oxidation products using gas- and high-performance liquid chromatography/mass spectrometry. Free Radic Biol Med 17: 397–409

Smith C, Mitchinson MJ, Aruoma OI, Halliwell B (1992) Stimulation of lipid peroxidation and hydroxyl-radical generation by the contents of human atherosclerotic lesions. Biochem J 286: 901–905

Sommer A, Prenner E, Gorges R, Stütz H, Grillhofer H, Kostner GM, Paltauf F, Hermetter A (1992) Organization of phosphatidyl choline and sphingomyelin in the surface monolayer of low density lipoprotein and lipoprotein(a) as determined by time-resolved fluorometry. J Biol Chem 267: 24217–24222

Sparrow CP, Olszewski J (1992) Cellular oxidative modification of low density lipoprotein does not require lipoxygenases. Proc Natl Acad Sci USA 89: 128–131

Sparrow CP, Olszewski J (1993) Cellular oxidation of low density lipoprotein is caused by thiol production in media containing transition metal ions. J Lipid Res 34: 1219–1228

Sparrow CP, Parthasarathy S, Steinberg D (1988) Enzymatic modification of low density lipoprotein by purified lipoxygenase and phospholipase A_2 mimics cell-mediated oxidative modification. J Lipid Res 29: 745–753

Stalenhoef AFH, Kleinveld HA, Kosmeijer-Schuil TG, Demacker PNM, Katan MB (1993) In vivo oxidised cholesterol in atheroscler o sis. Atherosclerosis 98: 113–114

Steinberg D, Parthasarathy S, Carew TE, Khoo JC, Witztum JL (1989) Beyond cholesterol. Modifications of low density lipoprotein that increase its atherogenicity. N Engl J Med 320: 915–924

Steinbrecher UP (1987) Oxidation of human low density lipoprotein results in derivatisation of lysine residues of apolipoprotein B by lipid peroxide decomposition products. J Biol Chem 262: 3603–3608

Steinbrecher UP, Witztum JL, Parthasarathy S, Steinberg D (1987) Decrease in reactive amino groups during oxidation or endothelial cell modification of LDL. Arteriosclerosis 7: 135–143

Steinbrecher UP, Zhang H, Lougheed M (1990) Role of oxidatively modified LDL in atherosclerosis. Free Radic Biol Med 9: 155–168

Stelmaszynska T, Kukovetz E, Egger G, Schaur RJ (1992) Possible involvement of myeloperoxidase in lipid peroxidation. Int J Biochem 24: 121–128

Stiko-Rahm A, Hultgardh-Nilsson A, Regnström J, Hamsten A, Nilsson J (1992) Native and oxidized LDL enhances production of PDGF AA and the surface expression of PDGF receptors in cultured human smooth muscle cells. Arterioscler Thromb 12: 1099–1109

Stocker R, Bowry VW, Frei B (1991) Ubiquinol-10 protects human low density lipoprotein more efficiently against lipid peroxidation than does α-tocopherol. Proc Natl Acad Sci USA 38: 1646–1650

Tamasawa N, Takebe K (1992) Analytical examination of oxidized free and esterified 7-ketocholesterol and related oxysterols in human plasma incubated with copper. Tohoku J Exp Med 168: 37–45

Terkeltaub R, Banka CL, Solan J, Santoro D, Brand K, Curtiss LK (1994) Oxidized LDL induces monocytic cell expression of interleukin-8, a chemokine with T-lymphocyte chemotactic activity. Arterioscler Thromb 14: 47–53

Thomas CE (1992) The influence of medium components on Cu^{2+}-dependent oxidation of low-density lipoproteins and its sensitivity to superoxide dismutase. Biochim Biophys Acta 1128: 50–57

Thomas CE, Jackson RL, Ohweiler DF, Ku G (1994a) Multiple lipid oxidation products in low density lipoproteins induce interleukin-1 beta release from human blood mononuclear cells. J Lipid Res 35: 417–427

Thomas MJ, Thornburg T, Manning J, Hooper K, Rudel LL (1994b) Fatty acid composition of low-density lipoprotein influences its susceptibility to autoxidation. Biochemistry 33: 1828–1834

Thomas JP, Kalyanaraman B, Girotti AW (1994c) Involvement of preexisting lipid hydroperoxides in Cu^{2+}-stimulated oxidation of low-density lipoprotein. Arch Biochem Biophys 315: 244–254

Tribble DL, Holl LG, Wood PD, Krauss RM (1992) Variations in oxidative susceptibility among six low density lipoprotein subfractions of differing density and particle size. Atherosclerosis 93: 189–199

Tribble DL, Van den Berg JJM, Motchnik PA, Ames BN, Lewis DM, Chait A, Krauss RM (1994) Oxidative susceptibillty of low density lipoprotein subfractions is related to their ubiquinol-10 and α-tocopherol content. Proc Natl Acad Sci USA 91: 1183–1187

Uchida K, Stadtman ER (1992) Modification of histidine residues in proteins by reaction with 4-hydroxynonenal. Proc Natl Acad Sci USA 89: 4544–4548

Uchida K, Stadtman ER (1993) Covalent attachment of 4-hydroxynonenal to glyceraldehyde-3-phosphate dehydrogenase. J Biol Chem 268: 6388–6393

Vance JE (1990) Lipoproteins secreted by cultured rat hepatocytes contain the antioxidant 1-alk-1-enyl-2-acylglycerophosphoethanolamine. Biochim Biophys Acta 1045: 128–134

Van Hinsbergh VWM, Van Scheffer M, Havekes L, Kempen HJM (1986) Role of endothelial cells and their products in the modification of low-density lipoproteins. Biochim Biophys Acta 878: 49–64

Wang T, Yu W, Powell WS (1992) Formation of monohydroxy derivatives of arachidonic acid, linoleic acid, and oleic acid during oxidation of low density lipoprotein by copper ions and endothelial cells. J Lipid Res 33: 525–537

Weis JR, Pitas RE, Wilson BD, Rodgers GM (1991) Oxidized low-density lipoprotein increases cultured human endothelial cell tissue factor activity and reduces protein C activation. FASEB J 5: 2459–2465

Wieland E, Parthasarathy S, Steinberg D (1993) Peroxidase-dependent metal-independent oxidation of low density lipoprotein in vitro: a model for in vivo oxidation? Proc Natl Acad Sci USA 90: 5929–5933

Yamaguchi M, Sato H, Bannai S (1993) Induction of stress proteins in mouse peritoneal macrophages by oxidized low density lipoprotein. Biochem Biophys Res Commun 193: 1198–1201

Yang C, Pownall HJ (1993) Structure and function of apolipoprotein B. In: Rosseneu M (ed) Structure and function of apolipoproteins. CRC Press, Boca Raton, pp 64–84

Ylä-Herttuala S, Rosenfeld ME, Parthasarathy S, Glass CK, Sigal E, Witztum JL, Steinberg D (1990) Colocalization of 15-lipoxygenase mRNA and protein with isotopes of oxidized low density lipoprotein in macrophage-rich areas of atherosclerotic lesions. Proc Natl Acad Sci USA 87: 6959–6963

Ylä-Herttuala S, Rosenfeld ME, Parthasarathy S, Glass CK, Sigal E, Särkioja T, Witztum JL, Steinberg D (1991) Gene expression in macrophage-rich human atherosclerotic lesions. 15-lipoxygenase and acetyl low density lipoprotein receptor messenger RNA colocalize with oxidation specific lipid-protein adducts. J Clin Invest 87: 1146–1152

Yoshida Y, Tsuchiya J, Niki E (1994) Interaction of α-tocopherol with copper and its effect on lipid peroxidation. Biochim Biophys Acta 1200: 85–92

Zhang H, Basra HJK, Steinbrecher UP (1990) Effects of oxidatively modified LDL on cholesterol esterification in cultured macrophages. J Lipid Res 31: 1361–1369

Zwijsen RML, Japenga SC, Heijen AMP, Van den Bos RC, Koeman JH (1992a) Induction of platelet-derived growth factor chain a gene expression in human smooth muscle cells by oxidized low density lipoproteins. Biochem Biophys Res Commun 186: 1410–1416

Zwijsen RML, Oudenhoven IMJ, De Haan LHJ (1992b) Effects of cholesterol and oxysterols on gap junctional communication between human smooth muscle cells. Eur J Pharmacol 228: 115–120

Editor-in-Charge: Professor H. Grunicke and Professor M. Schweiger

Rev. Physiol. Biochem. Pharmacol., Vol. 127

p53: DNA Damage, DNA Repair, and Apoptosis

C. GÖTZ and M. MONTENARH

Contents

1 Introduction

Multicellular organisms continually regulate growth and differentiation of their constituent cells. At various times during development and in adult life, cells must choose between growth, cytostasis, and death. Malignancy results when cells are exposed to DNA-damaging agents or radiation, with the result that these cells escape the normal mechanisms which regulate these decisions. Malignant cells acquire the ability to ignore intracellular and extracellular signals which normally regulate cell growth or differentiation or even programmed cell death.

Genes and gene products whose normal function is to inhibit uncontrolled growth might be expected to play key roles in cell cycle control, detection, and eventually repair of DNA damage and thereby to inhibit neoplasia. The growth-suppressor gene product p53 is known to play an efficient part in the cellular growth control machinery (Selter and Montenarh 1994; Donehower and Bradley 1993; Levine 1993). The importance of p53 as a general cellular regulator is evident in its high degree of evolutionary conservation from fish, amphibians, and rodents to man (for review see: Montenarh 1992; Soussi et al. 1990). Wild-type p53 blocks proliferation and transformation in a variety of different cellular systems (Finlay et al. 1989; Eliyahu et al. 1989; Mercer

Department of Medical Biochemistry, University of the Saarland, Building 44, 66421 Homburg, Germany et al. 1990a,b; Baker et al. 1990; Diller et al. 1990; Michalovitz et al. 1990). In addition, wild-type p53 can induce a pathway of programmed cell death in various cellular systems (Yonish-Rouach et al. 1991; Ryan et al. 1993; Shaw et al. 1992). Knockout animals lacking endogenous p53 genes develop normally but are susceptible to tumor formation early in life (Donehower et al. 1992), suggesting that p53 is, in general, dispensable for growth and development. However, the fact that knockout animals exhibit a greatly elevated predisposition to tumor development indicates that p53 is somehow involved in controlling the integrity of the genome. It is not yet known how p53 suppresses neoplastic development, how it inhibits growth, or how it directs cells into an apoptotic pathway. Wild-type p53 binds to double-stranded DNA in a sequence-specific manner (Kern et al. 1991b). It has the ability to repress or activate transcription on its own or by interacting with transcription factors (O'Rourke et al. 1990; Santhanam et al. 1991; Ginsberg et al. 1991; Farmer et al. 1992; Raycroft et al. 1990; Truant et al. 1993; Seto et al. 1992; Agoff et al. 1993). p53 mutations are found in a broad spectrum of human tumor types such as tumors of the breast, colon, lung, and liver and in a variety of leukemias (Caron de Fromentel and Soussi 1992; Hollstein et al. 1991; Greenblatt et al. 1994). In vitro, mutant p53 immortalizes primary cells and cooperates with *ras* or polyoma virus middle-T antigen in transformation of primary cells (Jenkins et al. 1984; Parada et al. 1984; Eliyahu et al. 1984; Reihsaus et al. 1992). Most of the missense mutations in human tumors occur in the highly conserved central region that encodes a domain which binds DNA in a sequence-specific manner (Bargonetti et al. 1993; Pavletich et al. 1993; Wang et al. 1993b; Cho et al. 1994). Mutant p53 is less able to bind to a DNA consensus sequence (El-Deiry et al. 1992; Funk et al. 1992) and fails to transactivate transcription. Most mutant p53 are more stable than wild-type p53; in addition, mutant p53 often have an altered conformation which can be detected with specific monoclonal antibodies (Montenarh 1992). Wild-type p53 has a short half-life, and its concentration in the nucleus of normal, nontransformed cells is low (Kraiss et al. 1991).

p53 is a phosphoprotein which binds to an increasing number of cellular proteins (Pietenpol and Vogelstein 1993). The amino-terminal domain of p53 is acidic and has been shown to interact with several viral and cellular proteins such as the adenovirus E1b 55-kDa protein (Levine 1993), mdm2, and TBP (J. Chen et al. 1993; X. Chen et al. 1993; De Benedetti et al. 1993; Seto et al. 1992). Furthermore, the amino-terminus of p53 functions as a transcriptional activator (Farmer et al. 1992; Zambetti et al. 1992) inducing several genes including mdm2 (Juven et al. 1993; Perry et al. 1993), GADD45 (Kastan et al. 1992; Zhan et al. 1993), and WAF1/CIP1 (El-Deiry et al. 1993). The WAF1 gene product is a potent inhibitor of cell cycle-dependent kinases (cdk) (Harper et al. 1994). C-terminal regions of p53 are involved in binding to cellular proteins such as the regulatory β-subunit of protein kinase CK2 (Herrmann et al. 1991; Wagner et al. 1994), tms1 of fission yeast (Wagner 1994), and a mammalian tms1-related 42-kDa protein (Appel et al. 1994). The internal core region of the polypeptide chain of p53 is involved in specific binding to DNA (Pavletich et al. 1993; Bargonetti et al. 1993; Wang et al. 1993b; Halazonetis et al. 1993; Cho et al. 1994) and binding to SV40 large T antigen (Jenkins et al. 1988; Tan et al. 1986). Mouse p53 has been shown to have the 5.8 S ribosomal RNA covalently bound to a C-terminal serine (Fontoura et al. 1992). Wild-type and mutant p53 are phosphorylated in vivo at several amino- and carboxy-terminal sites, and several sites can be phosphorylated in vitro by various protein kinases (Lees-Miller et al. 1992; Ullrich et al. 1993; Meek et al. 1990; Milne et al. 1992a, 1994; Herrmann et al. 1991; Stürzbecher et al. 1990; Bischoff et al. 1990). However, we know little so far about the in vivo function of p53 phosphorylation. Hupp et al. (1993) reported that the sequence-specific DNA binding of p53 could be activated via phosphorylation by protein kinase CK2. Milne et al. (1992b) reported that the CK2 site seemed to be somehow implicated in the inhibition of cell proliferation. However, it is not yet clear whether changes in phosphorylation also directly influence transcriptional activation by affecting the quaternary structure of p53. Moreover, interaction of p53 with itself (Kraiss et al. 1988) or with other cellular partners such as mdm2, the regulatory β subunit of protein kinase CK2, and a 42-kDa protein (Wu et al. 1993; Appel et al. 1994; Wagner et al. 1994) might affect sequence-specific binding to DNA or RNA.

p53 is a critical member of the G_1 cell cycle checkpoint pathway that arrests cell cycle progression in response to damaged DNA late in G_1 of the cell cycle (Mercer et al. 1990b) or induces programmed cell death, i.e., apoptosis (Oren 1992; Kastan et al. 1992; Kuerbitz et al. 1992; Nelson and Kastan 1994). DNA damage-inducing agents produce a transient increase in the concentration of p53 through post-transcriptional mechanisms that include temporary stabilization of the p53 protein (Maltzman and Czyzyk 1984; Nelson and Kastan 1994).

2 p53 and DNA Damage

Organisms are continuously exposed to a variety of potentially mutagenic or carcinogenic agents. Destruction of DNA by mutation or deletion is a potential step towards carcinogenesis. This was impressively demonstrated in the case of repair-deficient xeroderma pigmentosum cells, which are highly sensitive to UV irradiation-mediated malignant transformation (Cleaver 1989). However, there are various mechanisms known that cause DNA damage or at least base modifications, such as DNA alkylating agents like mitomycin C and cyclophosphamide (Cook and Brazell 1976; Reddy and Randerath 1987). Vinca alkaloids such as etoposide form complexes with DNA and topoisomerase II and eventually cause strand breaks and apoptosis (Kaufmann 1991; Walker et al. 1991). Nucleotide analogs such as 5-fluorouracil cause DNA fragmentation (Parker et al. 1987). DNA strand breaks appear to be the signal which induces p53 protein levels following DNA damage (Nelson and Kastan 1994; Zhan et al. 1994), although it is not clear whether there are single- or double-strand breaks. Derivatives of folate such as methotrexate inhibit purine nucleotide synthesis and thus deprive cells of purine nucleotides for DNA synthesis and DNA repair (Li and Kaminskas 1984). All of these agents or treatments induce high levels of wild-type p53 (Fritsche et al. 1993). Selenite is an essential nutrient (Neve 1991). At doses well above the physiological requirement it is proven anticarcinogenic in many animal systems (Medina and Morrison 1988), whereas at higher doses selenium is highly toxic (Buell 1983). In vitro, selenite is a strong inhibitor of cell growth (Medina and Oborn 1984). Very recently it was shown that selenodiglutathione, the initial metabolite of selenite, induces p53 protein levels in cells that contain wild-type p53 (Lanfear et al. 1994), indicating that selenodiglutathione induces the DNA damage-recognition pathway. In addition, this selenocompound induces apoptosis regardless of whether cells contain wild-type or mutant p53, indicating that DNA damage recognition and apoptosis may be two different pathways.

Retinoids act as therapeutic agents by reducing the incidence of second primary tumors in patients with squamous carcinomas of the head and neck, and they inhibit the growth of tumor cells in experimental systems (Lotan 1980; Hong et al. 1990). It was shown that treatment of non-small cell lung carcinoma cells with retinoids leads to growth inhibition which coincides with a considerable increase in the level of the p53 protein. This increase was specific for wild-type p53 and not detectable in cells carrying mutant p53 (Maxwell and Mukhopadhyay 1994).

Cells treated with the restriction enzyme *Pvu*II respond with a dramatic increase in p53 protein level (Lu and Lane 1993). *Pvu*II treatment is known to generate chromosomal damage (Bryant 1992). Nitric oxide (NO) is involved in several pathophysiological processes, because NO itself or the products of its redox chemistry can cause DNA damage. It was recently shown that NO treatment of cells leads to an accumulation of p53. The overexpression of p53 is later on correlated with apoptotic processes in these cells (Messmer et al. 1994).

Hypoxia and heat, agents that induce cellular stress primarily by inhibiting oxygen-dependent metabolism and denaturing proteins, induce accumulation of p53, although activation of the G_1 arrest by low oxygen conditions is independent of the p53 status. In addition, cells expressing the human papilloma virus E6 gene, which show increased degradation of p53 by ubiquitination and fail to accumulate p53 in response to DNA-damaging agents, do increase their p53 levels following heat and hypoxia (Graeber et al. 1994). Thus, it is tempting to speculate that the level of oxygen used in this study induces a G_1 arrest which is independent of the presence of wild-type p53.

It is well known that ionizing irradiation and UV irradiation induce different kinds of DNA damage, which is repaired through different pathways. Exposure of mammalian cells to γ-irradiation results in single- and double-strand breaks in DNA and in various forms of base alterations in the DNA (Ward 1975). Exposure to DNA-damaging agents probably contributes to the development of many human cancers. In addition to DNA damage, irradiation of cells leads to increased cell mortality and cell cycle arrest in G_1 or in G_2. This growth inhibition allows cells optimal repair of DNA damage before they start to replicate mutated or damaged DNA. Cells which are defective in this repair process may be prone to neoplastic development. Recently published papers indicate that p53 is a critical component of the cellular response to ionizing radiation. p53 seems to be involved in an irradiation-induced cell cycle arrest and in processes directing the cells to an apoptotic pathway of cell death. In general, the response in p53 accumulation seems to be more rapid in cells treated with ionizing irradiation than in cells treated with UV irradiation (Lu and Lane 1993). The elevated levels of p53 resulted in the stimulation of transcription of reporter genes that contain p53-binding sites in their promoter regions.

The level of p53 in transformed cells is regulated mainly by post-translational modification, resulting in prolonged half- life of the protein (Oren et al. 1981; Kraiss et al. 1991). High levels of p53 after DNA damage are due to an elevated stability of the p53 protein (Kastan et al. 1991). A similar observation was made by Maltzman and Cyzyzk (1984) in cells immediately after UV light irradiation. This post-transcriptional increase in p53 could result from changes in the phosphorylation of p53, from binding to other cellular proteins, or from self-oligomerization (for review see: Selter and Montenarh 1994). Both the increase in p53 protein levels and the decrease in DNA synthesis after DNA damage can be blocked by protein synthesis inhibitors such as cycloheximide and by phosphodiesterase inhibitors such as caffeine (Kastan et al. 1991).

Cells from mice after whole-body irradiation exhibit overexpression of wild-type p53 and corresponding growth arrest in G_1, or they go into apoptosis. Cells from mice lacking endogenous p53 have a reduced level of apoptosis compared with wild-type p53-expressing mice (Merritt et al. 1994). In general, cells which lack endogenous p53 or which express mutant p53 do not exhibit G_1 arrest after γ-irradiation. In mouse hepatoma cells expressing wild-type p53, γ-irradiation leads to a transient accumulation of p53 protein in the nuclei, whereas no such accumulation was observed in cells carrying mutant p53 protein (Unger et al. 1994). Crypt cells from small and large intestines of mice expressing wild-type p53 showed abundant apoptosis after γ-irradiation, whereas p53-deficient mice are completely resistant to the induction of apoptosis. Other responses of cells to γ-irradiation, namely G_2 block and reduction in DNA synthesis, were both intact in p53-deficient animals (Clarke et al. 1994). These data show that the reduction of DNA synthesis, DNA chain elongation, and G_2 arrest following γ-irradiation represent p53-independent cellular responses. Similar results were obtained in a comparable study using a different strain of p53-deficient mice (Merritt et al. 1994). These results might lead to the assumption that p53-deficient cells or cells that express mutant p53 have increased levels of DNA damage and therefore may develop tumors more rapidly. However, this hypothesis is not compatible with the finding that p53 deficiency does not lead, in general, to tumor formation in every tissue type. Furthermore, from comparison of the responses of fibroblasts (Lowe et al. 1993a; Slichenmyer et al. 1993), thymocytes (Clarke et al. 1993; Lowe et al. 1993b), hematopoietic cells (Lotem and Sachs 1993), and intestinal epithelium (Clarke et al. 1994), it is clear that not all cell types exhibit p53-dependent apoptosis following γ-irradiation-induced DNA damage.

Modern biochemistry and cell biology rely heavily on the use of radiolabeled tracer molecules. Pulse labeling of cells with ^3H thymidine leads to an increase in p53 mRNA and in p53 protein level. Both effects are accompanied by a reduction of cells in S-phase (Dover et al. 1994). In the light of these results, a great number of experimental results obtained by radiolabeling of cells have to be re-evaluated. When explants of human uroepithelium were exposed to acute doses of cobalt-60, cells that grew out to form primary cultures expressed elevated levels of p53 and *c-myc* (Mothersill et al. 1994b). Elevated levels of p53 are correlated with loss of contact inhibition and the expression of preneoplastic features. Exposure of mouse epidermis to high-dose β-radiation showed that before the appearance of a visible or histologically detectable tumor, overexpression of p53 was common at an exposed site and was accompanied by an overexpression of $p62^{c\text{-}fos}$ and $p21^{N\text{-}ras}$ (Leszczynski et al. 1994).

γ-Irradiation of human diploid fibroblasts in the G_1 phase caused a cell cycle arrest prior to S-phase. This cell cycle arrest is accompanied by a long-term induction of WAF1/CIP1 in normal human fibroblasts (Di Leonardo et al. 1994) and correlated with a lack of activation of both cyclin E/cdk2 and cyclin A/cdk2 kinases and dependent on the presence of wild-type p53. The intact cyclin/cdk2 complexes contain WAF1 (wild-type p53-

activated fragment), an inhibitor of cdk2 which is induced by wild-type p53 (El-Deiry et al. 1993, 1994). As a consequence of activation of the WAF1 inhibitor, phosphorylation of the Rb protein is dramatically inhibited in irradiated cells (Dulic et al. 1994). This block in Rb phosphorylation blocks cells from entering into S-phase.

In addition to that of p53 and WAF1, the expression of mdm2 was also elevated following γ-irradiation. The oncogene mdm2 encodes a 90-kDa protein that binds to the p53 protein and blocks p53-mediated transcriptional activation (Momand et al. 1992). A subset of human tumors contain amplified mdm2 protein and wild-type p53, suggesting that overexpression of mdm2 is another mechanism for altering p53 function in tumorigenesis (Oliner et al. 1992). The mdm2 protein binds to p53 and inactivates p53 wild-type function. Thus, binding of mdm2 to p53 resulted in a reduction of the ability of cells to arrest in G_1 after γ-irradiation, indicating a feedback loop (Chen et al. 1994). A p53-binding site was identified in an intronic sequence of the mdm2 gene (Wu et al. 1993), and expression of mdm2 protein is induced by elevated levels of p53 on the transcriptional level. On the other hand, overexpression of p53 is neutralized by binding of p53 to the mdm2 protein. p53 transcriptionally regulates another cellular gene, namely GADD45 (growth-arrested and DNA damage-inducible gene) (Kastan et al. 1992). GADD45 is a member of the cellular response machinery to radiation. The GADD45 gene harbors a conserved intronic sequence which comprises a p53-binding site, indicating that p53 transactivates GADD45 after γ-irradiation of cells. Although there is overexpression of p53 after radiation of cells, GADD45 transcription is reduced. These data suggest that a balance may exist between the levels of p53 protein and mdm2 protein and the GADD45 gene product. This balance may normally lead cells through the G_1 phase of the cell cycle and may also be involved in regulating DNA repair, apoptosis, and cell cycle progression.

Cells transfected or infected with high-risk human papilloma virus HPV 16 or 18 cannot arrest in G_1 after exposure to DNA-damaging agents (Gujuluva et al. 1994). Growth arrest by induction of p53 in DNA-damaged keratinocytes is also bypassed by human papilloma virus 16 E7 protein (Demers et al. 1994). Rapid degradation of wild-type p53 in HPV-infected or -transfected cells may be responsible for this phenomenon (Scheffner et al. 1990). In general, cells expressing HPV16 E7 proteins continue cell cycle progression following DNA damage-induced p53 activation (Hickman et al. 1994).

Further insight into the role of p53 in the response to radiation-induced DNA damage comes from studies with cells from patients with ataxia telangiectasia (AT) and xeroderma pigmentosum (XP). Ataxia telangiectasia, a human autosomal recessive disorder, is characterized by cerebellar ataxia, oculocutaneous telangiectasia, and sinopulmonary infection (Sedgwick and Boder 1991). Cancer predisposition and hypersensitivity to ionizing radiation are characteristic of this disease. AT cells are acutely sensitive to radiation. Cells of patients suffering from XP are acutely sensitive to UV light, but not to ionizing irradiation. Unlike other fibroblasts, AT cells do not exhibit an increase in p53 protein and fail to arrest in G_1 after ionizing irradiation. On the other hand, p53 induction is normal in AT cells after exposure to UV-B light, an agent to which these cells are not hypersensitive (Kum Kum Khanna and Lavin 1993). Agents that produce double-strand breaks in DNA cause an induction of p53 in the absence of radiation in control cells but not in AT cells (Kum Kum Khanna and Lavin 1993). Specific inhibitors of protein kinase C and serine/threonine phosphatases prevent the UV-radiation-based induction of the p53 protein, implicating these cell-signaling molecules in the p53 response of DNA damage. The failure to express wild-type p53 is associated with a failure to up-regulate the expression of GADD45 (Kastan et al. 1992). Other proteins which participate in the cellular response to γ-irradiation, including WAF1 and mdm2, are also only weakly induced in AT cells after ionizing irradiation (Gates et al. 1994). It has already been demonstrated that mdm2 mRNA levels do not increase in AT cells after ionizing irradiation as much as they do in cells from normal individuals (Price and Park 1994). Induction of p53 is also abnormal in AT cells following treatment with methylmethanesulfonate and bleomycin but appears relatively normal following treatment with UV-C irradiation or topoisomerase inhibitors (Gates et al. 1994). Further evidence for the implication of cell-signaling molecules comes from experiments with hepatocytes, spontaneously immortalized liver epithelial cells, and their H-*ras*-ransformed derivatives. Accumulation of wild-type p53 was observed in primary hepatocytes and immortalized cells, but the arrest was less evident in transformed cells in spite of the fact that these cells express wild-type p53 and accumulate p53. Thus, the signal transduction upstream of p53 may be retained in these transformed cells, although its downstream signals or pathways totally independent of this system could be altered (Kadohama et al. 1994).

Based on all of the above-mentioned results, wild-type p53 is thought to be a key factor in the G_1 arrest of the cell cycle after γ-irradiation. However, it was found that patients with Bloom's syndrome, who lack any detectable p53 protein, still have a G_1 arrest after irradiation with UV-C (Van Laar et al. 1994), indicating the presence of other cellular factors which are involved in the regulation of this cell cycle checkpoint. A complete failure of p53 accumulation in response to UV treatment was observed in two primary cultures obtained from individuals suffering from Bloom's syndrome, whereas in other cases of Bloom's syndrome p53 levels increase following exposure to UV irradiation or X-rays (Lu and Lane 1993).

All of the agents or treatments discussed so far induce high levels of wild-type p53 protein. It was therefore not surprising that p53 mutations are associated with some aspects of resistance to chemotherapy and short survival in various malignancies (Wattel et al. 1994; El Rouby et al. 1993). Although the mechanism whereby p53 might influence resistance to chemotherapy is unclear, it has been shown that wild-type p53 suppresses the multidrug resistance gene promoter, whereas mutated p53 can stimulate it (Chin et al. 1992; Andreeff et al. 1993; El Rouby et al. 1993; Preudhomme et al. 1993). Wild-type p53-expressing cells are radiosensitive, whereas expression of mutant p53 significantly increases the cellular restistance of a variety of hematopoietic cell lines to γ-irradiation. p53-deficient thymocytes remained viable following doses of up to 2000 cGy, while wild-type p53-expressing thymocytes were susceptible to treatment with as little as 100 cGy (Lowe et al. 1993b). Consistent with findings in other cell types, irradiation of thymocytes caused a dramatic increase in p53 levels. Mutation in the p53 gene increases cellular tolerance to DNA damage and abrogates the radiation-induced G_1 arrest (Lee and Bernstein 1993). This radioresistant phenotype has also been observed in fibroblasts from Li-Fraumeni patients, who are heterozygous for p53 mutation. McIlwrath et al. (1994) observed a significant correlation between the level of ionizing radiation-induced G_1 arrest and radiosensitivity; i.e., cells having an intact G_1 arrest are more radiosensitive than cells lacking the G_1 arrest. These results support the idea that wild-type p53 is required for sensitivity of tumor cells to DNA-damaging agents, and that the loss of p53 function in certain tumor cells can lead to resistance to ionizing radiation (McIlwrath et al. 1994). In contrast, Jung et al. demonstrated that mutated p53 was present in squamous cell carcinomas which are radiosensitive, as well as in those which are radioinsensitive. Thus, this study indicates that mutations in the p53 gene do not directly correlate with radioresistance (Jung et al. 1992). It has been shown in rat lung cells that the loss of both wild-type p53 alleles is accompanied by increased sensitivity to γ-irradiation (Biard et al. 1994), although other authors found no correlation between p53 status and radioresistance in a variety of head and neck cancer cell lines (Brachman et al. 1993). Expression of mutant p53 in cells leads to an increase in radiosensitivity of these cells, and this increase seems to correlate with the level of expression of mutant p53 in rat embryo fibroblasts (Bristow et al. 1994), whereas in human keratinocytes mutant p53 confers only a very mild radiosensitivity (Marchetti et al. 1994). Different doses of irradiation were used for these experiments and may account for these different findings.

Radiation therapy is an accepted treatment modality for many types of human tumors, and improvements in the local control of primary tumors may impact on patient survival. However, the potentially carcinogenic effect of therapeutic irradiation has been recognized for years. Radiation-induced sarcomas in most cases have genetic alterations in the p53 gene and in the Rb gene (Brachman et al. 1991). However, it was not clear from this study whether these mutations occurred during tumor progression or resulted directly from radiation exposure. The UV component of sunlight is the major carcinogen involved in the etiology of skin cancer. In skin cancer formation, UV acts as a mutagen for the p53 gene, and it seems that this event occurs at an early stage of multistep UV carcinogenesis (Nagano et al. 1993; Kanjilal et al. 1993). It was shown that 40% of XP skin tumors contain at least one point mutation in the p53 gene (Dumaz et al. 1993). All the mutations were located at dipyrimidine sites, although others reported no evidence for the presence of tandem CC→TT transitions (Amstad et al. 1994). Normal human mammary epithelial cells can be transformed in vitro by γ-irradiation at a clinically used dose (30 Gy). The transformed cells have lost p53 due to a deletion of the gene on one allele, and they have a further small deletion of some base pairs in the second allele (Wazer et al. 1994). The analysis of the p53 gene in radiation-induced osteosarcoma revealed a high incidence for p53 mutations, leading in some cases also to an elevated expression of mutant p53 protein (Strauss et al. 1992). In about 20% of cases of UV-B-induced epidermal tumor cells p53 mutations were found. Interestingly, these mutations were found in the nontranscribed DNA strand, indicating some site and strand specificity of UV-B-induced p53 mutations (Kress et al. 1992). This observation was further supported by data which showed that UV light-induced mutations in the p53 gene were removed from the transcribed strand more rapidly than from the nontranscribed strand of this gene, although the induction of mutations was similar for both strands (Ford et al. 1994; Dumaz et al. 1993). A novel type of inactivation of wild-type p53 was found in radiation-induced osteosarcomas, namely insertion of murine endogenous retrovirus-like elements into intron 4 of the p53 gene, which leads to a p53-deficient cell (Mitreiter et al. 1994). Normal human uroepithelium cells were treated in culture with ^{60}Co irradiation and with or without a chemical carcinogen (N-nitrosodiethanolamine). Four different types of cells were generated. One population of cells never showed abnormal p53 expression; one population was positive for abnormal p53 expression when treated with irradiation and carcinogen. Another population of cells expressed high levels of p53 after a single dose of radiation, and the remaining population of cells was initially negative for p53 and became positive after some days in culture in the presence of growth factor (Mothersill et al. 1994a). It is not clear from this study whether the genetic material from the patients differed in mutability or whether there were other factors involved. In contrast to these results, thymus tissue from mice treated with γ-irradiation had a low frequency (2/15) of p53 mutations, suggesting that p53 mutations do not play a major role in radiation-induced thymic lymphomas, although the types of genetic mutations observed in the p53 gene are similar to those reported in other human tumors (Brathwaite et al. 1992). However, the situation after irradiation is not clear at all. UV-irradiated p53-deficient mouse cells have an elevated number of sister chromatid exchanges compared with wild-type p53-expressing mouse cells. However, there was no deficiency in DNA repair in p53-deficient mouse cells compared with wild-type p53-expressing cells (Ishizaki et al. 1994).

In summary, the inconsistent finding of mutations in the p53 gene after exposure to DNA-damaging radiation may reflect the fact that other or additional factors contribute to the generation of mutated p53. Moreover, in addition to mutation, wild-type p53 can be inactivated by complex formation with cellular and viral proteins and by a subcellular translocation. The contribution of these latter factors to a radiation-induced inactivation of p53 still has to be elucidated.

3 p53 and DNA Repair

A variety of systems have evolved to protect cells from the consequences of DNA damage: DNA proofreading, base excision repair, photoreactivation, methyltransferases, and nucleotide excision repair. The nucleotide excision repair system protects DNA from the widest variety of lesions induced by diverse agents such as UV irradiation, chemical agents, and free radicals (Lindahl 1993). In addition to blocking the cell cycle progression and inducing programmed cell death, there is increasing evidence that p53 directly or indirectly stimulates the DNA repair machinery. The human ERCC-1 gene was the first mammalian DNA repair gene to be cloned (Westerveld et al. 1984). It was demonstrated that ERCC-1 deficiency in mice leads to elevated levels of p53 in kidney, brain, and liver (McWhir et al. 1993). In another approach it was found that wild-type p53 forms complexes with ERCC-3, a basic transcription factor that is involved in transcription-coupled repair, probably by its intrinsic helicase activity (Schaefer et al. 1993). Binding of p53 to the ERCC-3 protein was efficiently inhibited by the hepatitis B virus X protein (Wang et al. 1994). Thus, p53 could modulate repair of DNA by binding to ERCC-3. Inactivation of p53 by mutation or by binding to viral gene products could increase the mutation frequency and thus increase the probablity of neoplastic transformation of human cells.

The growth arrest and DNA damage-inducible gene GADD45 is a ubiquitously expressed mammalian gene that is induced by DNA damage and certain other stresses. The expression of GADD45 is transcriptionally regulated by wild-type p53, although this gene can also be induced by another pathway (Kastan et al. 1992). Although the precise mechanism by which GADD45 is involved in DNA repair is not yet known, by coimmunoprecipitation the proliferating cell nuclear antigen PCNA was found in complex with GADD45 (Smith et al. 1994). PCNA is a component of cyclin-cdk complexes known to be associated with restriction control points within the cell cycle. After DNA damage, PCNA changes its subcellular distribution from sites of DNA replication to sites of DNA damage (Hall et al. 1993). Furthermore, PCNA is an auxiliary factor for repair-associated DNA polymerases δ and ε and is required for nucleotide excision repair, at least in vitro (Shivji et al. 1992). Excision repair was enhanced in cells transfected with a GADD45 gene and reduced when GADD45 was depleted by GADD45-specific antibodies. The interaction of GADD45 with PCNA may have different and probably cumulative effects. It may displace PCNA from WAF1/cdk complexes, thus inhibiting cell cycle progression, or it may compete with cellular factors of the DNA replication machinery for binding to PCNA. This competition may inhibit DNA replication. p53 might therefore enhance DNA repair and inhibition of cellular DNA synthesis by inducing the transcription of the GADD45 gene.

p53 is known to bind single- and double-stranded DNA (Steinmeyer and Deppert 1988; Kern et al. 1991a,b). p53 binds preferentially to single-stranded DNA and catalyzes DNA renaturation and DNA strand transfer (Bakalkin et al. 1994). These findings suggest that p53 may play a direct role in the repair of DNA by binding to DNA.

Repair rates are highly variable and sequence dependent. Repair of p53 mutations due to UV-irradiation was very slow at seven of eight positions which are frequently mutated in skin cancer (Tornaletti and Pfeifer 1994). These results might indicate that the efficiency of DNA repair may somehow contribute to the mutational spectrum seen in cancer-associated genes. By analyzing the repair of p53, of other genes, and of nontranscribed regions of the DNA, the p53 gene was found to be repaired faster than the dihydrofolate reductase gene and much more efficiently than an inactive genomic region (M.K. Evans et al. 1993).

Although there seems to be a good correlation between the presence of wild-type p53 and DNA repair, another report shows that DNA repair can also be normal in p53-deficient cells (Ishizaki et al. 1994).

4 p53 and Apoptosis

Programmed cell death, apoptosis, is as necessary for the survival of organisms as cell proliferation. By apoptosis the organism tries to get rid of cells which have lost their function, which do not function properly (thymic education), which have completed their lifespan, or which are dangerous for the organism (Fesus 1993; Sen and D'Incalci 1992; Williams and Smith 1993). Another aspect of apoptosis is the production of dead cells for specific functions (cornification, lens cells). Thus, apoptosis is a process in the normal life of an organism, but one which must be severely regulated in order to protect cells from undesirable effects. Apoptosis of cells is quite different from necrosis (reviewed in: Sen and D'Incalci 1992), as it requires a genetically controlled sequence of morphological and biochemical events leading at least to death and elimination of the cell. It is a controlled process lacking any inflammatory events. With the onset of apoptosis, the cell shrinks and the chromatin becomes dense. DNA is fragmented by a cellular endonuclease; membrane architecture and membrane composition are drastically changed (membrane blebbing). Furthermore, cellular proteins are cross-linked by a transglutaminase to encapsulate the cell. The apoptotic cells are subsequently phagocytosed. Apoptosis occurs in cells undergoing DNA damage caused by irradiation, cytotoxic chemotherapy, or heating, and in cells suffering from hormone ablation or from growth factor or cytokine deprivation (reviewed in: Kerr et al. 1994). Another reason for the induction of apoptosis may be the inhibition of intercellular contact (Bates et al. 1994). As manifold as the reasons for the induction of cell death are the molecular events happening in response to the apoptosis-inducing signals. Some of the proteins or genes involved in apoptosis are known; some induce and some inhibit the programmed cell death (Williams and Smith 1993). Several cellular or viral proteins have been identified as inhibitors of apoptosis, such as bcl-2 (Hockenbery et al. 1990), adenoviral E1B (White et al. 1992), and bcr-abl (Evans et al. 1993). Other proteins represent the counterplayers of these apoptosis-inhibiting molecules. Bax, a protein with domains homologous to bcl-2, is a dominant negative inhibitor of bcl-2, and also adenoviral E1A (Lowe and Ruley 1993; Debbas and White 1993) and *c-myc* (Evan et al. 1992) have been identified as inducers of apoptosis. The tumor suppressor protein p53 displays a mediating function not only in G_1 arrest (reviewed in: Wagner and Montenarh 1995), but also in induction of apoptosis. Thus, it is a subtle task for a cell to find the balance between these positive and negative regulators of apoptosis.

Insight into the importance of p53 for the induction of apoptosis comes from a variety of cell lines carrying either mutant, inactivated or no p53. A metastasizing variant of human Saos-2 cells (osteogenic sarcoma) producing no endogenous p53 was transfected either with wild-type p53 or with a mutant p53 (p53143A) (Radinsky et al. 1994). After injection of cells into nude mice, metastases stemming from cells expressing p53 in the wild-type conformation consist of a differentiated osteoid matrix and apoptotic cells, whereas cells expressing the mutant conformation of p53 behave like the parental cells. Similar observations were made with murine erythroleukemia cells (Ryan et al. 1993), which also lack an endogenous p53. Transfected with a temperature-sensitive mutant of p53 (p53Val135), cells undergo apoptosis only when expressing the wild-type conformation. Cell lines carrying a mutated p53, such as a Burkitt's lymphoma line (Ramqvist et al. 1993) or a human colon carcinoma-derived cell line EB (Shaw et al. 1992), were drastically reduced in their viability or in their ability to form tumors in nude mice when wild-type p53 was introduced. The same effects were observed in cells where wild-type p53 was inactivated by the binding to SV40 large-T antigen. Both hepatocytes and kidney tubule cells from mice transformed with a temperature-sensitive T antigen were able to follow only the apoptotic pathway when grown at the nonpermissive temperature producing a nonfunctional T antigen (Yanai and Obinata 1994). A T antigen fragment providing the pRb-binding region but not the p53-binding domain is able to produce only slow-growing tumors when introduced into the brain chorioid plexus epithelial cells of transgenic mice (Symonds et al. 1994). Aggressive tumor development is exclusively possible with intact SV40 T antigen, thus inhibiting the p53-dependent apoptosis. Development of tumors occurs when the wild-type conformation of p53 is inactivated in some way, either by mutation or by interaction with viral proteins. From these experiments it is evident that wild-type p53 is implicated in the induction of apoptosis in many different cell lines, and this property also seems to be a critical regulator of oncogenesis. Whether p53 is engaged in inducing programmed cell death seems to depend on the kind of external or internal signals leading to apoptosis. Thus, p53 is not the universal executioner of a doomed cell.

In the programmed cell death of prostatic glandular cells following androgen ablation, p53 function is not required (Berges et al. 1993; Furuya and Isaacs 1993), because the same apoptotic effects were observed after castration of wild-type and p53-deficient mice. In murine proerythroblasts infected with an anemia-inducing strain of Friend virus, apoptosis is induced by the deprivation of erythropoietin (Kelley et al. 1994). Although p53-dependent pathways exist in those cells, the suicide in response to hormone ablation is not committed to conformational changes or stabilization of the wild-type p53 protein. The induction of cell suicide in thymocytes by Ca^{2+} or glucocorticoids does not make use of p53 as a mediating factor either (Clarke et al. 1993).

Most of the studies published so far deal with apoptosis induced by DNA damage. The majority of these experiments were performed with hematopoietic cells or thymocytes, where programmed cell death represents a natural event in the clonal selection of autoreactive thymocytes. Other cells which have been intensively studied in their response to DNA-damaging agents are cells from the intestinal mucosa. In some of these cells DNA damage is accompanied by a rapid increase in the expression of wild-type p53, followed by the apoptotic death of the cell. Cells from the small and large intestines of normal and p53-deficient mice were compared with regard to their response to γ-irradiation (Clarke et al. 1994; Merritt et al. 1994; Potten et al. 1994). Whereas there is no difference in the level of spontaneous apoptosis in small intestine and colon of normal and p53-deficient mice, the complete loss of p53 renders the cells clearly radiation resistant. The apoptosis-inducing function of p53 is more pronounced in small intestine than in colon cells. Furthermore, p53 is not able to produce an all-or-nothing response, as there is an obvious dose dependency for the induction of apoptosis (Jacquemier et al. 1994). Whereas mice deficient in p53 failed to show any response to γ-irradiation, mice heterozygous for p53 alleles exhibited a response intermediate between that of null and wild-type p53 mice. Similar observations concerning the p53-dependent apoptosis were made in hematopoietic cells of mice (Lotem and Sachs 1993) and in thymocytes. Immature thymocytes from mice lacking endogenous p53 (Lowe et al. 1993b), from mice expressing mutant p53 with a deletion (Clarke et al. 1993), or those in which p53 is inactivated by SV40 large-T antigen (McCarthy et al. 1994) are completely resistant to the lethal effects of ionizing radiation. When intact T antigen is substituted by a T antigen fragment unable to complex p53, the cells retain their sensitivity to γ-irradiation. As in intestinal mucosal cells, apoptotic effects showed a dose dependency (Clarke et al. 1993). Besides the p53-dependent pathways, apoptotic pathways exist in the same cell renouncing p53's cooperation. p53-deficient thymocytes which are insensitive to radiation are able to initiate apoptosis when exposed to compounds mimicking T-cell receptor engagement, or to glucocorticoids (Lowe et al. 1993b); they can also execute processes leading to the negative clonal selection of autoreactive thymocytes (McCarthy et al. 1994). Although all these findings concerning apoptosis in response to DNA damage are true for immature thymocytes, the induction of apoptosis in peripheral blood lymphocytes seems to be quite different (Seki et al. 1994). Human peripheral blood lymphocytes exposed to γ-irradiation undergo apoptosis but, depending on the subpopulation, the response differs. While B cells and the TcRα/β^+ subpopulations of T cells depend on p53, no identifiable levels of the tumor suppressor protein were induced in TcR γ/δ^+ T cells or natural killer cells. In subpopulations stemming from the same progenitor cell which carry out a p53-dependent apoptosis there has been a development of p53-dependent and -independent pathways, leading to radiation-induced apoptosis.

p53 is not involved in every apoptosis induction caused by DNA damage. Only DNA damage which results in strand breaks leads to the increased expression of p53 and the subsequent death of the cell (Nelson and Kastan 1994; Clarke et al. 1993). Agents rapidly inducing DNA damage such as ionizing radiation and bleomycin, and DNA topoisomerase-targeted drugs like etoposide (inhibitor of topoisomerase II) trigger the p53 elevation. Treatment of several human cell lines with the antimetabolite N-phosphono-acetyl-l-aspartate (PALA) resulted in delayed increases in p53 levels temporally correlated with the first appearance of DNA strand breaks. The camptothecin-stimulated trapping of DNA topoisomerase completely failed to increase p53 levels. A rapid increase in p53 level was also obtained by the introduction of nucleases into human cells (Nelson and Kastan 1994), confirming the assumption that p53 becomes active only when the DNA strand is broken. Thus, the tumor suppressor protein plays a major role in the induction of apoptosis caused by DNA damage. However, the responses to DNA-damaging agents differ strongly, depending on the cell type, the maturation grade, the kind of damage, and the p53 level in the cell.

Besides radiation-induced p53-dependent apoptosis, there are also other stimuli to release the tumor suppressor protein function. p53-dependent apoptosis is observed in response to survival or growth factor withdrawal or in response to certain cytokines. Hematopoietic cells usually die when they are deprived of cytokines like interleukin 3 (Gottlieb et al. 1994) or interleukin 6 (Yonish-Rouach et al. 1991, 1993). In the presence of these factors cells can survive even when p53 is overexpressed in the cell line; thus, the apoptotic effects are compensated by these effector substances. In factor-dependent acute myeloid leukemia blasts (AML) and human erythroleukemia cell line TF1 the induction of apoptosis could be attributed to wild-type p53 by microinjection of p53 antisense oligonucleotides (Zhu et al. 1994). Even in the absence of GM-CSF apoptosis of cells was suppressed by inhibition of p53. Another cytokine whose effectiveness in apoptosis has been demonstrated is tumor necrosis factor α (Gotlieb et al. 1994). Ovarian carcinoma cells exposed to TNF α showed a time- and dose-dependent up-regulation of p53 mRNA. Along with this up-regulation, cell suicide was indicated by the typical DNA fragmentation. As observed in the developing mouse lens, Rb deficiency leads to p53-dependent apoptosis (Morgenbesser et al. 1994). Under normal conditions, lens epithelial cells occasionally go into apoptosis, but lens fiber cells do not. Rb deficiency causes apoptosis in lens fiber cells which seems to depend on the presence of p53, as the hallmarks of apoptosis such as DNA fragmentation were absent in embryos from mice deficient in Rb and p53. The binding of pRb to the HPV E7 protein in the postnatal mouse lens led to very similar apoptotic effects, which could be compensated by the simultaneous expression of the p53-binding protein HPV E6 (Pan and Griep 1994). The expression of both viral proteins led to a reduction of apoptosis and to formation of lens tumors. The transcription factor E2F is possibly involved in this communicative pathway between Rb and p53 leading to apoptosis (Wu and Levine 1994). Coexpression of the two proteins in a murine cell line containing a temperature-sensitive p53 and E2F results in a loss of viability.

p53 expression and subsequent apoptosis were also induced by nitric oxide (Messmer et al. 1994). Nitric oxide is generated by a cytokine-induced NO synthase or by NO donors. In certain mouse macrophages and in a pancreatic β-cell line the NO generation stimulates the expression of p53 and the following cell death. By application of NO synthase inhibitors the effects initiated by the NO generation are suppressed.

Another field of p53 function in apoptosis has been described in response to cellular and viral oncogenes. A *v-myc*-induced T-cell lymphoma line which is heterozygous for p53 and whose one p53 allele is inactivated by the insertion of a Moloney murine leukemia provirus was transfected by a temperature-sensitive mutant of p53 (p53Val135). Shifted to its wild-type conformation, p53 rapidly induces loss of viability of the transformed cells (Wang et al. 1993a). The same seems to be true for the cellular counterpart of the oncogene. *c-myc* is involved in cell transformation and proliferation as well as in apoptosis. Constitutive *c-myc* expression can lead to cell death when cell proliferation is inhibited, due to the lack of growth-stimulation factors (Evan et al. 1992). *c-myc*-induced apoptosis seems to involve p53, as in quiescent mouse fibroblasts with wild-type p53 activation of *myc* was found to induce cell cycle reentry and apoptosis, preceded by a stabilization of p53 (Hermeking and Eick 1994). In p53-null fibroblasts only cell cycle reentry but not apoptosis was induced, thus suggesting that p53 takes over a protecting role to prevent cell proliferation induced by oncogene activation. Another viral oncogen which causes a p53-dependent apoptosis is the adenovirus E1A protein (Debbas and White 1993; Lowe and Ruley 1993; Lowe et al. 1994). Usually, together with the viral product E1B, E1A leads to transformation of primary rodent cells in a multistep process. By transfection of the E1A product alone, cell proliferation is counteracted by a stabilized p53, which turns on the apoptotic pathway. The E1B 19-kDa and the E1B 55-kDa protein can interact independently with p53, and both provide distinct mechanisms to inhibit apoptosis and thus make the transformation of cells a possible event (Shen and Shenk 1994).

An interesting reason for the initiation of apoptosis is the loss of intercellular contact in a colon carcinoma cell line (Bates et al. 1994). The inhibition of cell contact performed by an anti-integrin antibody drives the cells into apoptosis. In contact-inhibited cells two proteins with molecular weights of 68 and 72 kDa are produced and p53 is translocated from cytoplasma to the nucleus.

Only little is known about genes which are targeted by p53 and which, in the end, help to kill a cell. The wild-type p53-activated fragment WAF1/CIP1 seems to be involved not only in G_1 arrest but also in apoptosis of certain cells (El-Deiry et al. 1994). The *v-myc*-expressing T-cell lymphoma line J3D lacking endogenous p53 and a derivative transfected with a temperature-sensitive p53 (M3) were compared with regard to their ability to induce apoptosis. By shifting the cells to the permissive temperature, expression of WAF1 as well as the induction of apoptosis occurred only in the p53-expressing cell line M3. BAF3, a murine hematopoietic cell line, reacts upon ionizing radiation with G_1 arrest in the presence of Il-3 and with apoptosis after deprivation of IL-3. In both cases p53 and its downstream effector WAF1 are expressed. Thus, the decision to commit suicide or not depends on a second signal. In the same cell line WAF1 is not expressed during the induction of p53-independent pathways. Thus, WAF1 seems to be one of the specific effectors of p53-mediated apoptosis.

Two other genes are synergistic or antagonistic co-workers of p53 in the execution of apoptosis: bcl-2, an apoptosis-suppressing gene, and bax, a dominant inhibitor of bcl-2. bcl-2 is a 25-kDa integral protein of the inner mitochondrial membrane (Hockenbery et al. 1990) which was first detected because of its oncogenic function in the majority of non-Hodgkin's B-cell lymphomas (Tsujimoto et al. 1985) The localization in the mitochondrial membrane seems to be important for the apoptosis-inhibiting effects of bcl-2 (Nguyen et al. 1994). Overexpression of bcl-2 blocks apoptosis of a pre-B-lymphocyte line (Hockenbery et al. 1990; Hockenbery 1992). The counteracting role of bcl-2 concerning p53-dependent apoptosis is possibly due to the fact that in tumor cells often an inverse correlation or mutual exclusion between the expression of p53 and bcl-2 exists (Silvestrini et al. 1994; Haldar et al. 1994; Pezzella et al. 1993). Together with *c-myc,* bcl-2 is able to modulate the p53 function by altering the subcellular trafficking during the cell cycle (Ryan et al. 1994). The coexpression of two proto-oncogenes in a murine erythroleukemia cell line can totally overcome the p53-induced apoptosis by retaining the p53 protein in the cytoplasm during a critcal period in G_1. In a *v-myc*-induced T-cell lymphoma line wild-type p53-triggered apoptosis is inhibited by bcl-2 (Wang et al. 1993c). Thus, bcl-2 can prevent apoptosis in a cell line simultaneously exposed to the cell-proliferating signals of *c-myc* and the growth-arresting signals of wild-type p53. p53 turned out to be a regulator of bcl-2 and bax gene expression in vivo and in vitro (Miyashita et al. 1994). In a murine leukemia cell line transfected with a temperature-sensitive p53, a temperature-dependent decrease in the expression of bcl-2 was observed, accompanied by a simultaneous increase in the expression of the counteracting protein bax. In mice lacking p53 the reverse observation was made, thus confirming indirectly the results made in vitro. Two regions with a highly conserved homology (BH1 and BH2) are important for the inhibition of apoptosis and the heterodimerization with bax. Mutations in these regions abrogate the death-repressor activity and its ability for heterodimerization (Yin et al. 1994). A functional homolog of this apoptosis-inhibiting protein is the adenoviral E1B gene product (White et al. 1992; Chiou et al. 1994). Human bcl-2 expression completely prevented p53-dependent apoptosis in an adenovirus E1A-transformed rodent cell line and converted the cells into a growth-arrested status without affecting the levels or the localization of the p53 protein. Bax has been detected in association with bcl-2 (Oltvai et al. 1993). The 21-kDa protein shares a high sequence homology with its counteractor bcl-2 and can form heterodimers with this protein. Bax accelerates apoptotic cell death induced by cytokine deprivation (interleukin 3) and thus counteracts the repressor activity of bcl-2. The efficiency of the dominant inhibitor of bcl-2 is shown by comparison of p53-dependent and TGFβ1-dependent apoptosis (Selvakumaran et al. 1994). Both apoptotic pathways are based on the down-regulation of the bcl-2 expression, but only the p53-dependent pathway simultaneously induces an up-regulation of bax. Not only does the p53-dependent apoptosis occur more rapidly; even ectopic bcl-2 levels which inhibit the TGFβ1-induced apoptosis cause only a delay in the p53-dependent cell death. These differences can presumably be attributed to the efficacy of bax in the initiation of cell suicide. From all these experiments it becomes evident that the ratio of bcl-2 to bax determines the survival or the death of a cell.

There are also reports which deal with p53-dependent apoptosis in the absence of p53 target genes (Caelles et al. 1994). GHFT1 cells, SV40 T antigen-immortalized cells from a transgenic pituitary tumor transfected with a temperature-sensitive p53, undergo apoptosis in response to UV- and γ-irradiation, mitomycin C, and 4- nitroquinoline-1-oxide when p53 is present in its wild-type conformation. Apoptosis is triggered independent of RNA or protein synthesis, and apoptotic cleavage happens only while p53 is functional. The application of actinomycin D or cycloheximide inhibits the induction of potential p53 target genes like WAF1 or mdm-2, but the apoptotic program still takes place. According to these results, apoptosis does not depend on activation of p53 target genes; p53 rather represses genes necessary for cell survival or is a component of the enzymatic machinery for apoptotic cleavage or repair of DNA. The fact that p53 is able to interact with RPA (Dutta et al. 1993) would support this idea.

There is increasing evidence that cancer might be the consequence of the inability of cells to induce apoptosis at the right time and in the right place. Induction of cell death would presumably be an ingenious possibility of treating some kinds of cancer in order to achieve tumor regression. As p53 seems to be important for at least some forms of apoptosis, it would be a good tool in the fight against cancer. There are already some reports about the application of p53 alone or in combination with anticancer drugs in therapeutic trials.

The cytotoxicity of some anticancer agents like ionizing radiation, 5-fluorouracil, etoposide, and adriamycin is clearly modulated by a p53-dependent apoptosis (Lowe et al. 1993a). p53 is necessary for the efficient performance of the death program, as mouse embryonic fibroblasts deficient in p53 are resistant to this diverse group of chemotherapeutic drugs. The fact that different therapeutic agents seem to use a common mechanism to kill a cell also provides a possible explanation for the development of multidrug resistance.

Stimulation of apoptosis which might lead to tumor regression is turned on by the application of selenodiglutathione (SDG), for example (Lanfear et al. 1994). In murine erythroleukemia cells (MEL) and the ovarian cell line A2780, SDG induces rapid and irreversible changes in morphology characteristic for apoptosis. The apoptotic cells show a higher membrane permeability and the typical chromosomal fragmentation, but only the ovarian cell line which contains wild-type p53 shows an elevation in p53 level. As MEL cells contain mutant p53, the tumor suppressor is not necessarily connected to the induction of the apoptotic effects induced by the application of selenodiglutathione. The introduction of wild-type p53 itself is utilized to increase the chemosensitivity towards some kinds of drugs in human lung cancer cells (Fujiwara et al. 1994). The human lung cancer cell line H358, with a homozygous deletion of p53, shows a higher sensitivity towards cisplatin when transfected with a recombinant adenovirus construct containing wild-type p53. In nude mice implanted with H358 tumors the combined administration of cisplatin and the p53 construct led to a massive destruction of tumors by apoptosis. Also p53 alone turned out to be a successful therapeutic agent. Multicellular tumor spheroids of human non-small-cell lung cancer cell lines serve as a model for solid tumors (Fujiwara et al. 1993). H322a cells express a p53 gene which is homozygously mutated at codon 248; WT226 cells have an endogenous wild-type p53. A retroviral vector with a wild-type p53 can penetrate the tumor spheroids and cause apoptotic effects in the H322a cell line; the WT226 cell line is immune towards the apoptotic effects of the ectopic p53. The introduction of p53 in an adenoviral vector has also been successfully applied in squamous cell carcinoma of the head and neck (Liu et al. 1994). After the transduction of the plasmid, exogenous wild-type p53 was expressed at levels up to tenfold higher than the endogenous mutant p53. In vitro the expression resulted in growth arrest and morphological changes consistent with apoptosis. In nude mice with established carcinoma nodules regression of the tumors was observed after peritumoral infiltration of the vector. All these examples show the usefulness of p53 in cancer therapy by attenuating the replication rate of tumor cells in vitro and the growth of solid tumors in vivo.

5 Concluding Remarks

p53 plays a major role in some forms of apoptotic processes. In response to some external or internal signals like DNA damage, stress situations like growth factor withdrawal, or the presence of oncogenes, levels of wild-type p53 are elevated in the cell. p53 turns out to be a component of a genetically controlled cascade of events leading at least to cell suicide. In this process p53 does not take over the decision about life or death of a cell; rather, it acts as a mediator between the decisive institution and the executive tools. Thus, apoptosis proves to be a complex process in which p53 plays an important but not the decisive role; putting together all the pieces of this complex puzzle may help in fighting cancer.

Acknowledgements. This work was supported by grant W77/93/Mo2 from the *Deutsche Krebshilfe eV* and the *Deutsche Forschungsgemeinschaft* (SFB 246). The authors thank Katrin Appel for her critical reading of the manuscript and helpful suggestions and Marianne Buchholz for her editorial help.

References

Agoff SN, Hou J, Linzer DIH, Wu B (1993) Regulation of the human hsp70 promoter by p53. Science 259: 84–87

Amstad P, Hussain SP, Cerutti P (1994) Ultraviolet B light-induced mutagenesis of p53 hotspot codons 248 and 249 in human skin fibroblasts. Mol Carcinog 10: 181–188

Andreeff M, Zhao S, Drach D, Hegewisch-Becker S, Rees JHK, Liu Y, Hanania EG, Körbling M, Deisseroth AB, Drach J (1993) Expression of multidrug resistance (mdr-1) and p53 genes in hematologic cell systems: implications for biology and gene therapy. Cancer Bull 45: 131–138

Appel K, Schneider E, Wagner P, Höog J-O, Karlsson C, Montenarh M (1994) A new 42 KDa protein binding to the growth suppressor protein p53. Int J Oncol 5: 667–673

Bakalkin G, Yakovleva T, Selivanova G, Magnusson KP, Szekely L, Kiseleva E, Klein G, Terenius L, Wiman KG (1994) p53 Binds single-stranded DNA ends and catalyzes DNA renaturation and strand transfer. Proc Natl Acad Sci USA 91: 413–417

Baker SJ, Markowitz S, Fearon ER, Willson JKV, Vogelstein B (1990) Suppression of human colorectal carcinoma cell growth by wild-type p53. Science 249: 912–915

Bargonetti J, Manfredi JJ, Chen X, Marshak DR, Prives C (1993) A proteolytic fragment from the central region of p53 has marked sequence-specific DNA-binding activity when generated from wild-type but not from oncogenic mutant p53 protein. Genes Dev 7: 2565–2574

Bates RC, Buret A, van Helden DF, Horton MA, Burns GF (1994) Apoptosis induced by inhibition of intercellular contact. J Cell Biol 125: 403–415

Berges RR, Furuya Y, Remington L, English HF, Jacks T, Isaacs JT (1993) Cell proliferation, DNA repair, and p53 function are not required for programmed death of prostatic glandular cells induced by androgen ablation. Proc Natl Acad Sci USA 90: 8910–8914

Biard DSF, Martin M, Le Rhun Y, Duthu A, Lefaix JL, May E, May P (1994) Concomitant p53 gene mutation and increased radiosensitivity in rat lung embryo epithelial cells during neoplastic development. Cancer Res 54: 3361–3364

Bischoff JR, Friedman PN, Marshak DR, Prives C, Beach D (1990) Human p53 is phosphorylated by p60-cdc2 and cyclin B-cdc2. Proc Natl Acad Sci USA 87: 4766–4770

Brachman DG, Hallahan DE, Beckett MA, Yandell DW, Weichselbaum RR (1991) p53 gene mutations and abnormal retinoblastoma protein in radiation-induced human sarcomas. Cancer Res 51: 6393–6396

Brachman DG, Beckett M, Graves D, Haraf D, Vokes E, Weichselbaum RR (1993) p53 mutation does not correlate with radiosensitivity in 24 head and neck cancer cell lines. Cancer Res 53: 3667–3669

Brathwaite O, Bayona W, Newcomb EW (1992) p53 mutations in C57BL/6J murine thymic lymphomas induced by gamma-irradiation and N-methylnitrosourea. Cancer Res 52: 3791–3795

Bristow RG, Jang A, Peacock J, Chung S, Benchimol S, Hill RP (1994) Mutant p53 increases radioresistance in rat embryo fibroblasts simultaneously transfected with HPV16-E7 and/or activated H-*ras*. Oncogene 9: 1527–1536

Bryant PE (1992) Induction of chromosomal damage by restriction endonuclease in CHO cells porated with streptolysin O. Mutat Res 268: 27–34

Buell DN (1983) Potential hazards of selenium as a chemopreventive agent. Semin Oncol 10: 311–321

Caelles C, Helmberg A, Karin M (1994) p53-dependent apoptosis in the absence of transcriptional activation of p53-target genes. Nature 370: 220–223

Caron de Fromentel C, Soussi T (1992) p53 tumor suppressor gene: a model for investigating human mutagenesis. Genes Chrom Cancer 4: 1–15

Chen C-Y, Oliner JD, Zhan Q, Fornace A Jr, Vogelstein B, Kastan MB (1994) Interactions between p53 and MDM2 in a mammalian cell cycle checkpoint pathway. Proc Natl Acad Sci USA 91: 2684–2688

Chen J, Marechal V, Levine AJ (1993) Mapping of the p53 and mdm-2 interaction domains. Mol Cell Biol 13: 4107–4114

Chen X, Farmer G, Zhu H, Prywes R, Prives C (1993) Cooperative DNA binding of p53 with TFIID (TBP): a possible mechanism for transcriptional activation. Genes Dev 7: 1837–1849

Chin K-V, Ueda K, Pastan I, Gottesman MM (1992) Modulation of activity of the promoter of the human MDR1 gene by ras and p53. Science 255: 459–462

Chiou S-K, Rao L, White E (1994) Bcl-2 blocks p53-dependent apoptosis. Mol Cell Biol 14: 2556–2563

Cho Y, Gorina S, Jeffrey PD, Pavletich NP (1994) Crystal structure of a p53 tumor suppressor-DNA complex: understanding tumorigenic mutations. Science 265: 346–355

Clarke AR, Purdie CA, Harrison DJ, Morris RG, Bird CC, Hooper ML, Wyllie AH (1993) Thymocyte apoptosis induced by p53-dependent and -independent pathways. Nature 362: 849–852

Clarke AR, Gledhill S, Hooper ML, Bird CC, Wyllie AH (1994) p53 dependence of early apoptotic and proliferative responses within the mouse intestinal epithelium following gamma-irradiation. Oncogene 9: 1767–1773

Cleaver JE (1989) How many human genetic disorders affect cellular radiosensitivity? Cancer Cells 1: 108–110

Cook PR, Brazell IA (1976) Detection and repair of single-strand breaks in nuclear DNA. Nature 263: 679–682

De Benedetti L, Varesco L, Pellegata NS, Losi L, Gismondi V, Casarino L, Sciallero S, Bonelli L, Biticchi R, Bafico A, Masetti E, James R, Heouaine A, Ranzani GN, Aste H, Ferrara G (1993) Genetic events in sporadic colorectal adenomas: K-ras and p53 heterozygous mutations are not sufficient for malignant progression. Anticancer Res 13: 667–670

Debbas M, White E (1993) Wild-type p53 mediates apoptosis by E1A, which is inhibited by E1B. Genes Dev 7: 546–554

Demers GW, Foster SA, Halbert CL, Galloway DA (1994) Growth arrest by induction of p53 in DNA damaged keratinocytes is bypassed by human papillomavirus 16 E7. Proc Natl Acad Sci USA 91: 4382–4386

Di Leonardo A, Linke SP, Clarkin K, Wahl GM (1994) DNA damage triggers a prolonged p53-dependent G_1 arrest and long-term induction of Cip1 in normal human fibroblasts. Genes Dev 8: 2540–2551

Diller L, Kassel J, Nelson CE, Gryka MA, Litwak G, Gebhardt M, Bressac B, Ozturk M, Baker SJ, Vogelstein B, Friend SH (1990) p53 functions as a cell cycle control protein in osteosarcomas. Mol Cell Biol 10: 5772–5781

Donehower LA, Bradley A (1993) The tumor suppressor p53. Biochim Biophys Acta Rev Cancer 1155: 181–205

Donehower LA, Harvey M, Slagle BL, McArthur MJ, Montgomery CA Jr, Butel JS, Bradley A (1992) Mice deficient for p53 are developmentally normal but susceptible to spontaneous tumours. Nature 356: 215–221

Dover R, Jayaram Y, Patel K, Chinery R (1994) p53 expression in cultured cells following radioisotope labelling. J Cell Sci 107: 1181–1184

Dulic V, Kaufmann WK, Wilson SJ, Tlsty TD, Lees E, Harper JW, Elledge SJ, Reed SI (1994) p53-dependent inhibition of cyclin-dependent kinase activities in human fibroblasts during radiation-induced G1 arrest. Cell 76: 1013–1023

Dumaz N, Drougard C, Sarasin A, Daya-Grosjean L (1993) Specific UV-induced mutation spectrum in the p53 gene of skin tumors from DNA-repair-deficient xeroderma pigmentosum patients. Proc Natl Acad Sci USA 90: 10529–10533

Dutta A, Ruppert JM, Aster JC, Winchester E (1993) Inhibition of DNA replication factor RPA by p53. Nature 365: 79–82

El-Deiry WS, Kern SE, Pietenpol JA, Kinzler KW, Vogelstein B (1992) Definition of a consensus binding site for p53. Nat Genet 1: 45–49

El-Deiry WS, Tokino T, Velculescu VE, Levy DB, Parsons R, Trent JM, Lin D, Mercer WE, Kinzler KW, Vogelstein B (1993) WAF1, a potential mediator of p53 tumor suppression. Cell 75: 817–825

El-Deiry WS, Harper JW, O'Connor PM, Velculescu VE, Canman CE, Jackman J, Pietenpol JA, Burrell M, Hill DE, Wang Y, Wiman KG, Mercer WE, Kastan MB, Kohn KW, Elledge SJ, Kinzler KW, Vogelstein B (1994) WAF1/CIP1 is induced in p53-mediated G_1 arrest and apoptosis. Cancer Res 54: 1169–1174

El Rouby S, Thomas A, Costin D, Rosenberg CR, Potmesil M, Silber R, Newcomb EW (1993) p53 gene mutation in B-cell chronic lymphocytic leukemia is associated with drug resistance and is independent of MDR1/MDR3 gene expression. Blood 82: 3452–3459

Eliyahu D, Raz A, Gruss P, Oren M (1984) Participation of p53 cellular tumor antigen in transformation of normal embryonic cells. Nature 312: 646–649

Eliyahu D, Michalovitz D, Eliyahu S, Pinhasi-Kimhi O, Oren M (1989) Wild-type p53 can inhibit oncogene-mediated focus formation. Proc Natl Acad Sci USA 86: 8763–8767

Evan GI, Wyllie AH, Gilbert CS, Littlewood TD, Land H, Brooks M, Waters CM, Penn LZ, Hancock DC (1992) Induction of apoptosis in fibroblasts by *c-myc* protein. Cell 69: 119–128

Evans CA, Owen-Lynch PJ, Whetton AD, Dive C (1993) Activation of the Abelson tyrosine kinase activity is associated with suppression of apoptosis in hemopoietic cells. Cancer Res 53: 1735–1738

Evans MK, Taffe BG, Harris CC, Bohr VA (1993) DNA strand bias in the repair of the p53 gene in normal human and xeroderma pigmentosum group C fibroblasts. Cancer Res 53: 5377–5381

Farmer G, Bargonetti J, Zhu H, Friedman P, Prywes R, Prives C (1992) Wild-type p53 activates transcription in vitro. Nature 358: 83–86

Fesus L (1993) Biochemical events in naturally occurring forms of cell death. FEBS Lett 328: 1–5

Finlay CA, Hinds PW, Levine AJ (1989) The p53 proto-oncogene can act as a suppressor of transformation. Cell 57: 1083–1093

Fontoura BMA, Sorokina EA, David E, Carroll RB (1992) p53 is covalently linked to 5.8S rRNA. Mol Cell Biol 12: 5145–5151

Ford JM, Lommel L, Hanawalt PC (1994) Preferential repair of ultraviolet light-induced DNA damage in the transcribed strand of the human p53 gene. Mol Carcinog 10: 105–109

Fritsche M, Haessler C, Brandner G (1993) Induction of nuclear accumulation of the tumor-suppressor protein p53 by DNA-damaging agents. Oncogene 8: 307–318

Fujiwara T, Grimm EA, Mukhopadhyay T, Wei Cai D, Owen-Schaub LB, Roth JA (1993) A retroviral wild-type p53 expression vector penetrates human lung cancer spheroids and inhibits growth by inducing apoptosis. Cancer Res 53: 4129–4133

Fujiwara T, Grimm EA, Mukhopadhyay T, Zhang WW, Owen-Schaub LB, Roth JA (1994) Induction of chemosensitivity in human lung cancer cells in vivo by adenovirus-mediated transfer of the wild-type p53 gene. Cancer Res 54: 2287–2291

Funk WD, Pak DT, Karas RH, Wright WE, Shay JW (1992) A transcriptionally active DNA-binding site for human p53 protein complexes. Mol Cell Biol 12: 2866–2871

Furuya Y, Isaacs JT (1993) Differential gene regulation during programmed death (apoptosis) versus proliferation of prostatic glandular cells induced by androgen manipulation. Endocrinology 133: 2660–2666

Gates CE, Reed CE, Bromberg JS, Everett ET, Baron PL (1994) Prevalence of p53 mutations in patients with squamous cell carcinoma of the esophagus. J Thorac Cardiovasc Surg 108: 148–152

Ginsberg D, Mechta F, Yaniv M, Oren M (1991) Wild-type p53 can down-modulate the activity of various promoters. Proc Natl Acad Sci USA 88: 9979–9983

Gotlieb WH, Watson JM, Rezai A, Johnson M, Martínez-Maza O, Berek JS (1994) Cytokine-induced modulation of tumor suppressor gene expression in ovarian cancer cells: up-regulation of p53 gene expression and induction of apoptosis by tumor necrosis factor-a. Am J Obstet Gynecol 170: 1121–1130

Gottlieb E, Haffner R, Von Rüden T, Wagner EF, Oren M (1994) Down-regulation of wild-type p53 activity interferes with apoptosis of IL-3-dependent hematopoietic cells following IL-3 withdrawal. EMBO J 13: 1368–1374

Graeber TG, Peterson JF, Tsai M, Monica K, Fornace AJ, r., Giaccia AJ (1994) Hypoxia induces accumulation of p53 protein, but activation of a G_1-phase checkpoint by low-oxygen conditions is independent of p53 status. Mol Cell Biol 14: 6264–6277

Greenblatt MS, Bennett WP, Hollstein M, Harris CC (1994) Mutations in the p53 tumor suppressor gene: clues to cancer etiology and molecular pathogenesis. Cancer Res 54: 4855–4878

Gujuluva CN, Baek J-H, Shin K-H, Cherrick HM, Park N-H (1994) Effect of UV-irradiation on cell cycle, viability and the expression of p53, *gadd*153 and *gadd*45 genes in normal and HPV-immortalized human oral keratinocytes. Oncogene 9: 1819–1827

Halazonetis TD, Davis LJ, Kandil AN (1993) Wild-type p53 adopts a 'mutant'-like conformation when bound to DNA. EMBO J 12: 1021–1028

Haldar S, Negrini M, Monne M, Sabbioni S, Croce CM (1994) Down-regulation of bcl-2 by p53 in breast cancer cells. Cancer Res 54: 2095–2097

Hall PA, McKee PH, Du P.Menage H, Dover R, Lane DP (1993) High levels of p53 protein in UV-irradiated normal human skin. Oncogene 8: 203–207

Harper JW, Adami GR, Wei N, Keyomarsi K, Elledge SJ (1994) The p21 Cdk-interacting protein cip1 is a potent inhibitor of G1 cyclin-dependent kinases. Cell 75: 805–816

Hermeking H, Eick D (1994) Mediation of *c-myc*-induced apoptosis by p53. Science 265: 2091–2093

Herrmann CPE, Kraiss S, Montenarh M (1991) Association of casein kinase II with immunopurified p53. Oncogene 6: 877–884

Hickman ES, Picksley SM, Vousden KH (1994) Cells expressing HPV16 E7 continue cell cycle progression following DNA damage-induced p53 activation. Oncogene 9: 2177–2181

Hockenbery D, Nunez G, Milliman C, Schreiber RD, Korsmeyer SJ (1990) Bcl-2 is an inner mitochondrial membrane protein that blocks programmed cell death. Nature 348: 334–336

Hockenbery DM (1992) The bcl-2 oncogene and apoptosis. Semin Immunol 4: 413–420

Hollstein M, Sidransky D, Vogelstein B, Harris CC (1991) p53 mutations in human cancers. Science 253: 49–53

Hong WK, Lippman SM, Itri LM, Karp DD, Lee JS, Byers RM, Schantz RM, Kramer AM, Lotan R, Peters LJ, Dimery IW, Brown BW, Goepfert H (1990) Prevention of second primary tumors with isotretinoin in squamous-cell carcinoma of the head and neck. N Engl J Med 323: 795–801

Hupp TR, Meek DW, Midgley CA, Lane DP (1993) Activation of the cryptic DNA binding function of mutant forms of p53. Nucleic Acids Res 21: 3167–3174

Ishizaki K, Ejima Y, Matsunaga T, Hara R, Sakamoto A, Ikenaga M, Ikawa Y, Aizawa S (1994) Increased UV-induced SCEs but normal repair of DNA damage in p53-deficient mouse cells. Int J Cancer 58: 254–257

Jacquemier J, Molès JP, Penault-Llorca F, Adélaide J, Torrente M, Viens P, Birnbaum D, Theillet C (1994) p53 immunohistochemical analysis in breast cancer with four monoclonal antibodies: comparison of staining and PCR-SSCP results. Br J Cancer 69: 846–852

Jenkins JR, Rudge K, Currie GA (1984) Cellular immortalization by a cDNA clone encoding the transformation-associated phosphoprotein p53. Nature 312: 651–654

Jenkins JR, Chumakov P, Addison C, Stürzbecher H-W, Wade-Evans A (1988) Two distinct regions of the murine p53 primary amino acid sequence are implicated in stable complex formation with simian virus 40 T antigen. J Virol 62: 3903–3906

Jung M, Notario V, Dritschilo A (1992) Mutations in the p53 gene in radiation-sensitive and -resistant human squamous carcinoma cells. Cancer Res 52: 6390–6393

Juven T, Barak Y, Zauberman A, George DL, Oren M (1993) Wild-type p53 can mediate sequence-specific transactivation of an internal promoter within the mdm2 gene. Oncogene 8: 3411–3416

Kadohama T, Tsuji K, Ogawa K (1994) Indistinct cell cycle checkpoint after UV damage in H-*ras*-transformed mouse liver cells despite normal p53 gene expression. Oncogene 9: 2845–2852

Kanjilal S, Pierceall WE, Cummings KK, Kripke ML, Ananthaswamy HN (1993) High frequency of p53 mutations in ultraviolet radiation-induced murine skin tumors: evidence for strand bias and tumor heterogeneity. Cancer Res 53: 2961–2964

Kastan MB, Onyekwere O, Sidransky D, Vogelstein B, Craig RW (1991) Participation of p53 protein in the cellular response to DNA damage. Cancer Res 51: 6304–6311

Kastan MB, Zhan Q, El-Deiry WS, Carrier F, Jacks T, Walsh WV, Plunkett BS, Vogelstein B, Fornace AJ (1992) A mammalian cell cycle checkpoint pathway utilizing p53 and GADD45 is defective in ataxia-telangiectasia. Cell 71: 587–597

Kaufmann S (1991) DNA topoisomerases in chemotherapy. Cancer Cells 3: 24–27

Kelley LL, Green WF, Hicks GG, Bondurant MC, Koury MJ, Ruley HE (1994) Apoptosis in erythroid progenitors deprived of erythropoietin occurs during the G_1 and S phases of the cell cycle without growth arrest or stabilization of wild-type p53. Mol Cell Biol 14: 4183–4192

Kern SE, Kinzler KW, Baker SJ, Nigro JM, Rotter V, Levine AJ, Friedman P, Prives C, Vogelstein B (1991a) Mutant p53 proteins bind DNA abnormally in vitro. Oncogene 6: 131–136

Kern SE, Kinzler KW, Bruskin A, Jarosz D, Friedman P, Prives C, Vogelstein B (1991b) Identification of p53 as a sequence-specific DNA-binding protein. Science 252: 1708–1711

Kerr JF, Winterford CM, Harmon BV (1994) Apoptosis. Its significance in cancer and cancer therapy. Cancer 73: 2013–2026

Kraiss S, Quaiser A, Oren M, Montenarh M (1988) Oligomerization of oncoprotein p53. J Virol 62: 4737–4744

Kraiss S, Spiess S, Reihsaus E, Montenarh M (1991) Correlation of metabolic stability and altered quaternary structure of oncoprotein p53 with cell transformation. Exp Cell Res 192: 157–164

Kress S, Sutter C, Strickland PT, Mukhtar H, Schweizer J, Schwarz M (1992) Carcinogen-specific mutational pattern in the p53 gene in ultraviolet B radiation-induced squamous cell carcinomas of mouse skin. Cancer Res 52: 6400–6403

Kuerbitz SJ, Plunkett BS, Walsh WV, Kastan MB (1992) Wild-type p53 is a cell cycle checkpoint determinant following irradiation. Proc Natl Acad Sci USA 89: 7491–7495

Kum Kum Khanna, Lavin MF (1993) Ionizing radiation and UV induction of p53 protein by different pathways in ataxia-telangiectasia cells. Oncogene 8: 3307–3312

Lanfear J, Fleming J, Wu L, Webster G, Harrison PR (1994) The selenium metabolite selenodiglutathione induces p53 and apoptosis: relevance to the chemopreventive effects of selenium. Carcinogenesis 15: 1387–1392

Lee JM, Bernstein A (1993) p53 Mutations increase resistance to ionizing radiation. Proc Natl Acad Sci USA 90: 5742–5746

Lees-Miller SP, Sakaguchi K, Ullrich SJ, Appella E, Anderson CW (1992) Human DNA-activated protein kinase phosphorylates serines 15 and 37 in the amino-terminal transactivation domain of human p53. Mol Cell Biol 12: 5041–5049

Leszczynski D, Servomaa K, Lang S, Kosma V-M, Rytömaa T (1994) Radiation-induced concomitant overexpression of p53, $p62^{c\text{-}fos}$ and $p21^{N\text{-}ras}$ in mouse epidermis. Cell Prolif 27: 517–528

Levine AJ (1993) The tumor suppressor genes. Annu Rev Biochem 62: 623–651

Li JC, Kaminskas E (1984) Accumulation of DNA strand breaks and methothrexate cytotoxicity. Proc Natl Acad Sci USA 81: 5694–5698

Lindahl T (1993) Instability and decay of the primary structure of DNA. Nature 362: 709–715

Liu T-J, Zhang W-W, Taylor DL, Roth JA, Goepfert H, Clayman GL (1994) Growth suppression of human head and neck cancer cells by the introduction of a wild-type p53 gene via a recombinant adenovirus. Cancer Res 54: 3662–3667

Lotan R (1980) Effects of vitamin A and its analogs (retinoids) on normal and neoplastic cells. Biochim Biophys Acta 605: 33–91

Lotem J, Sachs L (1993) Hematopoietic cells from mice deficient in wild-type p53 are more resistant to induction of apoptosis by some agents. Blood 82: 1092–1096

Lowe SW, Ruley HE (1993) Stabilization of the p53 tumor suppressor is induced by adenovirus 5 E1A and accompanies apoptosis. Genes Dev 7: 535–545

Lowe SW, Ruley HE, Jacks T, Housman DE (1993a) p53-dependent apoptosis modulates the cytotoxicity of anticancer agents. Cell 74: 957–967

Lowe SW, Schmitt EM, Smith SW, Osborne BA, Jacks T (1993b) p53 is required for radiation-induced apoptosis in mouse thymocytes. Nature 362: 847–849

Lowe SW, Jacks T, Housman DE, Ruley HE (1994) Abrogation of oncogene-associated apoptosis allows transformation of p53-deficient cells. Proc Natl Acad Sci USA 91: 2026–2030

Lu X, Lane DP (1993) Differential induction of transcriptionally active p53 following UV or ionizing radiation: defects in chromosome instability syndromes. Cell 75: 765–778

Maltzman W, Czyzyk L (1984) UV irradiation stimulates levels of p53 cellular tumor antigen in nontransformed mouse cells. Mol Cell Biol 4: 1689–1694

Marchetti E, Romero J, Sanchez R, Vargas JA, Dominguez C, Lacal JC, Ramon y Cajal S (1994) Oncogenes and cellular sensitivity of radiotherapy. A study on murine keratinocytes transformed by v-H-ras, v-myc, v-neu, adenovirus E1a and mutant p53. Int J Oncol 5: 611–618

Maxwell SA, Mukhopadhyay T (1994) Transient stabilization of p53 in non-small cell lung carcinoma cultures arrested for growth by retinoic acid. Exp Cell Res 214: 67–74

McCarthy SA, Symonds HS, Van Dyke T (1994) Regulation of apoptosis in transgenic mice by simian virus 40 T antigen-mediated inactivation of p53. Proc Natl Acad Sci USA 91: 3979–3983

McIlwrath AJ, Vasey PA, Ross GM, Brown R (1994) Cell cycle arrests and radiosensitivity of human tumor cell lines: dependence on wild-type p53 for radiosensitivity. Cancer Res 54: 3718–3722

McWhir J, Selfridge J, Harrison DJ, Squires S, Melton DW (1993) Mice with DNA repair gene (ERCC-1) deficiency have elevated levels of p53, liver nuclear abnormalities and die before weaning. Nat Genet 5: 217–224
Medina D, Morrison DG (1988) Current ideas on selenium as a chemopreventive agent. Pathol Immunpathol 7: 187–199
Medina D, Oborn CJ (1984) Selenium inhibition of DNA synthesis in mouse mammary epithelial cell line YN-4. Cancer Res 44: 4361–4365
Meek DW, Simon S, Kikkawa U, Eckhart W (1990) The p53 tumour suppressor protein is phosphorylated at serine 389 by casein kinase II. EMBO J 9: 3253–3260
Mercer WE, Amin M, Sauve GJ, Appella E, Ullrich SJ, Romano JW (1990a) Wild-type human p53 is antiproliferative in SV40-transformed hamster cells. Oncogene 5: 973–980
Mercer WE, Shields MT, Amin M, Sauve GJ, Appella E, Romano JW, Ullrich SJ (1990b) Negative growth regulation in a glioblastoma tumor cell line that conditionally expresses human wild-type p53. Proc Natl Acad Sci USA 87: 6166–6170
Merritt AJ, Potten CS, Kemp CJ, Hickman JA, Balmain A, Lane DP, Hall PA (1994) The role of p53 in spontaneous and radiation-induced apoptosis in the gastrointestinal tract of normal and p53-deficient mice. Cancer Res 54: 614–617
Messmer UK, Ankarcrona M, Nicotera P, Brüne B (1994) p53 expression in nitric oxide-induced apoptosis. FEBS Lett 355: 23–26
Michalovitz D, Halevy O, Oren M (1990) Conditional inhibition of transformation and of cell proliferation by a temperature-sensitive mutant of p53. Cell 62: 671–680
Milne DM, Palmer RH, Campbell DG, Meek DW (1992a) Phosphorylation of the p53 tumour-suppressor protein at three N-terminal sites by a novel casein kinase I-like enzyme. Oncogene 7: 1361–1369
Milne DM, Palmer RH, Meek DW (1992b) Mutation of the casein kinase II phosphorylation site abolishes the anti-proliferative activity of p53. Nucleic Acids Res 20: 5565–5570
Milne DM, Campbell DG, Caudwell FB, Meek DW (1994) Phosphorylation of the tumor suppressor protein p53 by mitogen-activated protein kinases. J Biol Chem 269: 9253–9260
Mitreiter K, Schmidt J, Luz A, Atkinson MJ, Höfler H, Erfle V, Strauss PG (1994) Disruption of the murine p53 gene by insertion of an endogenous retrovirus-like element (ETn) in a cell line from radiation-induced osteosarcoma. Virology 200: 837–841
Miyashita T, Krajewski S, Krajewska M, Wang HG, Lin HK, Liebermann DA, Hoffman B, Reed JC (1994) Tumor suppressor p53 is a regulator of *bcl-2* and *bax* gene expression in vitro and in vivo. Oncogene 9: 1799–1805
Momand J, Zambetti GP, Olson DC, George D, Levine AJ (1992) The *mdm-2* oncogene product forms a complex with the p53 protein and inhibits p53-mediated transactivation. Cell 69: 1237–1245
Montenarh M (1992) Biochemical, immunological, and functional aspects of the growth-suppressor/oncoprotein p53. In: Pimentel E, Perucho M (eds) Critical reviews in oncogenesis. CRC Press, Boca Raton, pp 233–256
Morgenbesser SD, Williams BO, Jacks T, DePinho RA (1994) p53-dependent apoptosis produced by Rb deficiency in the developing mouse lens. Nature 371: 72–74
Mothersill C, Harney J, Seymour CB (1994a) Induction of stable p53 oncoprotein and of *c-myc* overexpression in cultured normal human uroepithelium by radiation and N-nitrosodiethanolamine. Radiat Res 138: 93–98
Mothersill C, Seymour CB, Harney J, Hennessy TP (1994b) High levels of stable p53 protein and the expression of *c-myc* in cultured human epithelial tissue after cobalt-60 irradiation. Radiat Res 137: 317–322
Nagano T, Ueda M, Ichihashi M (1993) Expression of p53 protein is an early event in ultraviolet light-induced cutaneous squamous cell carcinogenesis. Arch Dermatol 129: 1157–1161
Nelson WG, Kastan MB (1994) DNA strand breaks: the DNA template alterations that trigger p53-dependent DNA damage response pathways. Mol Cell Biol 14: 1815–1823
Neve J (1991) Physiological and nutrional importance of selenium. Experientia 47: 187–193

Nguyen M, Branton PE, Walton PA, Oltvai ZN, Korsmeyer SJ, Shore GC (1994) Role of membrane anchor domain of bcl-2 in suppression of apoptosis caused by E1B-defective adenovirus. J Biol Chem 269: 16521–16524
O'Rourke RW, Miller CW, Kato GJ, Simon KJ, Chen D-L, Dang CV, Koeffler HP (1990) A potential transcriptional activation element in the p53 protein. Oncogene 5: 1829–1832
Oliner JD, Kinzler KW, Meltzer PS, George DL, Vogelstein B (1992) Amplification of a gene encoding a p53-associated protein in human sarcomas. Nature 358: 80–83
Oltvai ZN, Milliman CL, Korsmeyer SJ (1993) Bcl-2 heterodimerizes in vivo with a conserved homolog, Bax, that accelerates programmed cell death. Cell 74: 609–619
Oren M (1992) The involvement of oncogenes and tumor suppressor genes in the control of apoptosis. Cancer Metastasis Rev 11: 141–148
Oren M, Maltzman W, Levine AJ (1981) Post-translational regulation of the 54 K cellular tumor antigen in normal and transformed cells. Mol Cell Biol 1: 101–110
Pan H, Griep AE (1994) Altered cell cycle regulation in the lens of HPV-16 E6 or E7 transgenic mice: implications for tumor suppressor gene function in development. Genes Dev 8: 1285–1299
Parada LF, Land H, Weinberg RA, Rotter V (1984) Cooperation between the gene encoding p53 tumor antigen and ras in cellular transformation. Nature 312: 649–651
Parker WB, Kennedy KA, Klubes P (1987) Dissociation of 5-fluorouracil-induced DNA fragmentation from either its incorporation into DNA or its cytotoxicity in murine T-lymphoma (S-49) cells. Cancer Res 47: 979–982
Pavletich NP, Chambers KA, Pabo CO (1993) The DNA-binding domain of p53 contains the four conserved regions and the major mutation hot spots. Genes Dev 7: 2556–2564
Perry ME, Piette J, Zawadzki JA, Harvey D, Levine AJ (1993) The mdm-2 gene is induced in response to UV light in a p53-dependent manner. Proc Natl Acad Sci USA 90: 11623–11627
Pezzella F, Morrison H, Jones M, Gatter KC, Lane D, Harris AL, Mason DY (1993) Immunohistochemical detection of p53 and *bcl-2* proteins in non-Hodgkin's lymphoma. Histopathology 22: 39–44
Pietenpol JA, Vogelstein B (1993) Tumour suppressor genes: no room at the p53 inn. Nature 365: 17–18
Potten CS, Merritt A, Hickman J, Hall P, Faranda A (1994) Characterization of radiation-induced apoptosis in the small intestine and its biological implications. Int J Radiat Biol 65: 71–78
Preudhomme C, Lepelley P, Vachee A, Soenen V, Quesnel B, Cosson A, Fenaux P (1993) Relationship between p53 gene mutations and multidrug resistance (mdr_1) gene expression in myelodysplastic syndromes. Leukemia 7: 1888–1890
Price BD, Park SJ (1994) DNA damage increases the levels of MDM2 messenger RNA in wtp53 human cells. Cancer Res 54: 896–899
Radinsky R, Fidler IJ, Price JE, Esumi N, Tsan R, Petty CM, Bucana CD, Bar-Eli M (1994) Terminal differentiation and apoptosis in experimental lung metastases of human osteogenic sarcoma cells by wild-type p53. Oncogene 9: 1877–1883
Ramqvist T, Magnusson KP, Wang Y, Szekely L, Klein G, Wiman KG (1993) Wild-type p53 induces apoptosis in a Burkitt lymphoma (BL) line that carries mutant p53. Oncogene 8: 1495–1500
Raycroft L, Wu H, Lozano G (1990) Transcriptional activation by wild-type but not transforming mutants of the p53 anti-oncogene. Science 248: 1048–1051
Reddy MV, Randerath K (1987) 32P-analysis of DNA adducts in somatic and reproductive tissues of rats treated with the anticancer antibiotic, mitomycin C. Mutat Res 179: 75–88
Reihsaus E, Kraiss S, Barnekow A, Montenarh M (1992) Cooperation of p53 and polyoma virus middle T antigen in the transformation of primary rat embryo fibroblasts. Exp Cell Res 199: 10–18
Ryan JJ, Danish R, Gottlieb CA, Clarke MF (1993) Cell cycle analysis of p53-induced cell death in murine erythroleukemia cells. Mol Cell Biol 13: 711–719

Ryan JJ, Prochownik E, Gottlieb CA, Apel IJ, Merino R, Nuñez G, Clarke MF (1994) *c-myc* and *bcl-2* modulate p53 function by altering p53 subcellular trafficking during the cell cycle. Proc Natl Acad Sci USA 91: 5878–5882

Santhanam U, Ray A, Sehgal PB (1991) Repression of the interleukin 6 gene promoter by p53 and the retinoblastoma susceptibility gene product. Proc Natl Acad Sci USA 88: 7605–7609

Schaefer L, Roy R, Humbert S, Moncollin V, Vermeulen W, Hoeijmakers JH, Chambon P, Egly JM (1993) DNA repair helicase: a component of BTf2 (TFIIH) basic transcription factor. Science 260: 58–63

Scheffner M, Werness BA, Huibregtse JM, Levine AJ, Howley PM (1990) The E6 oncoprotein encoded by human papillomavirus types 16 and 18 promotes the degradation of p53. Cell 63: 1129–1136

Sedgwick RP, Boder E (1991) Ataxia-telangiectasia. In: In: Vianney de Jong JMB (ed) Hereditary neuropathies and spinocerebellar atrophies. Elsevier, Amsterdam, (Handbook of clinical neurology, vol. 16), pp 347–423

Seki H, Kanegane H, Iwai K, Konno A, Ohta K, Yachie A, Taniguchi N, Miyawaki T (1994) Ionizing radiation induces apoptotic cell death in human TcR-gamma/d$^+$ T and natural killer cells without detectable p53 protein. Eur J Immunol 24: 2914–2917

Selter H, Montenarh M (1994) The emerging picture of p53. Int J Biochem 26: 145–154

Selvakumaran M, Lin H-K, Miyashita T, Wang HG, Krajewski S, Reed JC, Hoffman B, Liebermann D (1994) Immediate early up-regulation of *bax* expression by p53 but not TGFb1: a paradigm for distinct apoptotic pathways. Oncogene 9: 1791–1798

Sen S, D'Incalci M (1992) Apoptosis. Biochemical events and relevance to cancer chemotherapy. FEBS Lett 307: 122–127

Seto E, Usheva A, Zambetti GP, Momand J, Horikoshi N, Weinmann R, Levine AJ, Shenk T (1992) Wild-type p53 binds to the TATA-binding protein and represses transcription. Proc Natl Acad Sci USA 89: 12028–12032

Shaw P, Bovey R, Tardy S, Sahli R, Sordat B, Costa J (1992) Induction of apoptosis by wild-type p53 in a human colon tumor-derived cell line. Proc Natl Acad Sci USA 89: 4495–4499

Shen Y, Shenk T (1994) Relief of p53-mediated transcriptional repression by the adenovirus E1B 19-kDa protein or the cellular Bcl-2 protein. Proc Natl Acad Sci USA 91: 8940–8944

Shivji KK, Kenny MK, Wood RD (1992) Proliferating cell nuclear antigen is required for DNA excision repair. Cell 69: 367–374

Silvestrini R, Veneroni S, Daidone MG, Benini E, Boracchi P, Mezzetti M, Di Fronzo G, Rilke F, Veronesi U (1994) The bcl-2 protein: a prognostic indicator strongly related to p53 protein in lymph node-negative breast cancer patients. JNCI 86: 499–504

Slichenmyer WJ, Nelson WG, Slebos RJ, Kastan MB (1993) Loss of a p53-associated G1 checkpoint does not decrease cell survival following DNA damage. Cancer Res 53: 4164–4168

Smith ML, Chen I-T, Zhan Q, Bae I, Chen C-Y, Gilmer TM, Kastan MB, O'Connor PM, Fornace AJ (1994) Interaction of the p53-regulated protein Gadd45 with proliferating cell nuclear antigen. Science 266: 1376–1380

Soussi T, Caron de Fromentel C, May P (1990) Structural aspects of the p53 protein in relation to gene evolution. Oncogene 5: 945–952

Steinmeyer K, Deppert W (1988) DNA binding properties of murine p53. Oncogene 3: 501–507

Strauss PG, Mitreiter K, Zitzelsberger H, Luz A, Schmidt J, Erfle V, Höfler H (1992) Elevated p53 RNA expression correlates with incomplete osteogenic differentiation of radiation-induced murine osteosarcomas. Int J Cancer 50: 252–258

Stürzbecher H-W, Maimets T, Chumakov P, Brain R, Addison C, Simanis V, Rudge K, Philp R, Grimaldi M, Court W, Jenkins JR (1990) p53 interacts with p34^{cdc2} in mammalian cells: implications for cell cycle control and oncogenesis. Oncogene 5: 795–802

Symonds H, Krall L, Remington L, Saenz-Robles M, Lowe S, Jacks T, Van Dyke T (1994) p53-dependent apoptosis suppresses tumor growth and progression in vivo. Cell 78: 703–711

Tan T-H, Wallis J, Levine AJ (1986) Identification of the p53 protein domain involved in formation of the simian virus 40 large T-antigen-p53 protein complex. J Virol 59: 574–583

Tornaletti S, Pfeifer GP (1994) Slow repair of pyrimidine dimers at p53 mutation hotspots in skin cancer. Science 263: 1436–1438

Truant R, Xiao H, Ingles CJ, Greenblatt J (1993) Direct interaction between the transcriptional activation domain of human p53 and the TATA box-binding protein. J Biol Chem 268: 2284–2287

Tsujimoto Y, Cossman J, Jaffe E, Croce CM (1985) Involvement of the bcl-2 gene in human follicular lymphoma. Science 228: 1440–1443

Ullrich SJ, Sakaguchi K, Lees-Miller SP, Fiscella M, Mercer WE, Anderson CW, Appella E (1993) Phosphorylation at Ser-15 and Ser-392 in mutant p53 molecules from human tumors is altered compared to wild-type p53. Proc Natl Acad Sci USA 90: 5954–5958

Unger C, Kress S, Buchmann A, Schwarz M (1994) gamma-Irradiation-induced micronuclei from mouse hepatoma cells accumulate high levels of the tumor suppressor protein p53. Cancer Res 54: 3651–3655

Van Laar T, Steegenga WT, Jochemsen AG, Terleth C, Van der Eb AJ (1994) Bloom's syndrome cells GM1492 lack detectable p53 protein but exhibit normal G1 cell-cycle arrest after UV irradiation. Oncogene 9: 981–983

Wagner P (1994) p53 forms tight complexes with tms1 of fission yeast. Int J Oncol 4: 987–992

Wagner P, Montenarh M (1995) G1 control of the eukaryotic cell cycle. In: Spandidos D (ed) Current perspectives on molecular and cellular oncology. Jai, London (in press)

Wagner P, Appel K, Issinger OG, Montenarh M (1994) On the interaction of p53 with casein kinase II. Int J Oncol 4: 491–498

Walker PR, Smith C, Youdale T, Leblanc J, Whitfield JF, Sikorska M (1991) Topoisomerase II reactive chemotherapeutic drugs induce apoptosis in thymocytes. Cancer Res 51: 1078–1085

Wang XW, Forrester K, Yeh H, Feitelson MA, Gu J, Harris CC (1994) Hepatitis B virus X protein inhibits p53 sequence-specific DNA binding, transcriptional activity, and association with transcription factor ERCC3. Proc Natl Acad Sci USA 91: 2230–2234

Wang Y, Ramqvist T, Szekely L, Axelson H, Klein G, Wiman KG (1993a) Reconstitution of wild-type p53 expression triggers apoptosis in a p53-negative v-myc retrovirus-induced T-cell lymphoma line. Cell Growth Differ 4: 467–473

Wang Y, Reed M, Wang P, Stenger JE, Mayr G, Anderson ME, Schwedes JF, Tegtmeyer P (1993b) p53 domains: identification and characterization of two autonomous DNA-binding regions. Genes Dev 7: 2575–2586

Wang Y, Szekely L, Okan I, Klein G, Wiman KG (1993c) Wild-type p53-triggered apoptosis is inhibited by *bcl-2* in a v-*myc*-induced T-cell lymphoma line. Oncogene 8: 3427–3431

Ward JF (1975) Molecular mechanisms of radiation-induced damage to nucleic acids. Adv Radiat Biol 5: 181–239

Wattel E, Preudhomme C, Hecquet B, Vanrumbeke M, Quesnel B, Dervite I, Morel P, Fenaux P (1994) p53 mutations are associated with resistance to chemotherapy and short survival in hematologic malignancies. Blood 84: 3148–3157

Wazer DE, Chu Q, Liu X-L, Gao Q, Safaii H, Band V (1994) Loss of p53 protein during radiation transformation of primary human mammary epithelial cells. Mol Cell Biol 14: 2468–2478

Westerveld A, Hoeijmakers JH, Van Duin M, De Wit J, Odijk H, Pastink A, Wood RD, Bootsma D (1984) Molecular cloning of a human DNA repair gene. Nature 310: 425–429

White E, Sabbatini P, Debbas M, Wold WS, Kusher DI, Gooding LR (1992) The 19-kilodalton adenovirus E1b transforming protein inhibits programmed cell death and prevents cytolysis by tumor necrosis factor alpha. Mol Cell Biol 12: 2570–2580

Williams GT, Smith CA (1993) Molecular regulation of apoptosis: genetic controls on cell death. Cell 74: 777–779

Wu X, Levine AJ (1994) p53 and E2F-1 cooperate to mediate apoptosis. Proc Natl Acad Sci USA 91: 3602–3606

Wu X, Bayle JH, Olson D, Levine AJ (1993) The p53-mdm-2 autoregulatory feedback loop. Genes Dev 7: 1126–1132

Yanai N, Obinata M (1994) Apoptosis is induced at nonpermissive temperature by a transient increase in p53 in cell lines immortalized with temperature-sensitive SV40 large T-antigen gene. Exp Cell Res 211: 296–300

Yin X-M, Oltvai ZN, Korsmeyer SJ (1994) BH1 and BH2 domains of Bcl-2 are required for inhibition of apoptosis and heterodimerization with Bax. Nature 369: 321–323

Yonish-Rouach E, Resnitzky D, Lotem J, Sachs L, Kimchi A, Oren M (1991) Wild-type p53 induces apoptosis of myeloid leukaemic cells that is inhibited by interleukin-6. Nature 352: 345–347

Yonish-Rouach E, Grunwald D, Wilder S, Kimchi A, May E, Lawrence J-J, May P, Oren M (1993) p53-mediated cell death: relationship to cell cycle control. Mol Cell Biol 13: 1415–1423

Zambetti GP, Bargonetti J, Walker K, Prives C, Levine AJ (1992) Wild-type p53 mediates positive regulation of gene expression through a specific DNA sequence element. Genes Dev 6: 1143–1152

Zhan Q, Carrier F, Fornace AJ (1993) Induction of cellular p53 activity by DNA-damaging agents and growth arrest. Mol Cell Biol 13: 4242–4250

Zhan Q, Bae I, Kastan MB, Fornace AJ (1994) The p53-dependent gamma-ray response of *GADD45*. Cancer Res 54: 2755–2760

Zhu Y-M, Bradbury DA, Russell NH (1994) Wild-type p53 is required for apoptosis induced by growth factor deprivation in factor-dependent leukaemic cells. Br J Cancer 69: 468–472

Editor-in-Charge: Professor H. Grunicke and Professor M. Schweiger

Rev. Physiol. Biochem. Pharmacol., Vol. 127

Biosynthesis of Nitric Oxide: Dependence on Pteridine Metabolism

E.R. WERNER[1], G. WERNER-FELMAYER[1], H. WACHTER[1], and B. MAYER[2]

Contents

1 Introduction

Not very long ago, a nitric oxide synthase was purified (Bredt and Snyder 1990; Mayer et al. 1990), the identities of nitric oxide (NO) and endothelium-derived relaxing factor were demonstrated (Ignarro et al. 1987; Palmer et al. 1987), and synthesis of nitrite and nitrate from L-arginine by activated macrophages was observed (Stuehr and Marletta 1985). Research in nitric oxide biochemistry is developing rapidly. Nitric oxide synthesis is widespread, occurring not only in mammals (Knowles and Moncada 1994), but also in the chick (Holscher and Rose 1992; Sung et al. 1994), a terrestrial mollusk (Gelperin 1994), a fish (Ostholm et al. 1994), insects (Ribeiro and Nussenzweig 1993; Müller 1994), and a slime mold (Werner-Felmayer et al. 1994). Actions of nitric oxide in the organism are diverse (Nathan 1992),

[1]Institut für Medizinische Chemie und Biochemie der Universität Innsbruck, Fritz-Pregl-Str. 3, 6020 Innsbruck, Austria
[2]Institut für Pharmakologie und Toxikologie der Universität Graz, Universitätsplatz 2, 8010 Graz, Austria

including neurotransmission and vascular relaxation (Moncada et al. 1991), and antimicrobial, antiviral, and antiproliferative activities (Nathan and Hibbs 1991; Karupiah et al. 1993; Lepoivre et al. 1991), to mention just a few. Biochemically, nitric oxide can regulate transcription by action on a transcription factor (Peunova and Enikolopov 1993), alter translation by action on a protein binding to RNA (Drapier et al. 1993; Weiss et al. 1993a), and directly activate or inhibit several enzyme proteins.

A number of reviews on various aspects of nitric oxide biochemistry have been published. In 1994, reviews focused on NO as a physiological messenger molecule (Lowenstein et al. 1994; Bredt and Snyder 1994), mammalian NO synthases (Knowles and Moncada 1994), control of NO synthases (Nathan and Xie 1994), aspects concerning structure and catalysis (Marletta 1994a), target interactions (Stamler 1994), and functional roles (Schmidt and Walter 1994). Our review will put special emphasis on aspects related to our work, including the role of pteridines for NO synthesis by pure enzyme and in intact cells, as well as the action of NO on soluble guanylyl cyclase and iron metabolism.

2 Pteridine Metabolism

Pteridines are a class of of compounds sharing a common pyrazino [*2, 3-d*] pyrimidine structure. These comprise vitamins such as riboflavin and folic acid, as well as compounds also synthesized by mammals like neopterin, biopterin, and molybdopterin (Nichol et al. 1985). Neopterin derives from the first intermediate in tetrahydrobiopterin synthesis. It occurs in significant amounts only in the body fluids of humans and primates, where its concentration is increased by cytokines formed in response to a challenge of the cell-mediated immune system (Fuchs et al. 1988; Wachter et al. 1989). Tetrahydrobiopterin is one of the cofactors of nitric oxide synthases. Its role has thus far been studied in most detail in hydroxylation of aromatic amino acids (Kaufman 1993). Molybdopterin acts as a cofactor of xanthine and sulfite oxidases in animals and of several oxidoreductases in plants and bacteria. Molybdopterin is thought to be synthesized also in mammals, but the pathway is not yet understood in detail. Interestingly, molybdopterin occurs in some enzymes as a dinucleotide (Rajagopalan and Johnson 1992).

Pteridines were first characterized as pigments in butterfly wings (Hopkins 1889), which explains the designation of the compound class (the Greek *pteron* signifies wing). Apparently, all pteridines are synthesized from GTP by hydrolytic cleavage of the heterocycle and rearrangement including the ribose moiety (Fig. 1). The first step in riboflavin synthesis is catalyzed by GTP cyclohydrolase II (Brown and Williamson 1982; Richter et al. 1993).

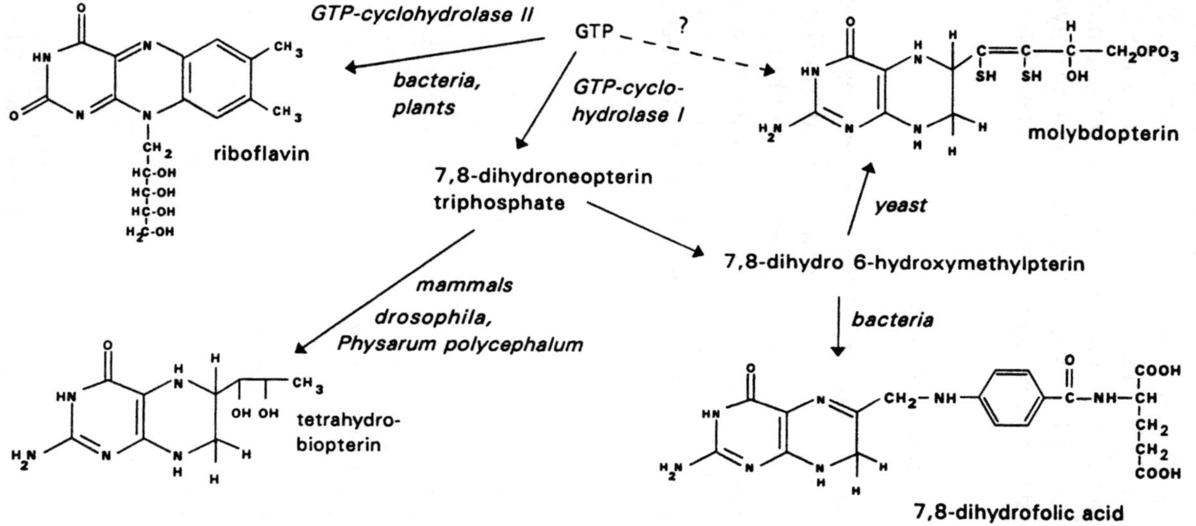

Fig. 1. Some pathways of pteridine synthesis. Pteridines are synthesized from GTP by hydrolytic cleavage by the action of GTP cyclohydrolases. Only some of the known pathways of pteridine synthesis are shown. Other pathways, like the formation of methanopterin, a cofactor used by methanogenic archaebacteria (Van Beelen et al. 1984; White 1990), or drosopterin, a pigment occurring in insects (Pfleiderer et al. 1992), have been omitted. The notation of species in which the synthesis occurs is exemplary rather than complete, since the occurrence of these pathways has not yet been studied in detail in many different species. It remains to be seen, for example, whether biosynthesis of molybdopterin proceeds in all species as it does in yeast, or may be formed via a different route from GTP in other species (*dashed line*)

Most other pteridines formed, however, derive from dihydroneopterin triphosphate, which is synthesized from GTP by the action of GTP cyclohydrolase I (Fig. 1). These include tetrahydrobiopterin in mammals, dihydrofolate in bacteria (Brown and Williamson 1982), and possibly also molybdopterin, which in yeast has recently been shown to be formed from 6-hydroxymethyl pterin and glycerinaldehyde 3-phosphate (Irby and Adair 1994).

2.1 Biosynthesis of Tetrahydrobiopterin

Biosynthesis of tetrahydrobiopterin from GTP requires three enzymatic steps (Nichol et al. 1985; Fig. 2). GTP is first cleaved by GTP cyclohydrolase I (EC 3.5.4.16) to yield dihydroneopterin triphosphate, which is then converted in an Mg(2+)-dependent reaction to 6-pyruvoyl tetrahydropterin by 6-pyruvoyl tetrahydropterin synthase (EC 4.6.1.10). Subsequently, the two keto-groups of the side chain are reduced in an NADPH-dependent way

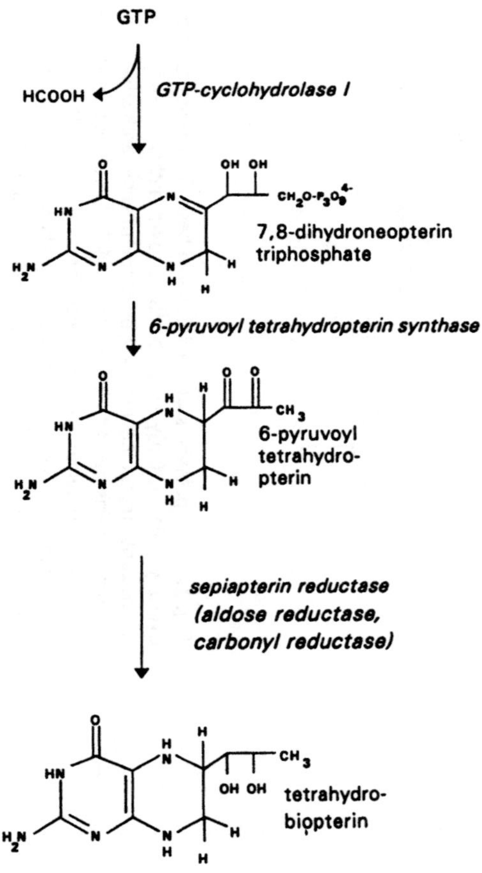

Fig. 2. Biosynthesis of tetrahydrobiopterin from GTP. GTP cyclohydrolase I, the major regulatory point of the pathway (see Fig. 4), does not require cofactors. 6-Pyruvoyl tetrahydropterin synthase needs Mg^{2+} and sepiapterin reductase needs NADPH to operate. In cases of sepiapterin reductase impairment or inhibition, the two keto groups of the side chain can also be reduced by aldose reductase and carbonyl reductase

by sepiapterin reductase (EC 1.1.1.153) to yield 6R 5,6,7,8-tetrahydrobiopterin, which, for the sake of brevity, will be called tetrahydrobiopterin in this review. The reductions are also catalyzed by action of carbonyl reductase and aldose reductase (Park et al.1991). These two additional enzymes may become important in circumstances of sepiapterin reductase impairment or inhibition. The three enzymes required for tetrahydrobiopterin synthesis in rat and human liver, GTP cyclohydrolase I, 6-pyruvoyl tetrahydropterin synthase, and sepiapterin reductase, have been cloned (Hatekeyama et al. 1991; Inoue et al. 1991; Citron et al. 1990; Gütlich et al. 1994; Thöny et al. 1992; Ashida et al. 1993; Ichinose et al. 1991). GTP cyclohydrolase I from *Escherichia coli* has been crystallized, and some structural data are known (Schmid et al. 1992). For 6-pyruvoyl tetrahydropterin synthase the crystal structure and human chromosomal location (11Q22.3–Q23.3) have recently been determined (Nar et al. 1994; Thöny et al. 1994).

Rare inherited diseases with partial defects in tetrahydrobiopterin biosynthesis or recycling lead to hyperphenylalaninemia caused by lack of sufficient cofactor for phenylalanine hydroxylase (Blau 1988). Strategies and guidelines for the screening of these defects have been developed (Dhondt 1991; Blau et al. 1992). Since the biosynthesis of 5-hydroxy tryptophan and 3,4-dihydroxy phenylalanine is also affected by tetrahydrobiopterin deficiency, patients are additionally treated with these compounds to cure the mental retardation associated with the deficiency (McInnes et al. 1984; Synderman et al. 1987). Attempts have been made to use tetrahydrobiopterin and sepiapterin (Curtius et al. 1979) or 6-methyl tetrahydropterin (McInnes et al. 1984) to treat the defects in tetrahydrobiopterin biosynthesis. Nitric oxide synthases require tetrahydrobiopterin and might therefore also be disturbed in inherited defects of tetrahydrobiopterin synthesis, and this disturbance might contribute to the neurological symptoms. In this context, it is important to note that only partial defects of tetrahydrobiopterin synthesis are known. Thus, total lack of tetrahydrobiopterin may be incompatible with life. While treatment with sepiapterin (see Sect. 2.2) or tetrahydrobiopterin would also stimulate nitric oxide synthase, 6-methyl tetrahydropterin might stimulate only aromatic amino acid hydroxylation, due to a lower selectivity of the latter enzymes for the natural cofactor tetrahydrobiopterin. It remains to be seen whether differences in the clincal outcome of treatment regimens are caused by different effects on nitric oxide synthesis.

In human liver, as well as in cultured human cells, 6-pyruvoyl tetrahydropterin synthase activity is found to be about two orders of magnitude lower than in tissues of other mammals (Hasler and Niederwieser 1986; Werner et al. 1990, 1991a, 1993). This enzyme activity is particularly low in human macrophages (Schoedon et al. 1987; Werner et al. 1990). As a result, neopterin derivatives accumulate in tissues and body fluids of human beings and are synthesized in high amounts in cytokine-stimulated human macrophages (Huber et al. 1984; Schoedon et al. 1986). Considerable amounts of neopterin were also found in tissues of a primate (Duch et al. 1984). Stimulation of GTP cyclohydrolase I by cytokines (see below) can therefore be monitored by measurement of neopterin levels in body fluids of human beings and primates (Fuchs et al. 1988; Wachter et al. 1989). The biochemical cause for the lower activity of 6-pyruvoyl tetrahydropterin synthase in human tissues and cells remains to be elucidated. It is also not clear at present whether neopterin derivatives accumulate to serve functional roles (Weiss et al. 1993b; Wöll et al. 1993; Überall et al. 1994) or indicate a partial degeneration of the pathway.

2.2 Pharmacological Manipulation of Intracellular Tetrahydrobiopterin Levels

Tetrahydrobiopterin biosynthesis is regulated by several mechanisms, which almost exclusively affect the first and key enzyme, GTP cyclohydrolase I. GTP cyclohydrolase I from rat liver shows a positive cooperativity for GTP (Hatakeyama et al. 1989) and is inhibited by dihydro- and tetrahydropterins, the end products of the pathway (Shen et al. 1988). Phenylalanine, which requires tetrahydrobiopterin to be metabolized, stimulates GTP cyclohydrolase I activity in rat liver by a regulatory protein binding to the enzyme (Harada et al. 1993).

In addition to these actions on GTP cyclohydrolase I protein, proinflammatory cytokines like interferon-gamma or tumor necrosis factor-alpha stimulate the activity of GTP cyclohydrolase I (Blau et al. 1985; Schoedon et al. 1986; Werner et al. 1989, 1993), presumably by causing an induction, i.e., an increased transcription and de novo synthesis of the enzyme. The extent of this induction is remarkable. In cultured human cells, for example, the enzyme activity is stimulated up to 100-fold by interferon-gamma (Werner et al. 1989). This stimulation is thought to be caused by induction of GTP cyclohydrolase I because it takes about 24 h to reach maximal activity and because the K_m for GTP remains unchanged (Werner et al. 1990). In addition, mRNA from cytokine-treated cells showed increased hybridization with a cDNA probe of rat liver GTP cyclohydrolase I (Geller et al. 1993; Schott et al. 1993; Scott-Burden et al. 1993; Nakayama et al. 1994). Indications for an increase of GTP cyclohydrolase I mRNA were also found with a polymerase-chain reaction using primers specific for rat liver GTP cyclohydrase I (Hattori and Gross 1993). Following treatment of rats with bacterial lipopolysaccharide, GTP cyclohydrolase I activities and tetrahydrobiopterin levels were found to be increased in several regions of the body, including

Table 1. Stimulation of pteridine synthesis in cultured cells. Although this has been demonstrated in only a few examples, it is assumed that the cytokines stimulate pteridine synthesis by increased transcription and following de novo synthesis of GTP cyclohydrolase I. Some of the actions of the mentioned cytokines may be exerted in an indirect way, i.e., by induction of other cytokines, in particular, when mixtures of cells like peripheral blood mononuclear cells are stimulated. Due to a low 6-pyruvoyl tetrahydropterin synthase activity in human cells, induction of GTP cyclohydrolase I leads to accumulation of neopterin derivatives in addition to tetrahydrobiopterin in the cells. In cells of rats and mice, however, only tetrahydrobiopterin is formed, due to an activity of 6-pyruvyol tetrahydropterin synthase that is two orders of magnitude higher

Cell	Stimuli	Reference
Human		
A431 epidermoid carcinoma	IFNγ	Werner et al. (1989)
A498 kidney carcinoma	IFNγ	Werner et al. (1989)
A549 lung carcinoma	IFNγ	Werner et al. (1989)
B lymphocytes (blood)	IFNγ, IL-2, PWM	Hofmann et al. (1992)
Endothelial cells, umbilical vein	IFNγ, TNFα, LPS	Werner-Felmayer et al. (1993b), Rosenkranz-Weiss et al. 1994)
Endothelial cells ea.hy926	IFNγ, TNFα, IL-1, LPS	Schoedon et al. (1993a)
Fibroblasts (dermis)	IFNγ, TNFα, IL-1[a]	Werner et al. (1989, 1990, 1993)
HUT102 T-lymphocytes	IFNγ, IL-2	Ziegler et al. (1990), Gütlich et al. (1992)
Macrophages, peripheral blood	IFNα,β,γ, TNFα, LPS	Huber et al. (1984), Schoedon et al. (1986), Troppmair et al. (1988), Werner-Felmayer et al. (1989)
Mononuclear cells, periph. blood	IFNα,β,γ, IL-2, LPS, PHA,PWM	Huber et al. (1983, 1984), Ziegler (1985), Ziegler et al. (1986, 1990), Blau et al. (1985), Troppamir et al. (1988), Werner-Felmayer et al. (1989), Gütlich et al. (1992)
ME-180 cervial carcinoma cells	IFNγ	Werner-Felmayer et al. (1993c)
MT-2 T cells (umbilical cord)	IFNγ , IL-2	Ziegler et al. (1990), Schott et al. (1993)
SK-HEP-1 hepatoma	IFNγ	Werner et al. (1989)
T lymphocytes (periph. blood)	PHA	Ziegler et al. (1990), Schott et al. (1993)
THP-1 monocytoma	IFNα,β,γ, TNFα, LPS	Werner-Felmayer et al. (1990)
T24 bladder carcinoma	IFNγ	Werner et al. (1989, 1990)
U138MG glioblastoma	IFNγ	Werner et al. (1989)
U373 gliobalstoma	IFNγ	Sakai et al. (1993a)
U937 monocytoma	IFNγ , PHA	Schott et al. (1993)

Table 1 (continued)

Cell	Stimuli	Reference
Murine		
Fibroblasts, dermal	IFNγ, TNFα	Werner et al. (1991a)
J774A1 macrophage line	constitutive	Werner et al. (1991a)
L929 fibroblasts	IFNγ, TNFα	Werner-Felmayer et al., unpublished
Macrophages, B10-BR, resident	LPS	Schoedon et al. (1987)
Macrophages, balb/c, resident	IFNγ, TNFα	Werner et al. (1991a)
Macrophages, C3H/HeN, elicited	LPS	Schoedon et al. (1993b)
OVA T-cell line	IL-2	Schoedon et al. (1987)
P388D1 macrophage line	constitutive	Werner et al. (1991a)
RAW264 macrophage line	constitutive, IFNγ, LPS	Tayeh and Marletta (1989), Kwon et al. (1989), Sakai et al. (1993a)
Rat		
Macrophages elicited	IFNγ + LPS	Sakai et al. (1993a)
Glomerular mesangial cells	IL-1	Mühl and Pfeilschifter (1994)
Smooth muscle cells	IL-1, LPS	Gross and Levi (1992), Scott-Burden et al. (1993), Hattori et al. (1993)

[a] Werner-Felmayer et al., unpublished.

spleen, liver, and brain (Werner-Felmayer et al. 1993a). Induction of pteridine synthesis has been observed in different cell types of human beings, rats and mice (Table 1). In addition to the cells mentioned in Table 1, increased GTP cyclohydrolase I transcription has been observed in rat and human hepatocytes (Geller et al. 1993; Nüssler, personal communication).

Hormones also stimulate tetrahydrobiopterin synthesis by induction of GTP cyclohydrolase I, as indicated by increased activities in adrenal medulla and cortex of rats treated with insulin or reserpine (Viveros et al. 1981) and by increased biopterin levels in glucocorticoid-treated rat hepatoma cells (Parniak and Pilkington 1989). Dexamethasone increased the stimulatory action of lipopolysaccharide on purified macrophages. In lipopolysaccharide-mediated stimulation of GTP cyclohydrolase I in peripheral blood mononuclear cells, however, glucocorticoids decrease GTP cyclohydrolase I induction by interfering with the production of proinflammatory cytokines like interferon-gamma (Werner-Felmayer et al. 1989). Antiinflammatory cytokines are able to inhibit induction of GTP cyclohydrolase I activities. These include transforming growth factor beta (Schoedon et al. 1993a; Scott-Burden et al. 1993), as well as interleukin 4 and interleukin 10 (Schoedon et al. 1993a).

Whereas cytokines regulate the activity of GTP cyclohydrolase I, the two subsequent enzymes, 6-pyruvoyl tetrahydropterin synthase and sepiapterin reductase, are constitutively expressed and remain unaffected by the cytokine treatment (Werner et al. 1990). Although the subsequent enzyme activity, 6-pyruvoyl tetrahydropterin synthase, is not sufficiently active in human cells to convert all 7,8-dihydroneopterin triphosphate to tetrahydrobiopterin (see Sect. 2.1), tetrahydrobiopterin levels in these cells are still under control of GTP cyclohydrolase I. The reason for this behavior is the K_m of 10 μM 7,8-dihydroneopterin triphosphate for 6 pyruvoyl-tetrahydropterin synthase (Takikawa et al. 1986; Werner et al. 1991b), which is at least one order of magnitude higher than the intracellular concentrations of 7,8-dihydroneopterin triphosphate reached in the cells.

The various ways to manipulate tetrahydrobiopterin levels in cultured cells are summarized in Fig. 3. In addition to stimulation of GTP cyclohydrolase I activity, intracellular tetrahydrobiopterin levels may be raised by treating cells with tetrahydrobiopterin or with sepiapterin. Sepiapterin is taken up by the cells and is reduced first by sepiapterin reductase to 7,8-dihydrobiopterin and then by dihydrofolate reductase to tetrahydrobiopterin (Nichol et al. 1985). Inhibitors of dihydrofolate reductase like methotrexate therefore efficiently inhibit the generation of tetrahydrobiopterin by this route. Depletion of GTP pools by IMP-dehydrogenase inhibitors (e.g., thiazofurin) decreases intracellular tetrahydrobiopterin concentrations (Hatakeyama et al. 1992). In PC-12 cells, an increase of GTP above control levels did not augment the tetrahydrobiopterin content, and GTP cyclohydrolase I seems to be almost saturated with GTP (Hatakeyama et al. 1992). Although GTP levels rise in cytokine-treated macrophages up to threefold (Schoedon et al. 1987), the increase in pteridine synthesis in this case appears to be mediated predominantly by induction of the enzyme, which results in up to 100-fold activity measured at saturating GTP levels (Blau et al. 1985; Schoedon et al. 1986; Werner et al. 1989, 1990). The next and key regulator step is GTP cyclohydrolase I. As outlined above, its activity is increased by proinflammatory cytokines, hormones, and, in the liver, by phenylalanine. Inhibition of GTP cyclohydrolase I is mediated by anti-inflammatory cytokines (Schoedon et al. 1993a; Scott-Burden et al. 1993), by di- and tetrahydropterin end products (Shen et al. 1988), and by certain pyrimidines like 2,4-diamino 6-hydroxy pyrimidine (DAHP, Gal et al. 1978). This inhibitor is required in millimolar concentrations and is therefore only marginally suited for in vivo work. 6-Pyruvoyl tetrahydropterin synthase, the sub-

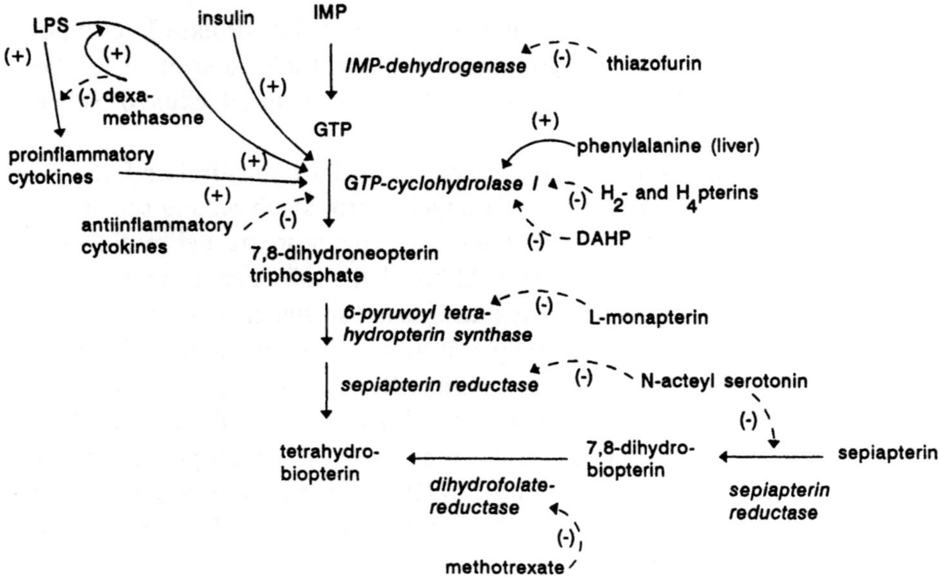

Fig. 3. Regulation of intracellular tetrahydrobiopterin by drugs. The indicated agents either stimulate (*solid-line arrows*) or inhibit (*dashed arrows*) the respective enzymes. These agents alter transcription (the cytokines and, presumably, also the glucocorticoids) or interact with the proteins to stimulate or inhibit the enzyme activity (the other agents). GTP cyclohydrolase I shows a positive cooperativity for GTP. The significance of this finding for the regulation of intracellular tetrahydrobiopterin levels is not entirely clear at present. Stimulatory action of phenylalanine on GTP cyclohydrolase I in the rat liver is mediated by a separate regulatory protein, which also has a binding site for feedback inhibition by tetrahydrobiopterin. The pathway from sepiapterin to tetrahydrobiopterin is not important for normal metabolism of the cell but provides a means of increasing intracellular tetrahydrobiopterin independently of its biosynthesis from GTP by administering synthetic compounds. *LPS,* lipopolysaccharide; *DAHP,* 2,4-diamino 6-hydroxy pyrimidine

sequent enzyme, is inhibited by comparatively high doses of monapterin (Blau et al. 1987). The next step for interference with tetrahydrobiopterin biosynthesis is sepiapterin reductase, which is efficiently inhibited by N-acetyl serotonin (Katoh et al. 1982), by phenprocoumon (Sueka and Katoh 1985), or by more recently devoloped sepiapterin reductase inhibitors (Smith et al. 1992).

2.3 Pteridine-Dependent Hydroxylases

The metabolic role of tetrahydrobiopterin has thus far been studied in most detail in aromatic amino acid hydroxylases (Kaufman 1993). In addition to NO synthase, four reactions are known to depend on tetrahydrobiopterin: phenylalanine 4-mono-oxygenase (EC 1.14.16.1; Kaufman 1963), tyrosine

3-mono-oxygenase (EC 1.14.16.2; Shiman et al. 1971), tryptophan 5-mono-oxygenase (EC 1.14.16.4; Friedman et al. 1972), and alkyl glycol-ether mono-oxygenase (EC 1.14.16.5; Tietz et al. 1964, Kaufman et al. 1990). Regarding the role of the tetrahydropterin cofactor, these enzymes share characteristics of the mechansim which are in clear contrast to nitric oxide synthase: (a) In these four enzymes not only tetrahydrobiopterin, but also a number of other tetrahydropterins are capable of stimulating the activity. Although there is a role for the stereochemistry of C6 of the pterin (Bailey et al. 1991), the differences are much less pronounced than with NO synthase, which for activity stimulation and binding shows a high specificity for the natural tetrahydrobiopterin isomer (Kwon et al. 1989; Giovanelli et al. 1991; Klatt et al. 1994a). (b) Compared with phenylalanine hydroxylase, which shows an apparent K_m for tetrahydrobiopterin of 21 μM (Bailey et al. 1991), NO synthases are stimulated half-maximally by much lower concentrations, i.e., 0.03–0.15 μM (Kwon et al. 1989; Mayer et al. 1990; Giovanelli et al. 1991). (c) In all four enzymes, in the absence of dihydropteridine reductase and NADPH, added tetrahydrobiopterin catalyzes a stoichiometric amount of product formation. In NO synthase, in contrast, no reaction occurs when tetrahydrobiopterin is added in the absence of NADPH. In the presence of NADPH, however, one equivalent of tetrahydrobiopterin added stimulates the formation of at least ten equivalents of product (Mayer et al. 1991; Giovanelli et al. 1991). (d) Whereas nitric oxide synthases are heme proteins (see Sect. 3.2), the pteridine-dependent hydroxylases contain non-heme iron, which in phenylalanine hydroxylase has been shown to be ligated by two specific histidine residues (Gibbs et al. 1993). The iron is essential for activity and is found in the vicinity of the hydroxylation site of the substrate in a distance compatible with a mechanism involving formation of an iron-bound oxo- or peroxo species as active agent (Martinez et al. 1993).

Following the hydroxylation of the substrate, tetrahydrobiopterin is recycled to the active cofactor by two enzymes, 4a-carbinolamine dehydratase and dihydropteridine reductase (Fig. 4; Kaufman 1993). The tetrahydropterin cofactor leaves the phenylalanine hydroxylase reaction as a 4a-hydroxy derivative. This is dehydrated by 4a-carbinolamine dehydratase, a protein that, surprisingly, was found to be identical to a nuclear factor involved in transcriptional regulation, DCoH (Citron et al. 1992; Thöny et al. 1993). If 4a-carbinolamine dehydratase is missing, a small part of the molecule rearranges to a derivative with the side chain in the 7- rather than in the 6-position of the pteridine ring (Fig. 4; Curtius et al. 1990; Davies et al. 1991). Most of the 4a-hydroxy derivative spontaneously dehydrates, though at a lower rate than in the presence of the enzyme. The product of the 4a-carbinolamine dehydratase reaction, the quinonoid 6,7(8H) dihydrobiopterin is then finally reduced by dihydropteridine reductase to tetrahydrobiop-

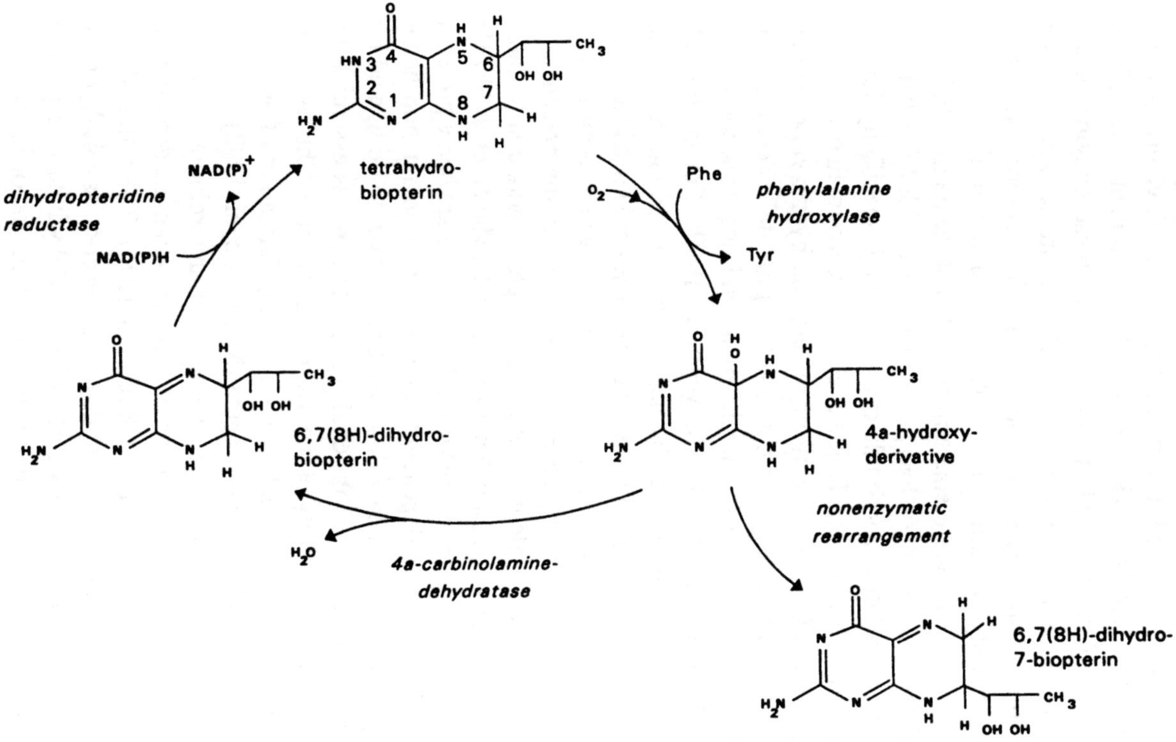

Fig. 4. Recycling of tetrahydrobiopterin in phenylalanine hydroxylation. The product formed from tetrahydrobiopterin in phenylalanine hydroxylation, the 4a-hydroxy derivative, spontaneously dehydrates to the quinonoid 6,7(8H)-dihydrobiopterin. This reaction is supported by 4a-carbinolamine dehydratase. In the absence or impairment of this dehydratase, some 4a-hydroxy derivative rearranges to 6,7(8H)-dihydro 7-biopterin, and the hydroxylation of phenylalanine cannot be carried out as efficiently. No similar recycling pathway has thus far been detected in the course of stimulation of NO synthase by tetrahydrobiopterin. *Phe,* Phenlylalanine; *Tyr,* tyrosine

terin. It is remarkable that up to now no indication of a similar cycle of tetrahydrobiopterin in the stimulation of NO synthases has been found. No dihydropteridine reductase activity (Giovanelli et al. 1991) and no 7-biopterin derivatives (Werner et al., unpublished) were found in investigations with pure NO synthase. Since there are indications that a redox reaction of tetrahydrobiopterin may contribute to product formation in NO synthase (see Sect. 3.2), it is conceivable that these reactions occur inside the active NO synthase dimer and are therefore not detected by the methods used thus far.

3 Biosynthesis of Nitric Oxide

3.1 Nitric Oxide Synthases

Soon after the first reports on mammalian NO synthesis appeared in 1987 (Ignarro et al. 1987; Palmer et al. 1987), cytosols of cytokine-activated macrophages were shown to contain an enzymatic activity catalyzing an NADPH-dependent conversion of the amino acid L-arginine to NO (Marletta 1988). Subsequently, enzymatic NO formation was detected in a variety of tissues, and, consistent with the apparent involvement of Ca^{2+} in NO-mediated signal transduction in blood vessels (Moncada et al. 1989) and neurons (Garthwaite et al. 1988), NO synthesis turned out to require micromolar concentrations of free Ca^{2+} in both endothelial cells (Mayer et al. 1989) and brain (Bredt and Snyder 1989; Knowles et al. 1989), whereas enzymatic NO formation occurred in a Ca^{2+}-independent manner in cytokine-activated macrophages (Marletta 1988). Meanwhile, three distinct NO synthase (EC 1.14.13.39) isozymes have been purified, cloned, and biochemically characterized. The different isozymes are biochemically similar, utilizing molecular oxygen and NADPH for the oxidation of L-arginine to L-citrulline and NO (see Fig. 5). As recently pointed out by Nathan and Xie (1994), no consensus on the nomenclature of NO synthase isozymes has been obtained so far (see also Förstermann et al. 1994). Here we use the terms "brain", "endothelial", and "inducible" NO synthase as follows: brain NO synthase for the soluble enzyme constitutively expressed in neurons, endothelial NO synthase for the membrane-associated isozyme constitutively expressed in endothelial cells, and inducible NO synthase for the enzyme which is expressed only upon activation of cells by cytokines and which does not require micromolar concentrations of Ca^{2+} for activity.

The first NO synthase isozyme was purified from rat cerebellum (Bredt and Snyder 1990). The enzyme was described as a soluble 160-kDa protein requiring Ca^{2+}/calmodulin for activity. The isolated enzyme was found to be labile, with a specific activity of about 1 mmol L-citrulline $\times$ min$^{-1}\times$ mg^{-1}

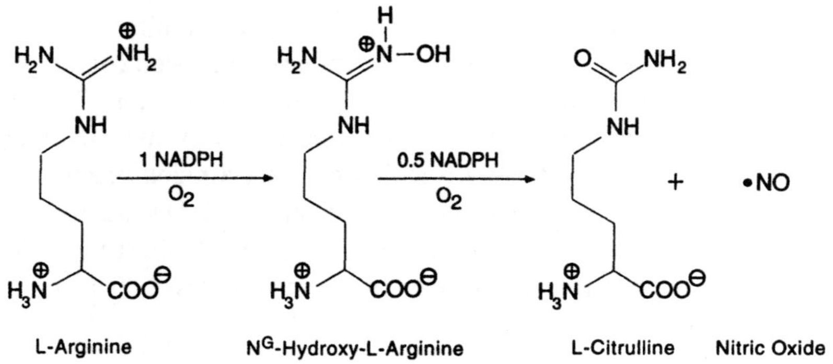

Fig. 5. The NO synthase reaction. NO synthases produce NO by catalyzing an NADPH-dependent 5-electron oxidation of L-arginine. In the course of two sequential oxidation steps, L-arginine is hydroxylated to N^G-hydroxy-L-arginine first. This intermediate remains bound to the enzyme and is oxidatively cleaved to L-citrulline and NO. The overall reaction involves reduction of two oxygen molecules requiring a total of eight electrons. Five reducing equivalents are derived from guanidino nitrogen oxidation; co-oxidation of NADPH provides three additional electrons

and a K_m for L-arginine of 1.5 μM. Subsequently, a similar protein was purified from porcine cerebellum on a larger scale and positively identified as NO synthase by means of NO chemiluminescence and functional reconstitution with purified soluble guanylyl cyclase (Mayer et al. 1990). The isolated porcine brain NO synthase was markedly stimulated by low concentrations of exogenous tetrahydrobiopterin, which at that time was believed to be a cofactor of inducible NO synthase only (see Sect. 3.2). Based on gel-filtration experiments, brain NO synthase was initially reported to be a 150- to 160-kDa monomer (Bredt and Snyder 1990; Mayer et al. 1990), whereas velocity sedimentation analysis of rat cerebellar NO synthase has suggested a homodimeric structure of the enzyme (Schmidt et al. 1991). Recent data suggest that purified rat brain NO synthase may exist in a monomer-dimer equilibrium, which is shifted towards the monomers at low protein concentration (Masters 1994). Although brain NO synthase was purified from soluble fractions, there is some evidence that the enzyme is to some degree attached to membrane structures in the cell. Thus, a 150-kDa protein with typical Ca^{2+}-dependent NO synthase activity was purified from detergent-solubilized particulate fractions of rat cerebellum (Hiki et al. 1992), and a considerable amount of NO synthase activity appears to be located in the endoplasmic reticulum of brain tissues (Hecker et al. 1994). Similar observations were made when the NADPH-diaphorase reaction (Bredt et al. 1991a; Hope et al. 1991) was used for histochemical localization of rat brain NO synthase by electron microscopy (Wolf et al. 1992). Cloning of rat brain NO synthase revealed sequence similarities to cytochrome P-450 reductase with consensus sites for binding of NADPH and the flavins FAD and FMN (Bredt et al. 1991b).

In contrast to the brain isozyme, endothelial NO synthase is present to 80% in membrane fractions (Förstermann et al. 1991; Mayer et al. 1993). It was purified from detergent-solubilized membranes of cultured bovine aortic endothelial cells and described as a 135-kDa protein. As observed with brain NO synthase, the endothelial enzyme required Ca^{2+}/calmodulin for activity and was stimulated by tetrahydrobiopterin (Pollock et al. 1991). Cloning of the enzyme revealed a consensus site for N-terminal myristoylation (Lamas et al. 1992; Sessa et al. 1992), and mutation of this motif resulted in expression of a soluble enzyme (Busconi and Michel 1993; Sessa et al. 1993), demonstrating that membrane association of endothelial NO synthase is due to post- or co-translational modification. Interestingly, agonist-induced phosphorylation was reported to trigger translocation of the enzyme from the membranes to the cytosol of cultured endothelial cells (Michel et al. 1993). Expression of this membrane-associated NO synthase isoform is apparently not confined to endothelial cells. Immunohistochemistry and in situ hybridization studies suggest that neurons in the CA1 region of the hippocampus express the endothelial isozyme (Dinerman et al. 1994). NO is a putative retrograde messenger mediating synaptic plasticity in this area of the brain, and studies with mice lacking the gene for neuronal NO synthase indicated that the endothelial, rather than the neuronal isozyme may be involved in long-term potentiation of synaptic transmission in the CA1 region (Odell et al. 1994).

Most of the studies on the properties of inducible NO synthase have been done with the enzyme obtained from cytokine-activated macrophages, but expression of this isozyme can be induced by various stimuli in many other mammalian cells (Nathan 1992; Nüssler and Billiar 1993). Inducible NO synthase has been purified from activated rat peritoneal macrophages (Yui et al. 1991) and the murine RAW 264.7 macrophage cell line (Hevel et al. 1991; Stuehr et al. 1991a), as well as from liver of endotoxin-treated rats (Evans et al. 1992). The native enzyme was reported to be a dimer with a subunit molecular mass of approximately 130 kDa; enzyme activity did not require the presence of micromolar concentrations of Ca^{2+} and was stimulated to various extents by tetrahydrobiopterin. Several lines of experimental evidence suggest that calmodulin is an essential cofactor of inducible NO synthase, although this isozyme exhibits maximal activity under nominally Ca^{2+}-free conditions. The hepatic enzyme isolated by Evans et al. (1992) was stimulated about fivefold by added calmodulin and only slightly blocked by up to 2 m*M* EGTA, and cloning of macrophage NO synthase revealed a consensus site for calmodulin binding (Lyons et al. 1992; Xie et al. 1992). Finally, calmodulin was found to copurify with macrophage NO synthase even in the presence of very low concentrations of Ca^{2+} (n*M*) (Cho et al. 1992), suggesting that calmodulin represents a tightly bound subunit of inducible NO synthase even at resting intracellular Ca^{2+} levels, whereas binding of calmodulin to the constitutive enzymes requires agonist-induced increases in free Ca^{2+} to the micromolar range.

Biochemical characterization of purified NO synthases obtained from various sources revealed a high complexity of this enzyme family. In accordance to the amino acid sequence similarities between NO synthase and flavin-containing cytochrome P-450 reductase (Bredt et al. 1991b), all three NO synthase isozymes were found to contain equimolar amounts of the reduced flavins FAD and FMN as prosthetic groups (Hevel et al. 1991; Mayer et al. 1991; Stuehr et al. 1991a; H.H.H.W. Schmidt et al. 1992; Pollock et al. 1993). As is well known for cytochrome P-450 reductase, NO synthase-bound flavins shuttle reducing equivalents from the nucleotide cofactor NADPH to an oxygenase domain, which is located in the active site of the enzyme and catalyzes reductive activation of molecular oxygen requisite for L-arginine oxidation. This electron transfer appears to operate independently of substrate metabolism, since both brain and macrophage NO synthase catalyze an NADPH-dependent reduction of cytochromes and low-molecular-mass electron acceptors in the absence of L-arginine (Klatt et al. 1992a; Abu-Soud and Stuehr 1993). This cytochrome reductase activity of NO synthase occurs at rates about 20-fold higher than those for L-arginine oxidation and is sensitive neither to cyanide nor to superoxide dismutase, indicating that this enzymatic activity of NO synthase is due to a direct transfer of electrons from the flavins to the acceptor molecules. The electron transfer from NADPH to the oxygenase domain appears to be the calmodulin-dependent step in NO synthesis, because the reductase activity of brain NO synthase requires the presence of Ca^{2+}/calmodulin.

Atomic absorption spectroscopic analysis showed that purified porcine brain NO synthase contained iron in a 1:1 stoichiometry and thus provided the first hint that reductive activation of molecular oxygen may be catalyzed by a redox-active iron in the catalytic site of the enzyme (Mayer et al. 1991). Subsequently, brain and macrophage NO synthase were identified as cytochrome P-450-like heme proteins, which are inhibited by agents such as CO, cyanide, or miconazole, which typically interfere with P-450-mediated hydroxylation reactions (Klatt et al. 1992c; McMillan et al. 1992; Stuehr and Ikeda-Saito 1992; White and Marletta 1992). More recently, properties and coordination of the heme site of NO synthase have been studied in detail using UV/VIS spectroscopy and other techniques (McMillan and Masters 1993; Matsuoka et al. 1994; Wang et al. 1993, 1994). Although the widely accepted consensus site for P-450s is not apparent in the sequence of NO synthases, amino acid residues believed to be critical for catalytic function of

P-450, including a cysteine presumably serving as thiolate ligand of the heme, are strongly conserved in the three isozymes known so far, suggesting that NO synthases belong to the cytochrome P-450 superfamily (Renaud et al. 1993).

In contrast to NO synthases, the well-established microsomal P-450-containing mono-oxygenase systems consist of two separate proteins: a cytochrome P-450 heme protein, which catalyzes substrate hydroxylation, and an FAD- and FMN-containing cytochrome P-450 reductase, responsible for electron transfer from NADPH to the cytochrome (Guengerich 1991; Degtyarenko and Archakov 1993). As outlined above, NO synthases contain the heme and the flavins within a single polypeptide and exhibit sequence similarities to both P-450 mono-oxygenases and the respective reductases (Bredt et al. 1991b; Renaud et al. 1993). Accordingly, NO synthases appear to have emerged from convergent evolution and combine two catalytic functions which have evolved separately, i.e., flavin-catalyzed cytochrome reductase and heme iron-catalyzed oxygenase activity. The bidomain structure of NO synthase was recently confirmed by separation of the reductase and oxygenase domains subsequent to tryptic cleavage of the rat brain enzyme (Sheta et al. 1994). As self-sufficient cytochromes P-450, NO synthases are unique within mammalian P-450 systems but resemble the soluble bacterial fatty acid hydroxylase cytochrome $P\text{-}450_{BM\text{-}3}$ which has been purified and cloned from *Bacillus megaterium* (Fulco 1991).

The precise mechanism of NO formation is not known. The reaction involves a 5-electron oxidation of one of the chemically equivalent guanidino-nitrogens of L-arginine, leading to the concomitant production of L-citrulline and NO. It is accompanied by an NADPH-dependent reduction of molecular oxygen (Mayer et al. 1991), which is incorporated into both reaction products (Kwon et al. 1990; Leone et al. 1991). Studies with purified macrophage and brain NO synthase suggest a two-step reaction involving formation of N^G-hydroxy-L-arginine as an intermediate, which is normally not released in considerable quantities from the enzyme but immediately undergoes oxidative cleavage to yield NO and L-citrulline (Stuehr et al. 1991b; Klatt et al. 1993; (see Fig. 5). Incubation of macrophage and brain NO synthase with substoichiometrical amounts of NADPH revealed that 1.5 mol of the nucleotide are consumed for the formation of 1.0 mol of L-citrulline (Mayer et al. 1991; Stuehr et al. 1991b). Two of the three NADPH-derived reducing equivalents appear to be utilized for the initial N-hydroxylation of L-arginine, and oxidative cleavage of N^G-hydroxy-L-arginine to NO and L-citrulline may require an additional electron. Since NADPH can transfer electrons only two at a time, it has been suggested that enzyme-bound flavins store the extra fourth electron for use in the next catalytic cycle. Accordingly, six electrons would be transferred to the heme from three molecules of NADPH in the course of two reaction cycles (Feldman et al. 1993).

Activation of brain NO synthase by Ca^{2+}/calmodulin triggers oxidation of NADPH and transfer of electrons to the prosthetic heme group for reductive activation of molecular oxygen. This enzymatic oxygen reduction also takes place in the absence of L-arginine and then gives rise to the generation of superoxide anions and hydrogen peroxide instead of NO (Mayer et al. 1991; Heinzel et al. 1992; Pou et al. 1992). Upon addition of L-arginine, oxygen activation couples to substrate metabolism, resulting in the formation of NO and L-citrulline. Inducible NO synthase appears to down-regulate this uncoupled oxygen reduction, as it does not oxidize appreciable quantities of NADPH without a ligand being bound to its substrate site (Abu-Soud and Stuehr 1993; Olken and Marletta 1993).

3.2 Role of Tetrahydrobiopterin as Cofactor of NO Synthases

It is now well established that tetrahydrobiopterin functions as an essential cofactor in NO biosynthesis. This novel role of the pteridine was uncovered in cytosolic fractions of LPS/interferon-gamma-activated macrophages (Kwon et al. 1989; Tayeh and Marletta 1989) and was subsequently confirmed with the constitutive brain (Mayer et al. 1990) and endothelial (Pollock et al. 1991) enzymes. Even after extensive purification, porcine brain NO synthase was found to exhibit about 20% of maximal enzyme activity in the absence of added tetrahydrobiopterin (Mayer et al. 1990). Analysis of purified NO synthase preparations for endogenous pteridines revealed that some amounts of tetrahydrobiopterin remained tightly bound during purification (Mayer et al. 1991). This has been confirmed with NO synthase isozymes isolated from various sources, although the amounts of copurified tetrahydrobiopterin apparently vary, depending on the tissue source and presumably also on the protocol applied for enzyme purification (Hevel and Marletta 1992; Klatt et al. 1992b; Schmidt et al. 1992; Baek et al. 1993; Pollock et al. 1993; Werner-Felmayer et al. 1994). Notwithstanding the reported differences in tetrahydrobiopterin content, recombinant rat brain NO synthase isolated from baculovirus-infected insect cells (Harteneck et al. 1994) was recently shown to contain essentially the same amount of tightly bound pteridines (0.45 mol per mol of NO synthase subunit) as previously found for the enzyme isolated from brain tissue (Schmidt et al. 1992). Moreover, although the amount of NO synthase-bound heme and flavins was markedly enhanced by feeding the cells with biosynthesis precursors, the amount of bound tetrahydrobiopterin was not increased when cellular tetrahydrobiopterin levels were increased by the addition of tetrahydrobiopterin or sepiapterin to the culture medium (List et al., unpublished). Thus, the substoichiometrical presence of tetrahydrobiopterin is apparently not due to limited intracellular levels of the pteridine. Together with previous findings demonstrating that the endogenous tetrahydrobiopterin is not displaced from the brain enzyme by high-affinity antagonists (Klatt et al. 1994a), these results possibly indicate that NO synthase dimers might contain two different pteridine-binding sites, one of them accounting for the 0.5 mol of tetrahydrobiopterin remaining irreversibly bound per mol of enzyme and the other being freely accessible to exogenously added ligands. However, this is only speculation at present and awaits confirmation from pteridine-binding studies with tetrahydrobiopterin-deficient NO synthase.

The role of tetrahydrobiopterin in NO synthesis is elusive thus far, but current experimental evidence suggests that it may act bifunctionally as a both redox-active and allosteric effector of NO synthase. Tetrahydrobiopterin is not involved directly in reductive activation of molecular oxygen (Mayer et al. 1992), but redox activity of the pteridine is suggested by experiments showing that the oxidized derivative dihydrobiopterin binds with fairly high affinity to NO synthase without supporting catalytic activity (Klatt et al. 1994a). NO synthase isozymes are unique as pteridine-dependent cytochromes P-450, and, as we have pointed out recently (Mayer and Werner 1995), evolution of tetrahydrobiopterin as a cofactor of these enzymes could be a consequence of the rather unique properties of the reaction product. NO binds with high affinity to ferrous heme and was described as a potent inhibitor of classical microsomal cytochromes P-450 (Khatsenko et al. 1993; Wink et al. 1993; Stadler et al. 1994). In fact, NO apparently blocks its own biosynthesis (Rogers and Ignarro 1992; Assreuy et al. 1993; Buga et al. 1993) through binding to the prosthetic heme group of NO synthase (Wang et al. 1994). Conceivably, tetrahydrobiopterin could be required to prevent the interference of NO with redox cycling of the heme iron. Recently, this hypothesis has gained some experimental support: (a) 7-Nitroindazole, a heme-site inhibitor of NO synthase, turned out to antagonize pteridine binding to NO synthase, suggesting that the pteridine-binding domain is located proximal to the heme (Mayer et al. 1994); (b) bound tetrahydrobiopterin was reported to affect the spectral properties of the heme, indicating a cross-talk between the two prosthetic groups (Marletta 1994b); (c) tetrahydrobiopterin was found to prevent feedback inhibition of NO synthase by NO (Griscavage et al. 1994); and (d) tetrahydrobiopterin chemically inactivates NO, presumably leading to formation of peroxynitrite (Mayer et al. 1995). Thus, enzyme-bound tetrahydrobiopterin appears to interact with the prosthetic heme group, and inactivation of NO by the free pteridine protects NO synthase from inhibition by its product. Future studies should clarify whether one of these effects may explain the crucial role of tetrahydrobiopterin as a redox-active cofactor in NO synthesis.

Several years ago it was proposed that tetrahydrobiopterin might act as an allosteric effector of NO synthase (Giovanelli et al. 1991). Taking into account our current knowledge about the presence of tightly bound tetrahydrobiopterin in purified NO synthase, these earlier data do not provide compelling evidence for the proposed role of the pteridine; nonetheless, this hypothesis is supported by recent experimental findings. First, it has been shown that tetrahydrobiopterin is required to keep macrophage NO synthase in its active dimeric state and that dissociated, inactive monomers reassociate only when heme, tetrahydrobiopterin, and L-arginine are present simultaneously (Baek et al. 1993). Although brain NO synthase appears to behave differently and remains a dimer under conditions that induce dissociation of macrophage NO synthase (Klatt et al. 1994b), there is evidence for a similar synergistic interaction between the L-arginine- and tetrahydrobiopterin-binding sites of the constitutive isozyme. We studied substrate and pteridine binding to brain NO synthase using tritium-labelled N^G-nitro-L-arginine and tetrahydrobiopterin as high-affinity ligands and found that L-arginine decreased the K_D of brain NO synthase for tetrahydrobiopterin from about 200 n*M* to 30 n*M* and that, vice versa, the affinity of the substrate site was increased in the presence of tetrahydrobiopterin (Klatt et al. 1994a). These data point to a positive cooperativity between substrate and pteridine binding to NO synthase, and, taking together all the experimental results available so far, one may conclude that tetrahydrobiopterin has a dual role in NO synthesis. Binding of the pteridine may convert NO synthase into an active high-affinity state, and the bound cofactor may participate in a redox-active manner in L-arginine oxidation.

3.3 Impact of Pteridine Metabolism on NO Formation by Intact Cells

As already outlined in Sect. 2.2., cytokines are strong stimulators of tetrahydrobiopterin synthesis in many cells. In investigating the physiological goal of this stimulation, research first focused on tryptophan degradation, which is induced in parallel to tetrahydrobiopterin synthesis (Werner et al. 1987, 1989). Although tetrahydrobiopterin is capable of stimulating indoleamine 2,3-dioxygenase (Nishikimi 1975; Ozaki et al. 1986), manipulation of intracellular tetrahydrobiopterin (see Sect. 2.2) had no effect on the amount of tryptophan degraded by the cells (Werner et al. 1991c; Sakai et al. 1993b).

Since NO synthase is induced by the same cytokines as GTP cyclohydrolase I, and since NO synthesis had been shown to require tetrahydrobiopterin, it was logical to assume that cytokines stimulate tetrahydrobiopterin synthesis to provide a cofactor essential for the formation of NO. To test this hypothesis, the tetrahydrobiopterin content of cytokine-treated murine fibroblasts was depleted by inhibition of GTP cyclohydrolase I with 2,4-diamino 6-hydroxypyrimidine (DAHP, see Sect. 2.2). This treatment decreased the amount of nitrite plus nitrate released to the culture medium. Tetrahydrobiopterin levels were restored in the presence of DAHP by adding sepiapterin, and nitrite plus nitrate formation by the cells was restored as well. Methotrexate abolished the stimulatory action of sepiapterin, underlining the fact that only the tetrahydroform is active in catalysis in intact cells. In this case, NO formation was lowest, presumably due to the inhibitory action of the accumulated 7,8-dihydrobiopterin. Addition of sepiapterin alone together with the cytokines even stimulated NO formation, indicating that NO synthase was not saturated with tetrahydrobiopterin. These data provided clear evidence that a goal of the induction of tetrahydrobiopterin synthesis is to provide a cofactor for the formation of nitric oxide by the cells (Werner-Felmayer et al. 1990).

Using a similar experimental approach, inhibition of NO formation by depletion of intracellular tetrahydrobiopterin was subsequently shown in several cultured cells (Werner-Felmayer et al. 1995; Table 2). These included not only cells with inducible NO synthase, but also cells with endothelial NO synthase such as porcine aortic (Schmidt et al. 1992) and human umbilical vein endothelial cells (HUVEC; Werner-Felmayer et al. 1993b; Rosenkranz-Weiss et al. 1994) as well as human ME-180 cells which express brain NO synthase (Werner-Felmayer et al. 1993c). In human cells with constitutive NO synthesis (HUVEC and ME-180), cytokines increase NO formation merely by increasing the tetrahydrobiopterin level without causing an induction of NO synthase itself (Werner-Felmayer et al. 1993a,b; Rosenkranz-Weiss et al. 1994). While most of the investigations confirmed the initial observations made with fibroblasts (Werner-Felmayer et al. 1990), rat alveolar macrophages behaved differently (Jorens et al. 1992). DAHP inhibited NO formation as in other cells. Pterins that should act as stimulators, like tetrahydrobiopterin or sepiapterin, however, inhibited NO formation. One possible interpretation is that these cells may contain a low dihydrofolate reductase activity. Since tetrahydrobiopterin added to culture medium is rapidly oxidized to 7,8-dihydrobiopterin, lack of sufficient dihydrofolate reductase activity would lead to accumulation of 7,8-dihydrobiopterin, which has been shown to inhibit the stimulatory effect of tetrahydrobiopterin on NO synthase (Klatt et al. 1994a). In addition, high concentrations of 7,8-dihydrobiopterin inhibit GTP cyclohydrolase I and hence de novo synthesis of tetrahydrobiopterin (see Sect. 2.2).

In addition to altering NO formation, manipulation of intracellular tetrahydrobiopterin also influenced NO-mediated effects on the cells. These

Table 2. Inhibition of NO formation in cultured cells by depletion of intracellular tetrahydrobiopterin. Depletion of tetrahydrobiopterin was achieved in most cases by inhibiting GTP cyclohydrolase I activity with 2,4-diamino 6-hydroxy pyrimidine. Some authors also used inhibition of sepiapterin reductase with N-acetyl serotonin or phenprocoumon, or a combination of both. In many of the cells studied, increasing intracellular tetrahydrobiopterin by the addition of sepiapterin or tetrahydrobiopterin effected an increase of NO formation above levels observed without such additions. In most cases, endothelial and brain NO synthase activities in the cells were measured indirectly by NO-mediated formation of cGMP after a Ca^{2+} stimulus

Cell	NO synthase	Reference
Murine dermal fibroblasts	Cytokine-inducible	Werner-Felmayer et al. (1990)
Murine brain endothelial cells	Cytokine-inducible	Gross et al. (1991)
Murine vascular endothelial, send1	Cytokine-inducible	Schoedon et al. (1994)
Murine macrophages	Cytokine-inducible	Sakai et al. (1993b), Schoedon et al. (1993b)
Porcine endothelial cells, aortic	Constitutive endothelial	K. Schmidt et al. (1992)
Human endothelial cells, umbilical vein	Constitutive endothelial	Werner-Felmayer et al. (1993b), Rosenkranz-Weiss et al. (1994)
Human endothelial, hybrid	Constitutive endothelial	Schoedon et al. (1993a)
Human cervial carinoma ME-180	Constitutive neuronal	Werner-Felmayer et al. (1993c)
Rat smooth muscle cells, aortic	Cytokine-inducible	Gross and Levi (1992), Nakayama et al. (1994)
Rat glomerular mesangila cells	Cytokine-inducible	Mühl and Pfeilschifter (1994)
Chicken macrophages	Cytokine-inducible	Sung et al. (1994)

include cytotoxicity in murine fibroblasts (Werner-Felmayer et al. 1990), cGMP formation in endothelial cells (Schmidt et al. 1992; Werner-Felmayer et al. 1993b; Schoedon et al. 1993a; Rosenkranz-Weiss et al. 1994), iron metabolism in murine macrophages (Weiss et al. 1993a), formation of iron-nitrosyl complexes in murine fibroblasts (Lancaster et al. 1994), and parasite killing in human and murine hepatocytes (Mellouk et al. 1994). When sepiapterin reductase inhibitors were administered to rats treated subsequently with endotoxin, a significant attenuation of the blood pressure drop caused by endotoxin was seen (Klemm et al. 1993). Although it is tempting to attribute this to inhibition of NO formation only, it cannot be excluded at present that other tetrahydrobiopterin dependent reactions also contributed to the observed effects.

4 Actions of Nitric Oxide

Actions of NO show a remarkable diversity. In target proteins, NO interacts with thiol groups and with iron as the main targets (Stamler 1994). Recent work makes clear that some effects like reaction with cysteine and glutathione (Wink et al. 1994), as well as inhibition of aconitase (Hausladen and Fridowich 1994; Castro et al. 1994), are mediated by reaction products of NO with superoxide or oxygen, rather than by NO itself. Some molecular aspects of the mechanism of stimulation of soluble guanylyl cyclase by NO have been characterized and will be detailed in Sect. 4.1. In addition to guanylyl cyclase, cyclo-oxygenases 1 and 2 are also stimulated by the action of NO (Salvemini et al. 1993; Corbett et al. 1993) via an unknown mechanism. Cytotoxic and anitproliferative actions of NO are mediated by inhibition of crucial enzymes: NO, or reaction products of NO with oxygen species (see above), inhibits mitochondrial aconitase and respiratory chain complexes I and II, i.e., NADH:ubiquinone oxidoreductase and NADH:succinate oxidoreductase (Hibbs et al. 1988; Drapier and Hibbs 1988; Stuehr and Nathan 1989; Stadler et al. 1991), glycerinaldehyde 3-phosphate dehydrogenase (Dimmeler et al. 1992), and ribonucleotide reductase (Kwon et al. 1991; Lepoivre et al. 1991, 1994). In addition, NO inhibits cytochrome P-450 enzymes (Khatsenko et al. 1993; Wink et al. 1993; Stadler et al. 1994), indoleamine 2,3-dioxygenase (Thomas et al. 1994), and even NO synthases themselves (see Sect. 3.2). Interaction of NO with signaling pathways includes AP-1 (Peunova and Enikolopov 1993; Tabuchi et al. 1994), protein kinase C (Gopalakrishna et al. 1993), P21-ras (Lander et al. 1993), and adenylyl cyclase (Duhe et al. 1994). Section 4.2 details the influence of NO on iron metabolism, which comprises an example of the action of NO on translational regulation by interacting with the RNA-binding iron-regulatory protein (Pantopoulos et al. 1994).

4.1 Soluble Guanylyl Cyclase

The soluble isoform of guanylyl cyclases [GTP pyrophosphate-lyase (cyclizing), EC 4.6.1.2] is the major physiological effector enzyme of NO in mammalian cells. In the presence of Mg^{2+} or Mn^{2+}, this enzyme catalyzes the cyclization of guanosine-5-triphosphate (GTP) to guanosine-3',5'-cyclic monophosphate (cGMP), which acts as an intracellular messenger affecting the function of specific protein kinases, phosphodiesterases, and ion channels (Walter 1989; Schmidt et al. 1993). As isolated from various tissues, soluble guanylyl cyclase is a heterodimer consisting of an α-subunit with a reported molecular mass of 73–83 kDa and α-subunit with a mass of 70 kDa

(Waldman and Murad 1987; Koesling et al. 1991). The enzyme exhibits a very low basal activity which is stimulated 100- to 300-fold in the presence of submicromolar concentrations of NO. Identification of the bovine lung enzyme as a hemeprotein containing 1 mol of ferroprotoporphyrin-IX per mol of heterodimer (Gerzer et al. 1981) has led to the proposal that enzyme stimulation is mediated through binding of NO to the heme (Ignarro 1991). This hypothesis has been substantiated by more recent work by Stone and Marletta (1994) involving spectral characterization of the bovine lung enzyme, showing that soluble guanylyl cyclase contains a pentacoordinated high-spin ferrous heme with histidine as the proximal ligand. As pointed out by the authors, soluble guanylyl cyclase differs from most other hemeproteins, such as hemoglobin or myoglobin, in that it does not bind oxygen to the sixth coordination position of the heme. This highly unusual feature of soluble guanylyl cyclase is important, because bound oxygen would prevent binding of NO under aerobic conditions and thus preclude formation of the ferrous-nitrosyl-heme complexes thought to be requisite for enzyme stimulation. With its exceptionally low affinity for oxygen, soluble guanylyl cyclase resembles the heme protein kinases FixL, representing sensors for oxygen, and possibly also for NO and CO, in nitrogen-fixing bacteria (Gilles-Gonzalez et al. 1994). Thus, both soluble guanylyl cyclase and FixL could belong to a novel class of heme proteins acting as sensors for gases, rather than as oxygen carriers or electron transporters. The spectral changes of soluble guanylyl cyclase induced by NO point to formation of a pentacoordinated ferrous-nitrosyl-heme complex, supporting the hypothesis that the bond of the proximal histidine ligand to the iron is broken upon binding of NO to the heme (Traylor and Sharma 1992). It is unclear at present how cleavage of this bond confers the change in protein conformation leading to stimulation of the cyclase. Another unsolved problem of considerable physiological significance is deactivation of the NO-stimulated enzyme. Deactivation should be fast in order to allow transduction of transient cellular signals, but ferrous-nitrosyl-heme complexes are known to be rather stable due to slow dissociation rates. Thus, additional mechanisms, e.g., redox modification of the nitrosyl ligand, may be involved in the rapid deactivation of soluble guanylyl cyclase.

In addition to NO, carbon monoxide, produced endogenously by heme oxygenase, has attracted much attention recently as a putative endogenous biological messenger acting through stimulation of soluble guanylyl cyclase (Maines 1993; Verma et al. 1993; Zhuo et al. 1993). However, the purified enzyme was reported to be not (Yu et al. 1994) or only marginally (Stone and Marletta 1994) stimulated by 100% CO. We have investigated this issue with the enzyme purified from bovine lung (Humbert et al. 1990) and observed an about tenfold increased activity in the presence of 100% CO, whereas the enzyme was stimulated approximately 200-fold by low concentrations of NO (Mayer et al., unpublished results). Also, the interaction of CO with the heme apparently differs from that seen upon binding of NO, as CO was found to form a hexacoordinated complex without inducing cleavage of the histidine-iron bond (Stone and Marletta 1994; Yu et al. 1994). Moreover, the metalloporphyrins used as inhibitors of heme oxygenase to demonstrate the involvement of endogenous CO in neurotransmission are unspecific and also block NO-induced accumulation of cGMP (Luo and Vincent 1994).

Both subunits of soluble guanylyl cyclase have been cloned, sequenced, and functionally expressed in various systems (Koesling et al. 1991). Two different α– (α_1 and α_2) and β- (β_1 and β_2) subunits have been identified to date, pointing to the possible existence of four different heterodimeric isoforms, although the $\alpha_1\beta_1$ dimer represents the most prominent isoform detected in most tissues so far. In their C-terminal part, the α- and β-subunits of soluble guanylyl cyclase show pronounced sequence similarities to the membrane-associated isoforms, which are stimulated by natriuretic peptides, as well as to the adenylyl cyclases, suggesting that these regions may contain the catalytic domains involved in GTP binding. The regulatory heme-binding domain may be located in the N-terminal half of the protein. In this region, the β-subunits contain a histidine residue (His-105 in β_1) with a leucine in the minus 4 position, a motif which has been conserved in the heme-binding sites of all hemoglobin and myoglobin side chains. Exchange of His-105 in the β_1-subunit of soluble guanylyl cyclase against a phenylalanine by site-directed mutagenesis yielded a heme-free apoenzyme that exhibited the same basal activity as the wild-type enzyme but was completely insensitive to NO (Wedel et al. 1994). Since mutagenesis of each of the other histidine residues in the α- and β-subunits had no effect on NO-induced enzyme stimulation, these data provide convincing evidence that the proximal imidazole ligand of the heme in soluble guanylyl cyclase is identical with the histidine in position 105 of the β-subunit.

4.2 Iron Metabolism

Interactions of NO with iron metabolism are complex. As already mentioned above, heme and nonheme iron centers are major targets of NO action. NO regulates iron metabolism by interaction with the iron-regulatory protein (Drapier et al. 1993; Weiss et al. 1993a), and iron in turn influences the expression of cytokine-induced nitric oxide synthase at the transcriptional level (Weiss et al. 1994). NO itself may cause removal of iron from ferritin stores (Reif and Simmons 1990).

The iron-regulatory protein (IRP), also called iron-regulatory factor (IRF) or iron-responsive element-binding protein (IRE-BP) is a protein binding to specific sequences in RNAs called iron-responsive elements (IREs). Surprisingly, IRP was found to be identical to cytoplasmatic aconitase (Constable et al. 1992; Haile et al. 1992). The protein contains a 4Fe-4S cluster, one of the four iron atoms that are labile and removed in conditions of low iron (Klausner et al. 1993; Pantopoulos et al. 1994). In this iron-depleted form, the factor has lost its aconitase activity and binds to the specific iron-responsive elements on RNA, which are located in the 5' untranslated region of ferritin and erythroid delta amino laevulinate synthase (eALAS), as well as in the 3' untranslated region of transferrin receptor. Thus, in conditions of low iron, translation of proteins ultimately leading to iron consumption, i.e., ferritin and erythroid delta amino laevulinate synthase, are blocked. In contrast, the mRNA of a protein increasing iron supply, the transferritin receptor, is protected from degradation, and more protein can therefore be translated.

NO produced by the cells upon cytokine induction, or added as a gas to the recombinant protein, converts the IRP to the same condition as low iron, causing a binding of the protein to mRNA (Drapier et al. 1993; Weiss et al. 1993a). This results in an increase of intracellular free iron by increased aquisition and decreased use or storage. As shown recently, intracellular levels of free iron modify the amount of NO formed by interacting with NO synthase expression. High levels of free iron reduce nuclear transcription of NO synthase, whereas in iron-depleted cells, NO synthase transcription rates are highly elevated (Weiss et al. 1994). This suggests the possibility of coordinated regulation of nitric oxide synthase induction and iron metabolism (Weiss et al. 1994).

5 Conclusions

All three – brain, endothelial, and inducible – NO synthase isozymes contain tightly bound tetrahydrobiopterin and require added tetrahydrobiopterin for full activity. The precise role of tetrahydrobiopterin is unknown, but several lines of evidence support the hypothesis of a role in redox catalysis in addition to allosteric activation. In cultured cells, depletion of intracellular tetrahydrobiopterin can limit NO formation by all three isozymes. Cytokine or drug-mediated increase in intracellular tetrahydrobiopterin, in contrast, stimulates NO formation and hence NO-mediated effects on the cells. Two of these effects have been detailed here, the activation of soluble guanylyl cyclase by binding of NO to its heme group and the regulation of iron metabolism by interaction with IRP.

Acknowledgement. Our experimental work presented here has been supported by the Austrian *Fonds zur Förderung der wissenschaftlichen Forschung*, projects 9685 (E.R.W.) and 10098 (B.M.)

References

Abu-Soud HM, Stuehr DJ (1993) Nitric oxide synthases reveal a role for calmodulin in controlling electron transfer. Proc Natl Acad Sci USA 90: 10769–10772

Ashida A, Hatakeyama K, Kagamiyama H (1993) cDNA cloning, expression in *Escherichia coli* and purification of human 6-pyruvoyl-tetrahydropterin synthase. Biochem Biophys Res Commun 195: 1386–1393

Assreuy J, Cunha FQ, Liew FY, Moncada S (1993) Feedback inhibition of nitric oxide synthase activity by nitric oxide. Br J Pharmacol 108: 833–837

Baek KJ, Thiel BA, Lucas S, Stuehr DJ (1993) Macrophage nitric oxide synthase subunits – purification, characterization, and role of prosthetic groups and substrate in regulating their association into a dimeric enzyme. J Biol Chem 268: 21120–21129

Bailey SW, Dillard SB, Ayling JE (1991) Role of C6 chirality of tetrahydropterin cofactor in catalysis and regulation of tyrosine and phenylalanine hydroxylases. Biochemistry 30: 10226–10235

Blau N (1988) Inborn errors of pterin metabolism. Annu Rev Nutr 8: 185–209

Blau N, Joller P, Atares M, Cardesa-Garcia J, Niederwieser A (1985) Increase of GTP cyclohydrolase I activity in mononuclear blood cells by stimulation: detection of heterozygotes of GTP cyclohydrolase I deficiency. Clin Chim Acta 148: 47–52

Blau N, Steinerstauch P, Redweik U et al (1987) Dihydromonapterin triphosphate: occurrence, analysis and effect on tetrahydrobiopterin synthesis in vitro and in vivo. In: Curtius HC, Blau N, Levine RA (eds) Unconjugated pterins and related biogenic amines. De Gruyter, Berlin, pp 105–116

Blau N, Kierat L, Heizmann CW, Endres W, Giudici T, Wang M (1992) Screening for tetrahydrobiopterin deficiency in newborns using dried urine on filter paper. J Inherited Metab Dis 15: 402–404

Bredt DS, Snyder SH (1989) Nitric oxide mediates glutamate-linked enhancement of cGMP levels in the cerebellum. Proc Natl Acad Sci USA 86: 9030–9033

Bredt DS, Snyder SH (1990) Isolation of nitric oxide synthetase, a calmodulin-requiring enzyme. Proc Natl Acad Sci USA 87: 682–685

Bredt DS, Snyder SH (1994) Nitric oxide: a physiologic messenger molecule. Annu Rev Biochem 63: 175–195

Bredt DS, Glatt CE, Hwang PM, Fotuhi M, Dawson TM, Snyder SH (1991a) Nitric oxide synthase protein and mRNA are discretely localized in neuronal populations of the mammalian CNS together with NADPH diaphorase. Neuron 7: 615–624

Bredt DS, Hwang PM, Glatt CE, Lowenstein C, Reed RR, Snyder SH (1991b) Cloned and expressed nitric oxide synthase structurally resembles cytochrome P-450 reductase. Nature 351: 714–718

Brown GM, Williamson JM (1982) Biosynthesis of riboflavin, folic acid, thiamin, and pantothenic acid. Adv Enzymol 53: 345–381

Buga GM, Griscavage JM, Rogers NE, Ignarro LJ (1993) Negative feedback regulation of endothelial cell function by nitric oxide. Circ Res 73: 808–812

Busconi L, Michel T (1993) Endothelial nitric oxide synthase – N-terminal myristoylation determines subcellular localization. J Biol Chem 268: 8410–8413

Castro L, Rodriguez M, Radi R (1994) Aconitase is readily inactivated by peroxynitrite, but not by its precursor, nitric oxide. J Biol Chem 269: 29409–29415

Cho HJ, Xie QW, Calaycay J, Mumford RA, Swiderek KM, Lee TD, Nathan C (1992) Calmodulin is a subunit of nitric oxide synthase from macrophages. J Exp Med 176: 599–604

Citron BA, Milstien S, Gutierrez JC, Levine RA, Yanak BL, Kaufman S (1990) Isolation and expression of rat liver sepiapterin reductase cDNA. Proc Natl Acad Sci USA 87: 6436–6440

Citron BA, Davis MD, Milstien S, Gutierrez J, Mendel DB, Crabtree GE, Kaufman S (1992) Identity of 4a-carbinolamine dehydratase, a component of the phenylalanine hydroxylation system, and DCoH, a transregulator of homeodomain proteins. Biochemistry 89: 11891–11894

Constable A, Quick S, Gray NK, Hentze MW (1992) Modulation of the RNA-binding of a regulatory protein by iron in vitro: switching between enzymatic and genetic function? Proc Natl Acad Sci USA 89: 4554–4558

Corbett JA, Kwon G, Turk J, McDaniel ML (1993) IL-1 beta induces the coexpression of both nitric oxide synthase and cyclooxygenase by islets of Langerhans – activation of cyclooxygenase by nitric oxide. Biochemistry 32: 13767–13770

Curtius HC, Niederwieser A, Viscontini M, Otten A, Schaub J, Scheibenreiter S, Schmidt H (1979) Atypical phenylketonuria due to tetrahydrobiopterin deficiency. Diagnosis and treatment with tetrahydrobiopterin and sepiapterin. Clin Chim Acta 93: 251–262

Curtius HC, Adler C, Rebrin I, Heizmann C, Ghisla S (1990) 7-Substituted pterins: formation during phenylalanine hydroxylation in the absence of dehydratase. Biochem Biophys Res Commun 172: 1060–1066

Davis MD, Kaufman S, Milstien S (1991) Conversion of 6-substituted tetrahydropterins to 7-isomers via phenylalanine hydroxylase-generated intermediates. Proc Natl Acad Sci USA 88: 385–389

Degtyarenko KN, Archakov AI (1993) Molecular evolution of P450 superfamily and P450-containing monooxygenase systems. FEBS Lett 332: 1–8

Dhondt JL (1991) Strategy for the screening of tetrahydrobiopterin deficiency among hyperphenylalaninaemic patients: 15-years experience. J Inherited Metab Dis 14: 117–127

Dimmeler S, Lottspeich F, Brune B (1992) Nitric oxide causes ADP-ribosylation and inhibition of glyceraldehyde-3-phosphate dehydrogenase. J Biol Chem 267: 16771–16774

Dinerman JL, Dawson TM, Schell MJ, Snowman A, Snyder SH (1994) Endothelial nitric oxide synthase localized to hippocampal pyramidal cells: implications for synaptic plasticity. Proc Natl Acad Sci USA 91: 4214–4218

Drapier JC, Hibbs JB (1988) Differentiation of murine macrophages to express nonspecific cytotoxicoty for tumour cells results in L-arginine-dependent inhibition of mitochondrial iron-sulfur enzymes in the macrophage effector cells. J Immunol 140: 2829–2838

Drapier JC, Hirling H, Wietzerbin J, Kaldy P, Kuhn LC (1993) Biosynthesis of nitric oxide activates iron regulatory factor in macrophages. EMBO J 12: 3643–3649

Duch DS, Bowers SW, Woolf JH, Nichol CA (1984) Biopterin cofactor biosynthesis: GTP cyclohydrolase, neopterin and biopterin in tissues and body fluids of mammalian species. Life Sci 35: 1895–1901

Duhe RJ, Nielsen MD, Dittman AH, Villacres EC, Choi EJ, Storm DR (1994) Oxidation of critical cysteine residues of type I adenylyl cyclase by o-iodosobenzoate or nitric oxide reversibly inhibits stimulation by calcium and calmodulin. J Biol Chem 269: 7290–7296

Evans T, Carpenter A, Cohen J (1992) Purification of a distinctive form of endotoxin-induced nitric oxide synthase from rat liver. Proc Natl Acad Sci USA 89: 5361–5365

Feldman PL, Griffith OW, Stuehr DJ (1993) The surprising life of nitric oxide. Chem Eng News 71: 26–38

Förstermann U, Pollock JS, Schmidt HH, Heller M, Murad F (1991) Calmodulin-dependent endothelium-derived relaxing factor/nitric oxide synthase activity is present in the particulate and cytosolic fractions of bovine aortic endothelial cells. Proc Natl Acad Sci USA 88: 1788–1792

Förstermann U, Closs EI, Pollock JS, Nakane M, Schwarz P, Gath I, Kleinert H (1994) Nitric oxide synthase isozymes – characterization, purification, molecular cloning, and functions. Hypertension 23: 1121–1131

Friedman PA, Kappelman AH, Kaufman S (1972) Partial purification and characterization of tryptophan hydroxylase from rabbit hindbrain. J Biol Chem 247: 4165–4173

Fuchs D, Hausen A, Reibnegger G, Werner ER, Dierich MP, Wachter H (1988) Neopterin as a marker for activated cell-mediated immunity: application in HIV infection. Immunol Today 9: 150–155
Fulco AJ (1991) P450BM-3 and other inducible bacterial P450 cytochromes: biochemistry and regulation. Annu Rev Pharmacol Toxicol 31: 177–203
Gal EM, Nelson JM, Sherman AD (1978) Biopterin III. Purification and characterization of enzymes involved in cerebral synthesis of 7,8-dihydrobiopterin. Neurochem Res 3: 69–88
Garthwaite J, Charles SL, Chess-Williams R (1988) Endothelium-derived relaxing factor release on activation of NMDA receptors suggests role as intercellular messenger in the brain. Nature 336: 385–388
Geller DA, Disilvio M, Nüssler AK, Freeswick PD, Nguyen D, Shah N, Simmons RL, Billiar TR (1993) Hepatocyte nitric oxide production during endotoxinemia requires expression of GTP-cyclohydrolase mRNA and synthesis of tetrahydrobiopterin. Surg Forum 44: 67
Gelperin A (1994) Nitric oxide mediates network oscillations of olfactory interneurons in a terrestrial mollusc. Nature 369: 61–63
Gerzer R, Böhme E, Hofmann F, Schultz G (1981) Soluble guanylate cyclase purified from bovine lung contains heme and copper. FEBS Lett 132: 71–74
Gibbs BS, Wojchowski D, Benkovic SJ (1993) Expression of rat liver phenylalanine hydroxylase in insect cells and site-directed mutagenesis of putative non-heme iron binding sites. J Biol Chem 268: 8046–8052
Gilles-Gonzales MA, Gonzales G, Perutz MF, Kiger L, Marden MC, Poyart C (1994) Heme-based sensors, exemplified by the kinase FixL, are a new class of heme protein with distictive ligand binding and autoxidation. Biochemistry 33: 8067–8073
Giovanelli J, Campos KL, Kaufman S (1991) Tetrahydrobiopterin, a cofactor for rat cerebellar nitric oxide synthase, does not function as a reactant in the oxygenation of arginine. Proc Natl Acad Sci USA 88: 7091–7095
Gopalakrishna R, Chen ZH, Gundimeda U (1993) Nitric oxide and nitric oxide-generating agents induce a reversible inactivation of protein kinase C activity and phorbol ester binding. J Biol Chem 268: 27180–27185
Griscavage JM, Fukuto JM, Komori Y, Ignarro LJ (1994) Nitric oxide inhibits neuronal nitric oxide synthase by interacting with the heme prosthetic group – role of tetrahydrobiopterin in modulating the inhibitory action of nitric oxide. J Biol Chem 269: 21644–21649
Gross SS, Levi R (1992) Tetrahydrobiopterin synthesis – an absolute requirement for cytokine-induced nitric oxide generation by vascular smooth muscle. J Biol Chem 267: 25722–25729
Gross SS, Jaffe EA, Levi R, Kilbourn RG (1991) Cytokine-activated endothelial cells express an isotype of nitric oxide synthase which is tetrahydrobiopterin dependent, calmodulin independent and inhibited by arginine analogs with a rank-order of potency characteristic of activated macrophages. Biochem Biophys Res Commun 178: 823–829
Guengrich FP (1991) Reactions and significance of cytochrome P450-enzymes. J Biol Chem 266: 10019–10022
Gütlich M, Schott K, Werner T, Bacher A, Ziegler I (1992) Species and tissue specificity of mammalian GTP cyclohydrolase-I messenger RNA. Biochim Biophys Acta 1171: 133–140
Gütlich M, Jaeger E, Rucknagel KP, Werner T, Rodl W, Ziegler I, Bacher A (1994) Human GTP cyclohydrolase I: only one out of three cDNA isoforms gives rise to the active enzyme. Biochem J 302: 215–221
Haile DJ, Rounalt TA, Harford JB, Kennedy MC, Blodin GA, Beinert H, Klausner RD (1992) Cellular regulation of the iron responsive element binding protein: disassembly of the cubane iron-sulfur cluster results in high-affinity RNA-binding. Proc Natl Acad Sci USA 89: 11735–11739
Harada T, Kagamiyama H, Hatakeyama K (1993) Feedback regulation mechanisms for the control of GTP cyclohydrolase-I activity. Science 260: 1507–1510

Harteneck C, Klatt P, Schmidt K, Mayer B (1994) Baculovirus-mediated expression of rat brain nitric oxide synthase and characterization of the purified enzyme. Biochem J 304: 683–686

Hasler T, Niederwieser A (1986) Tetrahydrobiopterin-producing enzyme activities in liver of animals and man. In: Cooper BA, Whitehead VM (eds) Chemistry and biology of pteridines. De Gruyter, Berlin, pp 319–322

Hatakeyama K, Harada T, Suzuki S, Watanabe Y, Kagamiyama H (1989) Purification and characterization of rat liver GTP cyclohydrolase I. Cooperative binding of GTP to the enzyme. J Biol Chem 264: 21660–21664

Hatakeyama K, Inoue Y, Harada T, Kagamiyama H (1991) Cloning and sequencing of cDNA encoding rat GTP cyclohydrolase I: the first enzyme of the tetrahydrobiopterin biosynthetic pathways. J Biol Chem 266: 765–769

Hatakeyama K, Harada T, Kagamiyama H (1992) IMP dehydrogenase inhibitors reduce intracellular tetrahydrobiopterin levels through reduction of intracellular GTP levels – indications of the regulation of GTP cyclohydrolase-I activity by restriction of GTP availability in the cells. J Biol Chem 267: 20734–20739

Hattori Y, Gross SS (1993) GTP cyclohydrolase I messenger RNA is induced by LPS in vascular smooth muscle – characterization, sequence and relationship to nitric oxide synthase. Biochem Biophys Res Commun 195: 435–441

Hausladen A, Fridovich I (1994) Superoxide and peroxynitrite inactivate aconitase, but not its precursor, nitric oxide. J Biol Chem 269: 29405–29408

Hecker M, Mülsch A, Busse R (1994) Subcellular localization and characterization of neuronal nitric oxide synthase. J Neurochem 62: 1524–1529

Heinzel B, John M, Klatt P, Böhme E, Mayer B (1992) Ca2+/calmodulin-dependent formation of hydrogen peroxide by brain nitric oxide synthase. Biochem J 281: 627–630

Hevel JM, Marletta MA (1992) Macrophage nitric oxide synthase: relationship between enzyme-bound tetrahydrobiopterin and synthase activity. Biochemistry 31: 7160–7165

Hevel JM, White KA, Marletta MA (1991) Purification of the inducible murine macrophage nitric oxide synthase. Identification as a flavoprotein. J Biol Chem 266: 22789–22791

Hibbs JBJ, Taintor RR, Vavrin Z, Rachlin EM (1988) Nitric oxide: a cytotoxic activated macrophage effector molecule. Biochem Biophys Res Commun 157: 87–94

Hiki K, Hattori R, Kawai C, Yui Y (1992) Purification of insoluble nitric oxide synthase from rat cerebellum. J Biochem (Tokyo) 111: 556–558

Hofmann B, Bass H, Nishanian P, Faisal M, Figlin RA, Sarna GP, Fahey JL (1992) Different lymphoid cell populations produce varied levels of neopterin beta-2-microglobulin and soluble sIL-2 receptor when stimulated with IL-2, interferon-gamma or tumour necrosis factor-alpha. Clin Exp Immunol 89: 1–7

Holscher C, Rose SP (1992) An inhibitor of nitric oxide synthesis prevents memory formation in the chick. Neurosci Lett 145: 165–167

Hope BT, Michael GJ, Knigge KM, Vincent SR (1991) Neuronal NADPH diaphorase is a nitric oxide synthase. Proc Natl Acad Sci USA 88: 2811–2814

Hopkins FG (1889) Note on a yellow pigment from butterflies. Nature 40: 335

Huber C, Fuchs D, Hausen A, Margreiter R, Reibnegger G, Spielberger M, Wachter H (1983) Pteridines as a new marker to detect human T cells activated by allogeneic or self-modified major histocompatibility complex (MHC) determinants. J Immunol 130: 1047–1050

Huber C, Batchelor JR, Fuchs D, Hausen A, Lang A, Niederwieser D, Reibnegger G, Swetly P, Troppmair J, Wachter H (1984) Immune response-associated production of neopterin – release from macrophages primarily under control of interferon-gamma. J Exp Med 160: 310–316

Humbert P, Niroomand F, Fischer G, Mayer B, Koesling D, Hinsch KD, Gausepohl H, Frank R, Schultz G, Böhme E (1990) Purification of soluble guanylyl cyclase from bovine lung by a new immunoaffinity chromatographic method. Eur J Biochem 190: 273–278

Ichinose H, Katoh S, Sueoka T, Titani K, Fujita K, Nagatsu T (1991) Cloning and sequencing of cDNA encoding human sepiapterin reductase. Biochem Biophys Res Commun 179: 183–189

Ignarro LJ (1991) Heme-dependent activation of guanylate cyclase by nitric oxide: a novel signal transduction mechanism. Blood Vessels 28: 67–73

Ignarro LJ, Buga GM, Wood KS, Byrns RE, Chaudhuri G (1987) Endothelium-derived relaxing factor produced and released from artery and vein is nitric oxide. Med Sci 84: 9265–9269

Inoue Y, Kawasaki Y, Harada T, Hatakeyama K, Kagamiyama H (1991) Purification and cDNA cloning of rat 6-pyruvoyl-tetrahydropterin synthase. J Biol Chem 266: 20791–20796

Irby RB, Adair WL (1994) Intermediates in the folic acid biosynthetic pathway are incorporated in the yeast *Pichia canadiensis*. J Biol Chem 269: 23981–23987

Jorens PG, Vanoverveld FJ, Bult H, Vermeire PA, Herman AG (1992) Pterins inhibit nitric oxide synthase activity in rat alveolar macrophages. Br J Pharmacol 107: 1088–1091

Karupiah G, Xie QW, Buller RML, Nathan C, Duarte C, Macmicking JD (1993) Inhibition of viral replication by interferon-gamma induced nitric oxide synthase. Science 261: 1445–1448

Katoh S, Sueka T, Yamada S (1982) Direct inhibition of brain sepiapterin reductase by a catecholamine and an indoleamine. Biochem Biophys Res Commun 105: 75–81

Kaufman S (1963) The structure of phenylalanine hydroxylation cofactor. Proc Natl Acad Sci USA 50: 1085–1093

Kaufman S (1993) New tetrahydrobiopterin-dependent systems. Annu Rev Nutr 13: 261-286

Kaufman S, Pollock RJ, Summer GK, Das AK, Hajra AK (1990) Dependence of an alkyl glycol-ether monooxygenase activity upon tetrahydropterins. Biochim Biophys Acta 1040: 19–27

Khatsenko OG, Gross SS, Rifkind AB, Vane JR (1993) Nitric oxide is a mediator of the decrease in cytochrome-P450-dependent metabolism caused by immunostimulants. Proc Natl Acad Sci USA 90: 11147–11151

Klatt P, Heinzel B, John M, Kastner M, Böhme E, Mayer B (1992a) Ca^{2+}/calmodulin-dependent cytochrome c reductase activity of brain nitric oxide synthase. J Biol Chem 267: 11374–11378

Klatt P, Heinzel B, Mayer B, Ambach E, Werner-Felmayer G, Wachter H, Werner ER (1992b) Stimulation of human nitric oxide synthase by tetrahydrobiopterin and selective binding of the cofactor. FEBS Lett 305: 160–162

Klatt P, Schmidt K, Mayer B (1992c) Brain nitric oxide synthase is a haemoprotein. Biochem J 288: 15–17

Klatt P, Schmidt K, Uray G, Mayer B (1993) Multiple catalytic functions of brain nitric oxide synthase – biochemical characterization, cofactor-requirement, and the role of N(G.W)-hydroxy-L-arginine as an intermediate. J Biol Chem 268: 14781–14787

Klatt P, Schmid M, Leopold E, Schmidt K, Werner ER, Mayer B (1994a) The pteridine binding site of brain nitric oxide synthase – tetrahydrobiopterin binding kinetics, specificity, and allosteric interaction with the substrate domain. J Biol Chem 269: 13861–13866

Klatt P, Schmidt K, Brunner F, Mayer B (1994b) Inhibitors of brain nitric oxide synthase – binding kinetics, metabolism, and enzyme inactivation. J Biol Chem 269: 1674–1680

Klausner RD, Rouault TA, Harford JB (1993) Regulating the fate of mRNA: the control of cellular iron metabolism. Cell 72: 19–28

Klemm P, Ostrowski J, Morath T, Gruber C, Martorana PA, Henning R (1993) N-Acetylserotonin prevents the hypotension induced by bacterial lipopolysaccharides in the rat. Eur J Pharmacol 250: R9–R10

Knowles RG, Moncada S (1994) Nitric oxide synthases in mammals. Biochem J 298: 249–258

Knowles RG, Palacios M, Palmer RM, Moncada S (1989) Formation of nitric oxide from L-arginine in the central nervous system: a transduction mechanism for stimulation of the soluble guanylate cyclase. Proc Natl Acad Sci USA 86: 5159–5162

Koesling D, Böhme E, Schultz G (1991) Guanylyl cyclases, a growing family of signal-transducing enzymes. FASEB J 5: 2785–2791

Kwon NS, Nathan CF, Stuehr DJ (1989) Reduced biopterin as a cofactor in the generation of nitrogen oxides by murine macrophages. J Biol Chem 264: 20496–20501

Kwon NS, Nathan CF, Gilker C, Griffith OW, Matthews DE, Stuehr DJ (1990) L-citrulline production from L-arginine by macrophage nitric oxide synthase. The ureido oxygen derives from dioxygen. J Biol Chem 265: 13442–13445

Kwon NS, Stuehr DJ, Nathan CF (1991) Inhibition of tumor cell ribonucleotide reductase by macrophage-derived nitric oxide. J Exp Med 174: 761–767

Lamas S, Marsden PA, Li GK, Tempst P, Michel T (1992) Endothelial nitric oxide synthase: molecular cloning and characterization of a distinct constitutive enzyme isoform. Proc Natl Acad Sci USA 89: 6348–6352

Lancaster JR, Werner-Felmayer G, Wachter H (1994) Coinduction of nitric oxide synthesis and intracellular nonheme iron-nitrosyl complexes in murine cytokine-treated fibroblasts. Free Radic Biol Med 16: 869–870

Lander HM, Sehajpal PK, Novogrodsky A (1993) Nitric oxide signaling – a possible role for G-proteins. J Immunol 151: 7182–7187

Leone AM, Palmer RM, Knowles RG, Francis PL, Ashton DS, Moncada S (1991) Constitutive and inducible nitric oxide synthases incorporate molecular oxygen into both nitric oxide and citrulline. J Biol Chem 266: 23790–23795

Lepoivre M, Fieschi F, Coves J, Thelander L, Fontecave M (1991) Inactivation of ribonucleotide reductase by nitric oxide. Biochem Biophys Res Commun 179: 442–448

Lepoivre M, Flaman JM, Bobe P, Lemaire G, Henry Y (1994) Quenching of the tyrosyl free radical of ribonucleotide reductase by nitric oxide – relationship to cytostasis induced in tumor cells by cytotoxic macrophages. J Biol Chem 269: 21891–21897

Lowenstein CJ, Dinerman JL, Snyder SH (1994) Nitric oxide – a physiologic messenger. Ann Intern Med 120: 227–237

Luo DS, Vincent SR (1994) Metalloporphyrins inhibit nitric oxide-dependent cGMP formation in vivo. Eur J Pharmacol Mol Pharmacol 267: 263–267

Lyons CR, Orloff GJ, Cunningham JM (1992) Molecular cloning and functional expression of an inducible nitric oxide synthase from a murine macrophage cell line. J Biol Chem 267: 6370-6374

Maines MD (1993) Carbon monoxide – an emerging regulator of cGMP in the brain. Mol Cell Neurosci 4: 389–397

Marletta MA (1988) Mammalian synthesis of nitrite, nitrate, nitric oxide, and N-nitrosating agents. Chem Res Toxicol 1: 249–257

Marletta MA (1994a) Nitric oxide synthase: aspects concerning structure and catalysis. Cell 78: 297–930

Marletta MA (1994b) Approaches toward selective inhibition of nitric oxide synthase. J Med Chem 37: 1899–1907

Martinez A, Abeygunawardana C, Haavik J, Flatmark T, Mildvan AS (1993) Conformation and interaction of phenylalanine with the divalent cation at the active site of human recombinant tyrosine hydroxylase as determined by proton NMR. Biochemistry 32: 6381–6390

Masters BSS (1994) Nitric oxide synthases: why so complex? Annu Rev Nutr 14: 131–145

Matsuoka A, Stuehr DJ, Olson JS, Clark P, Ikedasaito M (1994) L-arginine and calmodulin regulation of the heme iron reactivity in neuronal nitric oxide synthase. J Biol Chem 269: 20335–20339

Mayer B, Werner ER (1995) Why tetrahydrobiopterin? In: Ignarro L, Murad F (eds) Nitric oxide: biochemistry, molecular biology and therapeutic implications. Academic, San Diego (in press)

Mayer B, Schmidt K, Humbert P, Böhme E (1989) Biosynthesis of endothelium-derived relaxing factor: a cytosolic enzyme in porcine aortic endothelial cells Ca2+-dependently converts L-arginine into an activator of soluble guanylyl cyclase. Biochem Biophys Res Commun 164: 678–685

Mayer B, John M, Böhme E (1990) Purification of a Ca2+/calmodulin-dependent nitric oxide synthase from porcine cerebellum. Cofactor-role of tetrahydrobiopterin. FEBS Lett 277: 215–219

Mayer B, John M, Heinzel B, Werner ER, Wachter H, Schultz G, Böhme E (1991) Brain nitric oxide synthase is a biopterin- and flavin-containing multi-functional oxido-reductase. FEBS Lett 288: 187–191

Mayer B, Heinzel B, Klatt P, John M, Schmidt K, Böhme E (1992) Nitric oxide synthase-catalyzed activation of oxygen and reduction of cytochromes – reaction mechanisms and possible physiological implications. J Cardiovasc Pharmacol 20: S54–S56

Mayer B, Schmid M, Klatt P, Schmidt K (1993) Reversible inactivation of endothelial nitric oxide synthase by N(G)-nitro-L-arginine. FEBS Lett 333: 203–206

Mayer B, Klatt P, Werner ER, Schmidt K (1994) Molecular mechanism of inhibition of porcine brain nitric oxide synthase by the antinociceptive drug 7-nitroindazole. Neuropharmacology 33: 1253–1259

Mayer B, Klatt P, Werner ER, Schmidt K (1995) Kinetics and mechanism of tetrahydrobiopterin-induced oxidation of nitric oxide. J Biol Chem 270: 655–659

McMillan K, Masters BSS (1993) Optical difference spectrophotometry as a probe of rat brain nitric oxide synthase heme-substrate interaction. Biochemistry 32: 9875–9880

McInnes RR, Kaufman S, Warsh JJ, Van Loon GR, Milstien S, Kapatos G, Soldin S, Walsh P, MacGregor D, Hanley WB (1984) Biopterin synthesis defect treatment with L-dopa and 5-hydroxytryptophan compared with therapy with a tetrahydropterin. J Clin Invest 73: 458–469

McMillan K, Bredt DS, Hirsch DJ, Snyder SH, Clark JE, Masters BSS (1992) Cloned, expressed rat cerebellar nitric oxide synthase contains stoichiometric amounts of heme, which binds carbon monoxide. Proc Natl Acad Sci USA 89: 11141–11145

Mellouk S, Hoffman SL, Liu ZZ, Delavega P, Billiar TR, Nüssler AK (1994) Nitric oxide-mediated antiplasmodial activity in human and murine hepatocytes induced by gamma interferon and the parasite itself: enhancement by exogenous tetrahydrobiopterin. Infect Immun 62: 4043–4046

Michel T, Li GK, Busconi L (1993) Phosphorylation and subcellular translocation of endothelial nitric oxide synthase. Proc Natl Acad Sci USA 90: 6252–6256

Moncada S, Palmer RM, Higgs EA (1989) Biosynthesis of nitric oxide from L-arginine. A pathway for the regulation of cell function and communication. Biochem Pharmacol 38: 1709–1715

Moncada S, Palmer RM, Higgs EA (1991) Nitric oxide: physiology, pathophysiology, and pharmacology. Pharmacol Rev 43: 109–142

Mühl H, Pfeilschifter J (1994) Tetrahydrobiopterin is a limiting cofactor of nitric oxide generation in interleukin-1 beta stimulated rat glomerular mesangial cells. Kidney Int 46: 1302–1306

Müller U (1994) Ca2+/calmodulin-dependent nitric oxide synthase in *Apis mellifera* and *Drosophila melanogaster*. Eur J Neurosci 6: 1362–1370

Nakayama DK, Geller DA, Disilvio M, Bloomgarden G, Davies P, Pitt BR, Hatakeyama K, Kagamiyama H, Simmons RL, Billiar TR (1994) Tetrahydrobiopterin synthesis and inducible nitric oxide production in pulmonary artery smooth muscle. Am J Physiol 266: L455-L460

Nar H, Huber R, Heizmann CW, Thöny B, Burgisser D (1994) Three-dimensional structure of 6-pyruvoyl tetrahydropterin synthase, an enzyme involved in tetrahydrobiopterin biosynthesis. EMBO J 13: 1255–1262

Nathan C (1992) Nitric oxide as a secretory product of mammalian cells. FASEB J 6: 3051–3064

Nathan CF, Hibbs JBJ (1991) Role of nitric oxide synthesis in macrophage antimicrobial activity. Curr Opin Immunol 3: 65–70

Nathan C, Xie QW (1994) Nitric oxide synthases: roles, tolls, and controls. Cell 78: 915–918

Nichol CA, Smith GK, Duch DS (1985) Biosynthesis and metabolism of tetrahydrobiopterin and molybdopterin. Annu Rev Biochem 54: 729–764

Nishikimi M (1975) A function of tetrahydropteridines as cofactor for indoleamine 2, 3-dioxygenase. Biochem Biophys Res Commun 63: 92–98
Nüssler AK, Billiar TR (1993) Review – inflammation, immunoregulation, and inducible nitric oxide synthase. J Leukoc Biol 54: 171–178
Odell TJ, Huang PL, Dawson TM, Dinerman JL, Snyder SH, Kandel ER, Fishman MC (1994) Endothelial NO synthase and the blockade of LTP by NO synthase inhibitors in mice lacking neuronal NO synthase. Science 265: 542–546
Olken NM, Marletta MA (1993) N^0-Methyl-L-arginine functions as an alternate substrate and mechanism-based inhibitor of nitric oxide synthase. Biochemistry 32: 9677–9685
Ostholm T, Holmqvist BI, Alm P, Ekstrom P (1994) Nitric oxide synthase in the CNS of the Atlantic salmon. Neurosci Lett 168: 233–237
Ozaki Y, Reinhard J, Nichol CA (1986) Cofactor activity of dihydroflavin mononucleotide and tetrahydrobiopterin for murine epididymal indoleamine 2,3-dioxygenase. Biochem Biophys Res Commun 137: 1106–1111
Palmer RMJ, Ferrige AG, Moncada S (1987) Nitric oxide release accounts for the biological activity of endothelium-derived relaxing factor. Nature 327: 524–526
Pantopoulos K, Weiss G, Hentze MW (1994) Nitric oxide and posttranscriptional control of cellular iron traffic. Trends Cell Biol 4: 82–86
Park YS, Heizmann CW, Wermuth B, Levine RA, Steinerstauch P, Guzman J, Blau N (1991) Human carbonyl and aldose reductases: new catalytic functions in tetrahydrobiopterin biosynthesis. Biochem Biophys Res Commun 175: 738–744
Parniak MA, Pilkington J (1989) Glucocorticoid stimulation of tetrahydrobiopterin levels and phenylalanine hydroxylase activity in rat hepatoma cells. Biochem Cell Biol 67: 293–296
Peunova N, Enikolopov G (1993) Amplification of calcium-induced gene transcription by nitric oxide in neuronal cells. Nature 364: 450–453
Pfleiderer W, Kim SJ, Yim JJ (1992) The nonenzymatic synthesis of drosopterins. In: Blau N, Curtius HC, Levine R et al. (eds) Pteridines and related biogenic amines and folates. Hanrim, Seoul, pp 54–60
Pollock JS, Förstermann U, Mitchell JA, Warner TD, Schmidt HHHW, Nakane M, Murad F (1991) Purification and characterization of particulate endothelium-derived relaxing factor synthase from cultured and native bovine aortic endothelial cells. Proc Natl Acad Sci USA 88: 10480–10484
Pollock JS, Werner ER, Mitchell JA, Förstermann U (1993) Particulate endothelial nitric oxide synthase: requirement and content of tetrahydrobiopterin, FAD, and FMN. Endothelium 1: 147–152
Pou S, Pou WS, Bredt DS, Snyder SH, Rosen GM (1992) Generation of superoxide by purified brain nitric oxide synthase. J Biol Chem 267: 24173–24176
Rajagopalan KV, Johnson JL (1992) The pterin molybdenum cofactors. J Biol Chem 267: 10199–10202
Reif DW, Simmons RD (1990) Nitric oxide mediates iron release from ferritin. Arch Biochem Biophys 283: 537–541
Renaud JP, Boucher JL, Vadon S, Delaforge M, Mansuy D (1993) Particular ability of liver P450s3A to catalyze the oxidation of N(omega)-hydroxyarginine to citrulline and nitrogen oxides and occurrence in NO synthases of a sequence very similar to the heme-binding sequence in P450s. Biochem Biophys Res Commun 192: 53–60
Ribeiro JMC, Nussenzweig RH (1993) Nitric oxide synthase activity from a hematophagous insect salivary gland. FEBS Lett 330: 165–168
Richter G, Ritz H, Katzenmeier G, Volk R, Kohnle A, Lottspeich F, Allendorf D, Bacher A (1993) Biosynthesis of riboflavin – cloning, sequencing, mapping, and expression of the gene coding for GTP cyclohydrolase-II in *Escherichia coli*. J Bacteriol 175: 4045–4051
Rogers NE, Ignarro LJ (1992) Constitutive nitric oxide synthase from cerebellum is reversibly inhibited by nitric oxide formed from L-arginine. Biochem Biophys Res Commun 189: 242–249

Rosenkranz-Weiss P, Sessa WC, Milstien S, Kaufman S, Watson CA, Pober JS (1994) Regulation of nitric oxide synthesis by proinflammatory cytokines in human umbilical vein endothelial cells – elevations in tetrahydrobiopterin levels enhance endothelial nitric oxide synthase specific activity. J Clin Invest 93: 2236–2243

Sakai N, Kaufman S, Milstien S (1993a) Tetrahydrobiopterin is required for cytokine-induced nitric oxide production in a murine macrophage cell line (RAW-264). Mol Pharmacol 43: 6–10

Sakai N, Saito K, Kaufman S, Heyes MP, Milstien S (1993b) Induction of pterin synthesis is not required for cytokine-stimulated tryptophan metabolism. Biochem J 295: 543–547

Salvemini D, Misko TP, Masferrer JL, Seibert K, Currie MG, Needleman P (1993) Nitric oxide activates cyclooxygenase enzymes. Proc Natl Acad Sci USA 90: 7240–7244

Schmid C, Ladenstein R, Luecke H, Huber R, Bacher A (1992) Crystallization and preliminary crystallographic characterization of GTP cyclohydrolase I from *Escherichia coli.* J Mol Biol 226: 1279–1281

Schmidt HHHW, Walter U (1994) NO at work. Cell 78: 919–925

Schmidt HHHW, Pollock JS, Nakane M, Gorsky LD, Förstermann U, Murad F (1991) Purification of a soluble isoform of guanylyl cyclase-activating-factor synthase. Proc Natl Acad Sci USA 88: 365–369

Schmidt HHHW, Smith RM, Nakane M, Murad F (1992) Ca^{2+}/calmodulin-dependent NO synthase type I: a biopteroflavoprotein with Ca2+/calmodulin-independent diaphorase and reductase activities. Biochemistry 31: 3243–3249

Schmidt HHHW, Lohmann SM, Walter U (1993) The nitric oxide and cGMP signal transduction system – regulation and mechanism of action. Biochim Biophys Acta 1178: 153–175

Schmidt K, Werner ER, Mayer B, Wachter H, Kukovetz WR (1992) Tetrahydrobiopterin-dependent formation of endothelium-derived relaxing factor (nitric oxide) in aortic endothelial cells. Biochem J 281: 297–300

Schoedon G, Troppmair J, Adolf G, Huber C, Niederwieser A (1986) Interferon-gamma enhances biosynthesis of pterins in peripheral blood mononuclear cells by induction of GTP-cyclohydrolase I activity. J IFN Res 6: 697–703

Schoedon G, Troppmair J, Fontana A, Huber C, Curtius HC, Niederwieser A (1987) Biosynthesis and metabolism of pterins in peripheral blood mononuclear cells and leukemia lines of man and mouse. Eur J Biochem 166: 303–310

Schoedon G, Schneemann M, Blau N, Edgell CJS, Schaffner A (1993a) Modulation of human endothelial cell tetrahydrobiopterin synthesis by activating and deactivating cytokines – new perspectives on endothelium-derived relaxing factor. Biochem Biophys Res Commun 196: 1343–1348

Schoedon G, Schneemann M, Hofer S, Guerrero L, Blau N, Schaffner A (1993b) Regulation of the L-arginine-dependent and tetrahydrobiopterin-dependent biosynthesis of nitric oxide in murine macrophages. Eur J Biochem 213: 833–839

Schoedon G, Blau N, Schneemann M, Flury G, Schaffner A (1994) Nitric oxide production depends on preceding tetrahydrobiopterin synthesis by endothelial cells – selective suppression of induced nitric oxide production by sepiapterin reductase inhibitors. Biochem Biophys Res Commun 199: 504–510

Schott K, Gütlich M, Ziegler I (1993) Induction of GTP-cyclohydrolase-I messenger RNA expression by lectin activation and interferon-gamma treatment in human cells associated with the immune response. J Cell Physiol 156: 12–16

Scott-Burden T, Elizondo E, Ge T, Boulanger CM, Vanhoutte PM (1993) Growth factor regulation of interleukin-L beta-induced nitric oxide synthase and GTP – cyclohydrolase expression in cultured smooth muscle cells. Biochem Biophys Res Commun 196: 1261–1266

Sessa WC, Harrison JK, Barber CM, Zeng D, Durieux ME, D'Angelo DD, Lynch KR, Peach MJ (1992) Molecular cloning and expression of a cDNA encoding endothelial cell nitric oxide synthase. J Biol Chem 267: 15274–15276

Sessa WC, Barber CM, Lynch KR (1993) Mutation of N-myristoylation site converts endothelial cell nitric oxide synthase from a membrane to a cytosolic protein. Circ Res 72: 921–924

Shen RS, Alam A, Zhang YX (1988) Inhibition of GTP cyclohydrolase I by pterins. Biochem Biophys Acta 965: 9–15

Sheta EA, McMillan K, Masters BSS (1994) Evidence for a bidomain structure of constitutive cerebellar nitric oxide synthase. J Biol Chem 269: 15147–15153

Shiman R, Akino M, Kaufman S (1971) Solubilization and partial purification of tyrosine hydroxylase from bovine adrenal medulla. J Biol Chem 246: 1330–1340

Smith GK, Duch DS, Edelstein MP, Bigham EC (1992) New inhibitors of sepiapterin reductase. Lack of an effect of intracellular tetrahydrobiopterin depletion upon in vitro proliferation of two human cell lines. J Biol Chem 267: 5599–5607

Snyderman SE, Sansaricq C, Pulmones MT (1987) Successful long-term therapy of biopterin deficiency. J Inherited Metab Dis 10: 260–266

Stamler JS (1994) Redox signaling: nitrosylation and related target interactions of nitric oxide. Cell 78: 931–936

Stadler J, Billiar TR, Curran RD, Stuehr DJ, Ochoa JB, Simmons RL (1991) Effect of exogenous and endogenous nitric oxide on mitochondrial respiration of rat hepatocytes. Am J Physiol 260: C910–C916

Stadler J, Trockfeld J, Schmalix WA, Brill T, Siewert JR, Greim H, Doehmer J (1994) Inhibition of cytochromes P4501A by nitric oxide. Proc Natl Acad Sci USA 91: 3559–3563

Stone JR, Marletta MA (1994) Soluble guanylate cyclase from bovine lung – activation with nitric oxide and carbon monoxide and spectral characterization of the ferrous and ferric states. Biochemistry 33: 5636–5640

Stuehr DJ, Ikedasaito M (1992) Spectral characterization of brain and macrophage nitric oxide synthases – cytochrome-P-450-like hemeproteins that contain a flavin semiquinone radical. J Biol Chem 267: 20547–20550

Stuehr DJ, Marletta MM (1985) Mammalian nitrate biosynthesis: mouse macrophages produce nitrite in response to *Escherichia coli* lipopolysaccharide. Med Sci 82: 7738–7742

Stuehr DJ, Nathan CF (1989) Nitric oxide. A macrophage product responsible for cytostasis and respiratory inhibition in tumor target cells. J Exp Med 169: 1543–1555

Stuehr DJ, Cho HJ, Kwon NS, Weise MF, Nathan CF (1991a) Purification and characterization of the cytokine-induced macrophage nitric oxide synthase: an FAD- and FMN-containing flavoprotein. Proc Natl Acad Sci USA 88: 7773–7777

Stuehr DJ, Kwon NS, Nathan CF, Griffith OW, Feldman PL, Wiseman J (1991b) N omega-hydroxy-L-arginine is an intermediate in the biosynthesis of nitric oxide from L-arginine. J Biol Chem 266: 6259–6263

Sueka T, Katoh S (1985) Carbonyl reductase activites of sepiapterin reductase from rat erythrocytes. Biochim Biophys Acta 843: 193–198

Sung YJ, Hotchkiss JH, Dietert RR (1994) 2,4-Diamino-6-hydroxypyrimidine, an inhibitor of GTP cyclohydrolase I, suppresses nitric oxide production by chicken macrophages. Int J Immunopharmacol 16: 101–108

Tabuchi A, Sano K, Oh E, Tsuchiya T, Tsuda M (1994) Modulation of AP-1 activity by nitric oxide (NO) in vitro: NO-mediated modulation of AP-1. FEBS Lett 351: 123–127

Takikawa S, Curtius HC, Redweik U, Ghisla S (1986) Purification of 6-pyruvoyl-tetrahydropterin synthase from human liver. Biochem Biophys Res Commun 134: 646–651

Tayeh MA, Marletta MA (1989) Macrophage oxidation of L-arginine to nitric oxide, nitrite, and nitrate. Tetrahydrobiopterin is required as a cofactor. J Biol Chem 264: 19654–19658

Thomas SR, Mohr D, Stocker R (1994) Nitric oxide inhibits indoleamine 2,3-dioxygenase activity in interferon-gamma primed mononuclear phagocytes. J Biol Chem 269: 14457–14464

Thöny B, Leimbacher W, Burgisser D, Heizmann CW (1992) Human 6-pyruvoyltetrahydropterin synthase – cDNA cloning and heterologous expression of the recombinant enzyme. Biochem Biophys Res Commun 189: 1437–1443

Thöny B, Neuheiser CR, Hauer CR, Heizmann CW (1993) Molecular cloning and expression of the human liver phenylalanine hydroxylase stimulating factor revealed structural and functional identity with the dimerization cofactor for the nuclear transcription factor HNF-1alpha. Adv Exp Med Biol 336: 103–106

Thöny B, Heizmann CW, Mattei MG (1994) Chromosomal location of two human genes encoding tetrahydrobiopterin-metabolizing enzymes – 6-pyruvoyl-tetrahydropterin synthase maps to 11Q22.3-Q23.3, and pterin-4 alpha-carbinolamine dehydratase maps to 10Q22. Genomics 19: 365–368

Tietz A, Lindberg M, Kennedy EP (1964) A new pteridine-requiring enzyme system for the oxidation of glyceryl ethers. J Biol Chem 239: 4081–4090

Traylor TG, Sharma VS (1992) Why NO? Biochemistry 31: 2847–2849

Troppmair J, Nachbaur K, Herold M, Aulitzky W, Tilg H, Gaiul G, Bieling P, Kotlan B, Flener R, Mull B, Aulitzky WO, Rokosu H, Huber C (1988) In-vitro and in-vivo studies on the induction of neopterin biosynthesis by cytokines, alloantigen and lipopolysaccharide (LPS). Clin Exp Immunol 74: 392-397

Überall F, Werner–Felmayer G, Schubert C, Grunicke HH, Wachter H, Fuchs D (1994) Neopterin derivatives together with cyclic guanosine monophosphate induce c–fos gene expression. FEBS Lett 352: 11–14

Van Beelen P, Stassen APM, Bosch JWG, Vogels GD, Guijt W, Haasnoot CAG (1984) Elucidation of the structure of methanopterin, a coenzyme from *Methanobacterium thermoautotrophicum*, using two-dimensional nuclear-magnetic-resonance techniques. Eur J Biochem 138: 563–571

Verma A, Hirsch DJ, Glatt CE, Ronnett GV, Snyder SH (1993) Carbon monoxide – a putative neural messenger. Science 259: 381–384

Viveros OH, Lee CL, Abou-Donia MM, Nixon JC, Nichol CA (1981) Biopterin cofactor biosynthesis: independent regulation of GTP cyclohydrolase in adrenal medulla and cortex. Science 213: 349–350

Wachter H, Fuchs D, Hausen A, Reibnegger G, Werner ER (1989) Neopterin as marker for activation of cellular immunity: immunologic basis and clinical application. Adv Clin Chem 27: 81–141

Waldman S, Murad F (1987) Cyclic GMP synthesis and function. Pharmacol Rev 39: 163–195

Walter U (1989) Physiological role of cGMP and cGMP-dependent protein kinase in the cardiovascular system. Rev Physiol Biochem Pharmacol 113: 42–88

Wang JL, Stuehr DJ, Ikedasaito M, Rousseau DL (1993) Heme coordination and structure of the catalytic site in nitric oxide synthase. J Biol Chem 268: 22255–22258

Wang JL, Rousseau DL, Abu-Soud HM, Stuehr DJ (1994) Heme coordination of NO in NO synthase. Proc Natl Acad Sci USA 91: 10512–10516

Wedel B, Humbert P, Harteneck C, Foerster J, Malkewitz J, Böhme E, Schultz G, Koesling D (1994) Mutation of His-105 in the beta(1)-subunit yields a nitric oxide-insensitive form of soluble guanylyl cyclase. Proc Natl Acad Sci USA 91: 2592–2596

Weiss G, Goossen B, Doppler W, Fuchs D, Pantopoulos K, Werner-Felmayer G, Wachter H, Hentze MW (1993a) Translational regulation via iron-responsive elements by the nitric oxide/NO-synthase pathway. EMBO J 12: 3651–3657

Weiss G, Fuchs D, Hausen A, Reibnegger G, Werner ER, Werner-Felmayer G, Semenitz E, Dierich MP, Wachter H (1993b) Neopterin modulates toxicity mediated by reactive oxygen and chloride species. FEBS Lett 321: 89–92

Weiss G, Werner-Felmayer G, Werner ER, Grünewald K, Wachter H, Hentze MW (1994) Iron regulates nitric oxide synthase activity by controlling nuclear transcription. J Exp Med 180: 969–976

Werner ER, Bitterlich G, Fuchs D, Hausen A, Reibnegger G, Szabo G, Dierich MP, Wachter H (1987) Human macrophages degrade tryptophan upon induction by interferon-gamma. Life Sci 41: 273–280

Werner ER, Werner-Felmayer G, Fuchs D, Hausen A, Reibnegger G, Wachter H (1989) Parallel induction of tetrahydrobiopterin biosynthesis and indoleamine 2,3-dioxygenase activity in human cells and cell lines by interferon-gamma. Biochem J 262: 861–866

Werner ER, Werner-Felmayer G, Fuchs D, Hausen A, Reibnegger G, Yim JJ, Pfleiderer W, Wachter H (1990) Tetrahydrobiopterin biosynthetic activities in human macrophages, fibroblasts, THP-1, and T 24 cells. GTP-cyclohydrolase I is stimulated by interferon-gamma, and 6-pyruvoyl tetrahydropterin synthase and sepiapterin reductase are constitutively present. J Biol Chem 265: 3189–3192

Werner ER, Werner-Felmayer G, Fuchs D, Hausen A, Reibnegger G, Yim JJ, Wachter H (1991a) Impact of tumour necrosis factor-alpha and interferon-gamma on tetrahydrobiopterin synthesis in murine fibroblasts and macrophages. Biochem J 280: 709–714

Werner ER, Werner-Felmayer G, Fuchs D, Hausen A, Reibnegger G, Wels G, Yim JJ, Pfleiderer W, Wachter H (1991b) 6-Pyruvoyl tetrahydropterin synthase assay in extracts of cultured human cells using high-performance liquid chromatography with fluorescence detection of biopterin. J Chromatogr 570: 43–50

Werner ER, Werner-Felmayer G, Fuchs D et al. (1991c) Pteridine synthesis in cytokine-treated cells. In: Blau N, Curtius HC, Levine R et al (eds) Pterins and biogenic amines in neurology, pediatrics and immunology. Lakeshore, Grosse Pointe, MI, pp 213–224

Werner ER, Werner-Felmayer G, Wachter H (1993) Tetrahydrobiopterin and cytokines. Proc Soc Exp Biol Med 203: 1–12

Werner-Felmayer G, Werner ER, Fuchs D, Hausen A, Reibnegger G, Wachter H (1989) Tumour necrosis factor-alpha and lipopolysaccharide enhance interferon-induced tryptophan degradation and pteridine synthesis in human cells. Biol Chem Hoppe Seyler 370: 1063–1069

Werner-Felmayer G, Werner ER, Fuchs D, Hausen A, Reibnegger G, Wachter H (1990) Tetrahydrobiopterin-dependent formation of nitrite and nitrate in murine fibroblasts. J Exp Med 172: 1599–1607

Werner-Felmayer G, Prast H, Werner ER, Philippu A, Wachter H (1993a) Induction of GTP cyclohydrolase I by bacterial lipopolysaccharide in the rat. FEBS Lett 322: 223–226

Werner-Felmayer G, Werner ER, Fuchs D, Hausen A, Reibnegger G, Schmidt K, Weiss G, Wachter H (1993b) Pteridine biosynthesis in human endothelial cells. Impact on nitric oxide-mediated formation of cyclic GMP. J Biol Chem 268: 1842–1846

Werner-Felmayer G, Werner ER, Fuchs D, Hausen A, Mayer B, Reibnegger G, Weiss G, Wachter H (1993c) Ca(2+)/calmodulin-dependent nitric oxide synthase activity in the human cervix carcinoma cell line ME-180. Biochem J 289: 357–361

Werner-Felmayer G, Golderer G, Werner ER, Wachter H (1994) Pteridine biosynthesis and nitric oxide synthase in *Physarum polycephalum.* Biochem J 304: 105–111

Werner-Felmayer G, Werner ER, Wachter H, Gross SS (1995) Analysis of tetrahydrobiopterin and its role in nitric oxide synthesis. In: Feelisch M, Stamler JS (eds) Methods in nitric oxide research. Wiley, Chichester, UK (in press)

White KA, Marletta MA (1992) Nitric oxide synthase is a cytochrome P-450 type hemoprotein. Biochem 31: 6627–6631

White RH (1990) Biosynthesis of methanopterin. Biochemistry 29: 5397–5404

Wink DA, Osawa Y, Darbyshire JF, Jones CR, Eshenaur SC, Nims RW (1993) Inhibition of cytochromes-P450 by nitric oxide and a nitric oxide-releasing agent. Arch Biochem Biophys 300: 115–123

Wink DA, Nims RW, Darbyshire JF, Christodoulou D, Hanbauer I, Cox GW, Laval F, Laval J, Cook JA, Krishna MC, Degraff WG, Mitchell JB (1994) Reaction kinetics for nitrosation of cysteine and glutathione in aerobic nitric oxide solutions at neutral pH. Insights into the fate and physiological effects of intermediates generated in the NO/O-2 reaction. Chem Res Toxicol 7: 519–525

Wolf G, Wurdig S, Schunzel G (1992) Nitric oxide synthase in rat brain is predominantly located at neuronal endoplasmic reticulum – an electron microscopic demonstration of NADPH-diaphorase activity. Neurosci Lett 147: 63–66

Wöll E, Weiss G, Fuchs D, Lang F, Wachter H (1993) Effect of pteridine derivatives on intracellular calcium concentration in human monocytic cells. FEBS Lett 318: 249–252
Xie QW, Cho HJ, Calaycay J, Mumford RA, Swiderek KM, Lee TD, Ding A, Troso T, Nathan C (1992) Cloning and characterization of inducible nitric oxide synthase from mouse macrophages. Science 256: 225–228
Yu AE, Hu SZ, Spiro TG, Burstyn JN (1994) Resonance raman spectroscopy of soluble guanylyl cyclase reveals displacement of distal and proximal heme ligands by NO. J Am Chem Soc 116: 4117–4118
Yui Y, Hattori R, Kosuga K, Eizawa H, Hiki K, Kawai C (1991) Purification of nitric oxide synthase from rat macrophages. J Biol Chem 266: 12544–12547
Zhuo M, Small SA, Kandel ER, Hawkins RD (1993) Nitric oxide and carbon monoxide produce activity-dependent long-term synaptic enhancement in hippocampus. Science 260: 1946–1950
Ziegler I (1985) Synthesis and interferon-gamma controlled release of pteridines during activation of human peripheral blood mononuclear cells. Biochem Biophys Res Commun 132: 404–411
Ziegler I, Schwulera U, Ellwart J (1986) Pteridines are produced during interleukin 2-induced T-cell proliferation and modulate transmission of this signal. Exp Cell Res 167: 531–538
Ziegler I, Schott K, Lubbert M, Herrmann F, Schwulera U, Bacher A (1990) Control of tetrahydrobiopterin synthesis in T lymphocytes by synergistic action of interferon-gamma and interleukin-2. J Biol Chem 265: 17026–17030

Editor-in-Charge: Professor H. Grunicke and Professor M. Schweiger

Rev. Physiol. Biochem. Pharmacol., Vol. 127

Cytochrome P450: Structure, Function, and Generation of Reactive Oxygen Species

R. BERNHARDT

Contents

Max-Delbrück-Centrum für Molekulare Medizin, Robert-Rössle-Strasse 10, 13122 Berlin, Germany and Freie Universität Berlin, Institut für Biochemie, Thielallee 63, 14195 Berlin, Germany

1 Historical Background

Cytochromes P450 were discovered about 40 years ago. At that time, pharmacologists were intrigued by the observation that the "hexobarbital sleeping time" of animals could be used as an indicator of the effectiveness of various compounds to alter the rate of drug metabolism. The availability of high-speed centrifuges in the late 1940s, allowing the separation of subcellular fractions from tissue homogenates, led to the intracellular localization of many enzyme systems. Although it is impossible to trace back the discovery of cytochromes P450 to one person, Brodie's laboratory was one of the first to use in vitro methods to study the metabolism of many chemicals and to relate these findings to in vivo observations. It was shown that the TPNH-dependent enzyme system (later identified as NADPH-cytochrome P450 reductase), participating in the oxidation of many drugs, was associated with the microsomal fraction of liver (La Du et al. 1953, 1955; Cooper and Brodie 1954; Axelrod 1954, 1955, see also Axelrod 1982). The pioneering work of this laboratory in both in vivo and in vitro studies of drug metabolism (reviewed by Brodie 1956) provided evidence that the activity of the enzyme system involved in the metabolism of different drugs could be readily altered and that there were profound differences between species with regard to the extent and way in which they metabolized the same substance. From such studies, as well as from work of Miller and co-workers (Müller and Miller 1953; Conney et al. 1956), who discovered the formation of carcinogenic electrophilic compounds by liver microsomes, it became clear that liver microsomes contain oxidative enzymes capable of metabolizing drugs and xenobiotics. It was established that the reaction proceeds only in the presence of molecular oxygen. Since then, more than 800 different xenobiotics have been identified to be substrates of liver microsomal cytochromes P450.

The inducibility of the liver microsomal drug-metabolizing enzyme systems was discovered in Miller's laboratory in the 1950s (Miller et al. 1952; Brown et al. 1954; Conney et al. 1956, 1960, 1967). They observed that administration of polycyclic hydrocarbons to rats increased dramatically the activity of liver microsomes to N-demethylate methylated azo dyes. Independently, Remmer made similar observations when studying the mechanism by which tolerance to a drug developed (Remmer 1959).

In the late1950s and the 1960s, two general classes of enzymes involved in oxygen metabolism were discovered (Fig. 1): oxidases, transferring electrons from a substrate to oxygen, and oxygenases, transferring oxygen to a substrate after reductive splitting of molecular oxygen. Oxygenases can be divided into dioxygenases and mono-oxygenases. Mono-oxygenases (mixed-function oxidases) catalyze the incorporation of a single atom of

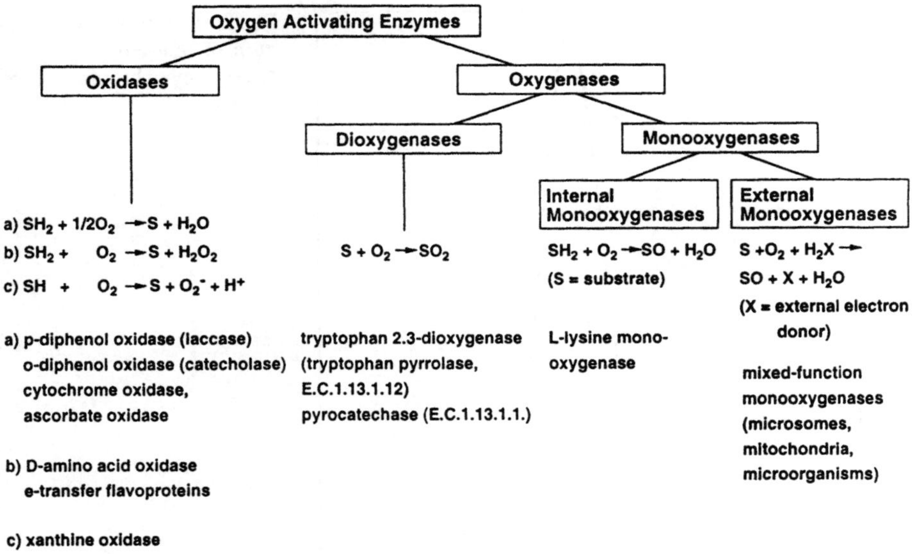

Fig. 1. Oxygen-activating enzymes. (From Ruckpaul et al. 1989)

molecular oxygen into a substrate with the concomitant reduction of the other atom to water (Mason et al. 1955; Mason 1957; Hayaishi and Nozaki 1969).

The mono-oxygenases were divided by Hayaishi and Nozaki (1969) into two classes, the internal and the external mono-oxygenases (see Fig. 1). Internal mono-oxygenases extract two reducing equivalents from the substrate to reduce one atom of dioxygen to water, whereas external mono-oxygenases utilize an external reductant. While the microsomal drug and xenobiotic-metabolizing enzymes were initially referred to as mixed-function oxidases, in recent years the term mono-oxygenase has become the more accepted one.

In 1958 Garfinkel and Klingenberg detected a CO-binding pigment in liver microsomes of pigs and rats, which was reducible by either NADPH or dithionite and displayed a typical absorption maximum of the reduced CO-bound complex at 450 nm (Fig. 2; Garfinkel 1958; Klingenberg 1958). It was not until 1964 that the hemoprotein nature of this CO-binding pigment was proven by Omura and Sato (1964). The identification of this pigment as a b-type cytochrome with an atypical absorption maximum (Soret band) of the CO complex at 450 nm led the authors to designate it cytochrome P450. The ability of reduced P450 to induce absorption at 450 nm upon CO binding is currently used for estimating the P450 content.

These developments were closely followed by studies in which it was experimentally substantiated that P450 is the key enzyme responsible for the

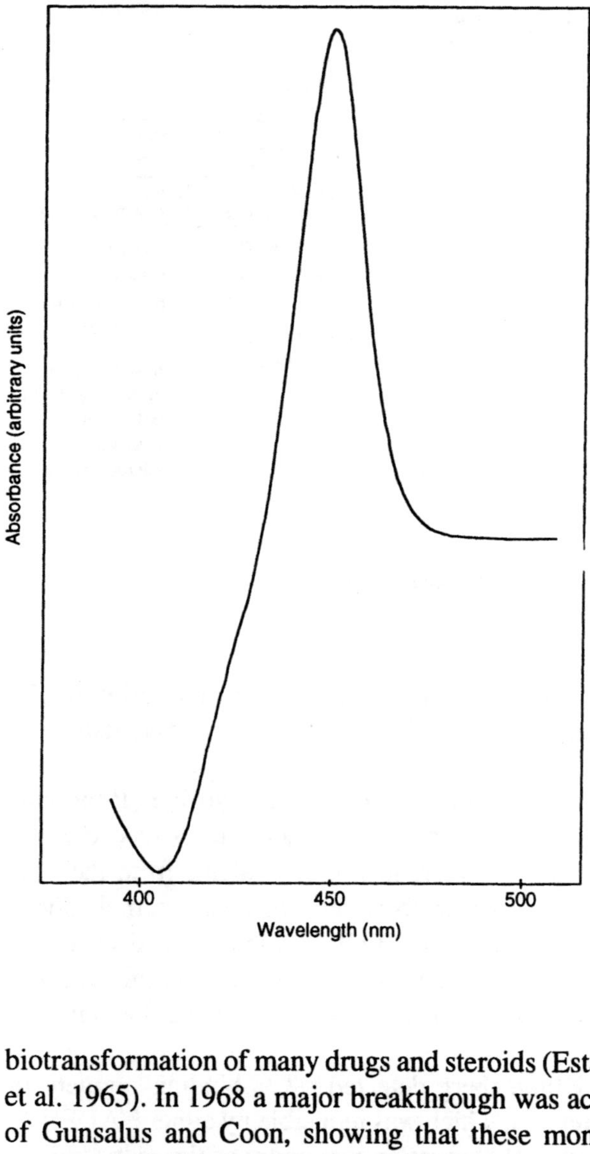

Fig. 2. CO-difference spectrum of reduced versus oxidized cytochrome P450 (CYP2B4)

biotransformation of many drugs and steroids (Estabrook et al. 1963; Cooper et al. 1965). In 1968 a major breakthrough was achieved by the laboratories of Gunsalus and Coon, showing that these mono-oxygenase systems are composed of different proteins (Katagiri et al. 1968; Lu and Coon 1968). This discovery started a series of studies leading to the purification and characterization of the components of cytochrome P450 systems and to their reconstitution in an in vitro system. Different isoforms of cytochromes P450 have been isolated from various species and tissues. In 1982 the primary sequences of two isoforms were solved, one by amino acid sequencing (Haniu et al. 1982) and one by nucleotide sequencing of a cloned cDNA (Fujii-Kuriyama et al. 1982). It was not until 1985 that the first three-dimensional structure of a P450 was reported (Poulos et al. 1985). This was the crystal structure of the soluble, bacterial camphor-hydroxylating P450cam

(CYP101). Unfortunately, due to tremendous difficulties with the crystallization of membrane proteins, no three-dimensional structure is so far available for a membrane-bound P450. During the 1980s, heterologous expression of cytochromes P450 in *Escherichia coli*, yeast, and cell cultures became available, inducing a major breakthrough in studies on the mechanism of P450 action and the regulation of the gene expression of these enzymes.

2 Description of Cytochrome P450 Systems

2.1 The Cytochrome P450 Supergene Family

Cytochrome P450 systems catalyze the following reaction:

$$RH + O_2 \longrightarrow ROH + H_2O \quad (1)$$

They are involved in the metabolism of many drugs and xenobiotics. They catalyze reactions as diverse as hydroxylation, N-, O,- and S-dealkylation, sulfoxidation, epoxidation, deamination, desulfuration, dehalogenation, peroxidation, and N-oxide reduction. Their substrates include fatty acids, steroids, and prostaglandins, as well as a multitude of foreign compounds such as drugs, anesthetics, organic solvents, ethanol, alkylaryl hydrocarbon products, pesticides, and carcinogens. Examples of cytochrome P450-dependent substrate conversions are shown in Table 1.

It is obvious that this diversity of substrates and catalyzed reactions cannot be managed by only a few different isoforms. When the first classification of the family members was made in 1991, 154 different P450 genes were described, grouped into 27 gene families. Ten of these families, including 18 subfamilies, were found in mammals (Nebert et al. 1991). By 1993, the number of P450 genes had already grown to 221, existing in 36 gene families. Twelve of the gene families, being divided into 22 subfamilies, were shown to exist in mammals (Fig. 3). For 17 P450 forms the localization on a distinct chromosome was identified (Nelson et al. 1993). To date, more than 300 different isoforms have been characterized, making an update of the classification necessary again. Nebert et al. (1991) also recommended a new nomenclature of the P450s to minimize confusion introduced by the use of different names by different groups for the same enzyme. CYP is used to characterize the respective P450 as a hemoprotein. The first arabic number defines the gene family, the following letter the subfamily, and the second number the individual enzyme, e.g., CYP1A1 for cytochrome P4501A1 (previously P450c).

Table 1. Cytochrome P450-dependent substrate conversions

	Substrates
Oxidation	
Aromatic hydroxylation	aniline, benzene, acetanilide, phenobarbital, 3,4-benzo[a]pyrene, steroids, bile acids
Aliphatic hydroxylation	fatty acids, alkanes, tolbutamide
Side-chain hydroxylation	cholesterol, barbiturates
Epoxidation	benzo[a]pyrene, vinylchloride, biphenyl compounds
Peroxidation	lipids
N-oxidation	nicotine, morphine, dimethylamphetamine, trimethylamine
S-oxidation	chlorpromazine, parathion, phenothiazine
N-dealkylation	benzphetamine, ephedrine, methadone
O-dealkylation	phenacetine, papaverine, griseofulvin, codeine
S-dealkylation	6-methylmercaptopurine
Deamination	amphetamine, ephedrine, mescaline
Desulfuration	parathion
Dehydrogenation	hexachlorocyclohexane, androstenedione, testosterone
Denitrification	2-nitropropane
Dehalogenation	halothane, chloroform
Reduction	
Azoreduction	prontosil
Nitro reduction	chloramphenicol
Carbonyl reduction	acetophenone
Hydrolysis	
Ester	procaine
Amide	phenacetin

Members of the same gene family are defined as usually having ≤ 40% sequence identity with a P450 protein from any other family. This definition was made arbitrarily but has turned out to be very useful. Mammalian sequences within the same subfamily are always > 55% identical. The main characteristics of the most important cytochrome P450 families will be discussed in brief.

Fig. 3. Evolutionary tree of the cytochrome P450 superfamily. Unweighted pair group method of analysis has been used. Of the 154 cDNAs or genes that were characterized in various species as of October 20, 1990, 147 are considered. (From Nebert et al. 1991)

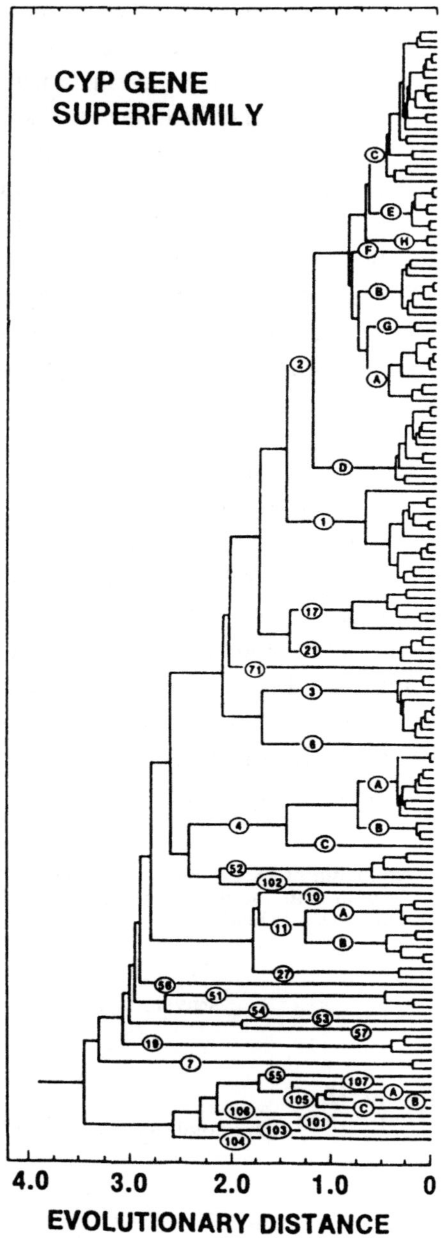

The **CYP1** family consists so far of only one subfamily with two members, CYP1A1 and CYP1A2. While CYP1A1 is detected only after treatment with inducers such as 3-methylcholanthrene (MC) and 2,3,7,8-tetrachlorodibenzo-p-dioxin (TCDD), CYP1A2 is constitutively expressed in liver but can also be induced by MC and TCDD. CYP1A1 has a high catalytic activity towards the polycyclic hydrocarbon benzo[a]pyrene. CYP1A2 is involved in the conversion of arylamines and thus can activate several heterocyclic amine promutagens derived from pyrolysates of proteins (see Gonzalez 1990 and papers cited therein; Gonzalez and Gelboin 1994). These enzymes may therefore play an essential role in chemical carcinogenesis. The **CYP2** family consists of ten subfamilies. Cytochromes P450 belonging to this family are involved mainly in the bioconversion of drugs (especially subfamilies CYP2D and CYP2C), in testosterone hydroxylation (CYP2A subfamily), and in the metabolism of ethanol, acetone, acetoacetone, and acetol (CYP2E subfamily; for details see Gonzalez et al. 1990; Nelson et al. 1993). The **CYP3** family consists of one subfamily with 12 individual enzymes. The most important form is CYP3A4, which is involved in the metabolism of a variety of very different drugs and xenobiotics such as nifedipine, cyclosporine, erythromycin, gestodene, and aflatoxins (Guengerich et al. 1993). This variety of substrates makes CYP3A4 one of the most important enzymes for drug metabolism. Members of the **CYP11** family are involved in steroid hormone biosynthesis. To this group belongs the side-chain cleavage enzyme (CYP11A1), catalyzing the conversion of cholesterol to pregnenolone, the initial and rate-limiting step of steroid hormone biosynthesis (reviewed by Lambeth 1990; Usanov et al. 1990; Vickery 1993) . CYP11B1 is involved in the biosynthesis of cortisol from 11-deoxycortisol and of corticosterone from 11-deoxycorticosterone, whereas CYP11B2 catalyzes the conversion of 11-deoxycorticosterone to aldosterone (reviewed by White et al. 1993). Family **CYP19** so far consists of only one member, aromatase,

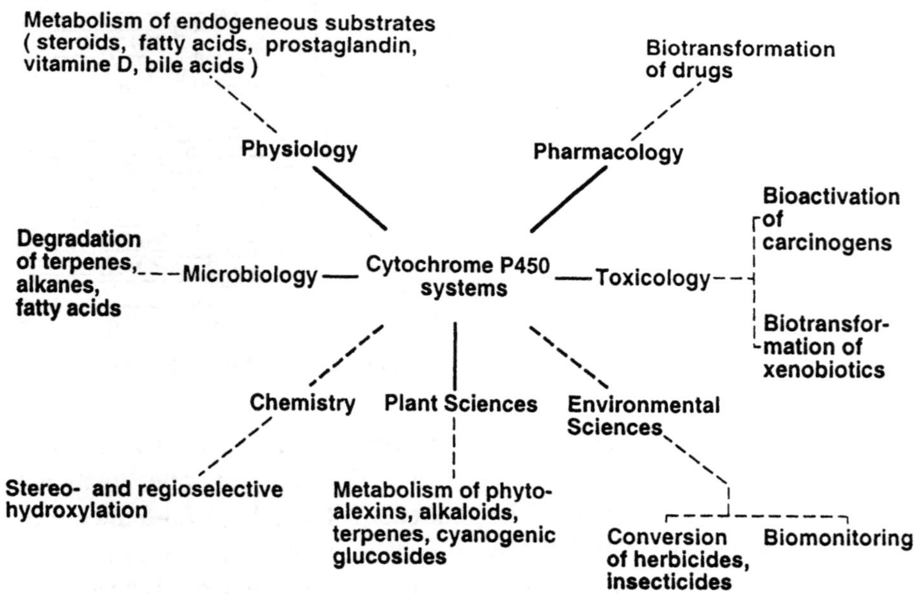

Fig. 4. Cytochrome P450 research and application fields

catalyzing the conversion of androgens to estrogens by aromatization of the A ring of androgens (reviewed by Simpson et al. 1993). The best-studied cytochrome P450, however, is **CYP101** (formerly called P450cam). This is a soluble, bacterial P450 involved in the catabolism of camphor by *Pseudomonas putida.* CYP101 was the first P450 for which a three-dimensional structure became available (Poulos et al. 1985), so that numerous structural and functional studies have been performed with it.

The diversity of cytochromes P450 and of catalyzed reactions has attracted the interest of scientists from very different fields in P450 research. Besides pharmacologists and toxicologists, endocrinologists, physiologists, microbiologists, organic chemists, plant biologists, and environmental scientists are working on diverse aspects of P450 function and regulation (Fig. 4). The growing variety of organisms in which P450 genes and enzymes are detected and the discovery of new P450 forms will probably lead to further applications and new research fields.

2.2 Structural Organization of Cytochrome P450 Systems

As mentioned in Sect. 1, cytochromes P450 belong to external mono-oxygenases (Fig. 1). This implies that they need an external electron donor which transfers the electrons necessary for oxygen activation and the subsequent substrate hydroxylation. Two main classes of cytochromes P450 principally different with respect to their electron-supporting system can be defined, (a) the microsomal type and (b) the mitochondrial/bacterial type (Fig. 5). Microsomal cytochromes P450 are membrane bound and accept electrons from a microsomal NADPH-cytochrome P450 reductase, containing FAD and FMN. All drug and xenobiotica metabolizing cytochromes P450 isolated so far belong to this class. In addition, CYP102 (P450BM-3) isolated from *Bacillus megaterium* was shown to belong to this class. This P450 system consists of a polypeptide chain with two different domains, one containing the hemoprotein and the other a FAD-reductase (Narhi and Fulco 1987; Boddupalli et al. 1992). The P450 domain of this protein was recently crystallized (Ravichandran et al. 1993).

Most of the other bacterial cytochromes P450 belong to the second class. They are soluble and obtain the electrons necessary for the reaction mechanism from an NADH-dependent FAD-containing reductase via an iron-sulfur protein of the /2Fe-2S/type. The camphor (Gunsalus and Wagner 1978), linalool (Ullah et al. 1990), and terpineol (Peterson et al. 1992) hydroxylating cytochromes P450 (CYP101, CYP111, and CYP108), as well as the 15-alpha-hydroxylase from *Bacillus megaterium* (Rauschenbach et al. 1993) belong to this class. Mitochondrial cytochromes P450 being involved in the

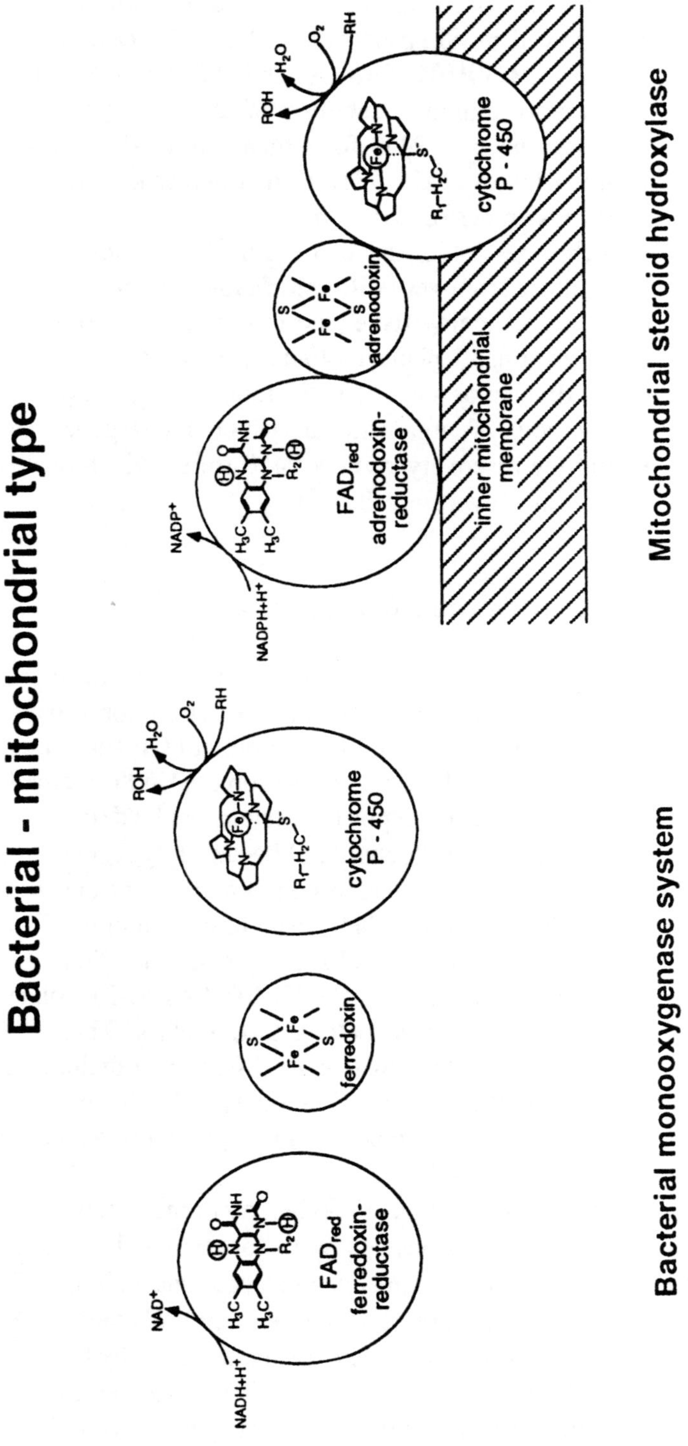
Bacterial - mitochondrial type
NADH+H+
NAD+
FADred
ferredoxin-reductase
ferredoxin
cytochrome P - 450
ROH
H2O
O2
RH
Bacterial monooxygenase system
NADPH+H+
NADP+
FADred
adrenodoxin-reductase
adrenodoxin
cytochrome P - 450
inner mitochondrial membrane
Mitochondrial steroid hydroxylase

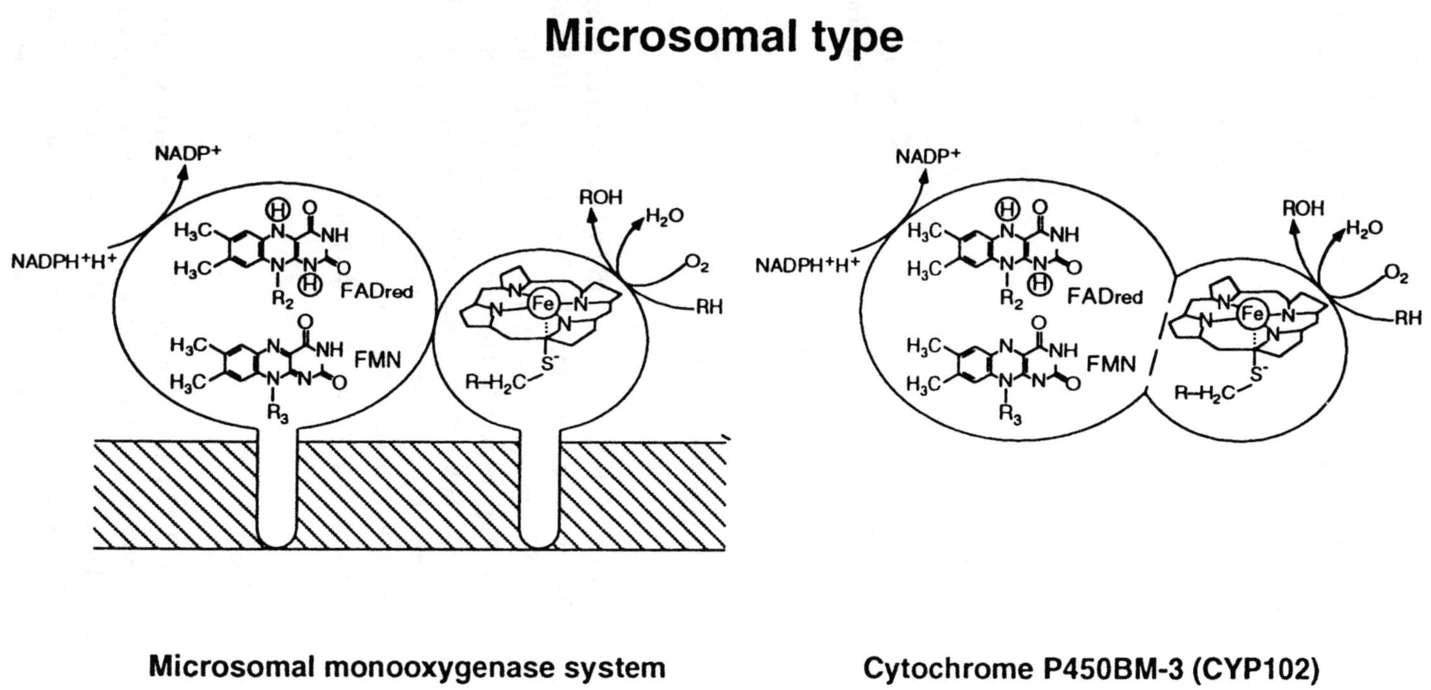

Fig. 5. Model for the organization of cytochrome P450 systems

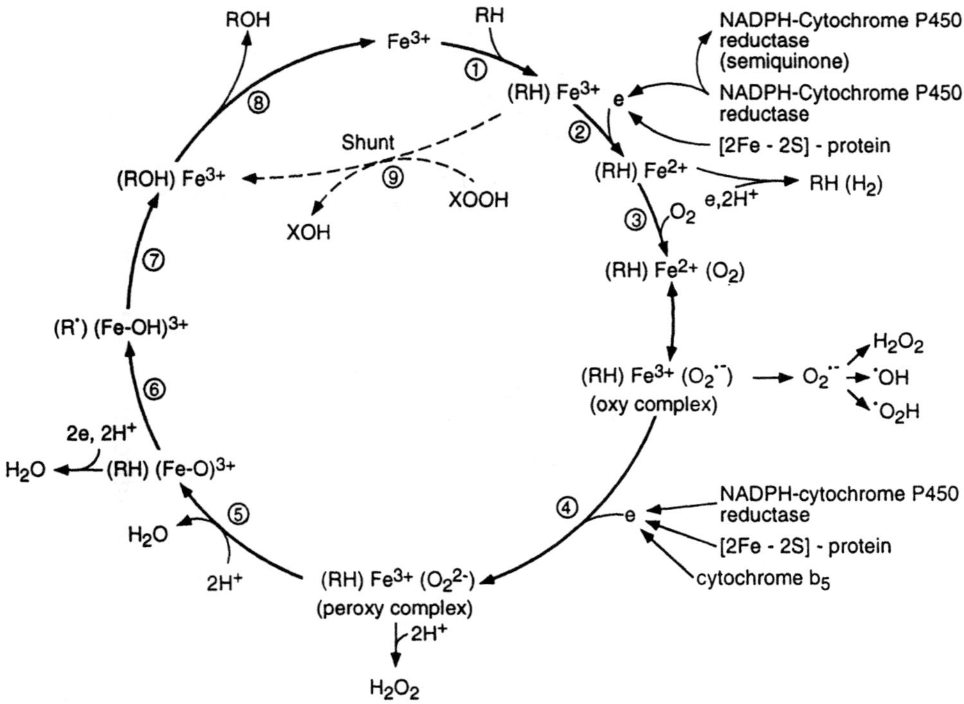

Fig. 6. Reaction cycle of cytochrome P450

side-chain cleavage of cholesterol, the 11β-hydroxylation of 11-deoxycortisol and the production of aldosterone also belong to the latter class. These cytochromes P450 are localized in the inner mitochondrial membrane, whereas the /2Fe-2S/protein, called adrenodoxin in the case of adrenal steroid hydroxylase systems, is a soluble protein of the matrix. The FAD-containing reductase, adrenodoxin reductase, is associated with the inner mitochondrial membrane. Recognition of the cytochromes P450 and the corresponding electron donor is a necessary prerequisite to the catalytic cycle.

2.3 Reaction Cycle

After cytochrome P450 had been identified as a hemoprotein, a series of studies on its hemoprotein nature began. A sulfhydryl group was identified in a variety of physicochemical studies (for review see Rein et al. 1984) and finally by resolution of the three-dimensional structure of a P450 (Poulos et al. 1985) as the fifth ligand of the heme iron. The generally accepted mechanism of cytochrome P450-dependent substrate conversion is depicted in the overall scheme presented in Fig. 6. Due to the fact that P450 is a one-electron acceptor (Peterson et al. 1977), radicals must be involved in the mechanism of mono-oxygenases (Ortiz de Montellano 1989; White 1991). The first step in the reaction cycle is the formation of the substrate-enzyme complex. The hemoprotein in its ferric state was shown to be an equilibrium mixture of low- and high- spin states of the heme iron ($S=5/2 \leftrightarrow S=1/2$). The spin-state equilibrium can be shifted by substrate binding as well as by temperature (reviewed by Schenkman 1982; Rein and Jung 1993). Substrate binding induces structural changes of the P450 which may result not only in a spin shift, but also in changes of the redox potential and in binding affinities between interacting components of the P450 system. These effects seem to be differently interrelated in different P450 systems. The binding of the substrate can be followed by observing spectral changes of the P450 in the near UV spectral region (Soret band) and in the visible spectral region caused by spin-state changes of the protein (Mitani and Horie 1969; Whysner et al. 1970). A classification of these small, but characteristic spectral changes was given by Schenkman et al. (1967). The majority of substrates being metabolized by cytochromes P450 produces so-called type-I difference spectra, characterized by a minimum at 417 nm and a maximum at 387 nm. Many inhibitors which bind directly to the heme iron produce type-II difference spectra with absorption maxima between 425 and 445 nm and minima between 390 and 420 nm. Finally, some compounds such as alcohols and ketones give rise to so-called inverse type-I difference spectra, characterized by an absorption maximum at about 420 nm and a minimum at 385–390 nm. The high-spin form of the P450 seems to be more readily reducible (Blanck et al. 1983; Fisher and Sligar 1985, 1987). The correlation between spin equilibrium and ability of the P450 to be reduced is not always valid, however (Lambeth and Kriengsiri 1985).

The second step of the reaction cycle is the introduction of the first electron, either by NADPH-dependent reductase or via a ferredoxin (Fig. 5). Iron (3+) is reduced to iron (2+), thereby oxidizing the microsomal reductase to a stable semiquinone free radical. This step is still not fully understood. The reason for this lies in the fact that the reduction of P450 in most cases is not monophasic or first order. Alternative explanations have been suggested to describe this phenomenon. The cluster model suggests the presence of aggregates of cytochrome P450 surrounding the NADPH-cytochrome P450 reductase and the fast phase to be due to reduction of the closest P450 molecules, whereas the slow phase is proposed to arise from exchange and subsequent reduction of the P450 molecules not initially in direct contact with the reductase (Peterson et al. 1976). The model is supported by kinetic (Blanck et al. 1983, 1989; Schwarze et al. 1985; Fisher and Sligar 1985, 1987) and structural investigations (Schwarz 1991, references therein). The sequential spin-state model (Backes et al. 1980, 1982; Tamburini et al. 1984) explains the biphasic cytochrome P450 reduction with differences in the reduction of high-spin and low-spin conformers. Although there is some evidence for a favored electron transfer to the high-spin conformer (Backes et al. 1985), the rapid relaxation of the spin equilibrium as shown for CYP101 and CYP2B4 is in disagreement with this model. In addition, formation of active (fast-phase) and inactive complexes between reductase and cytochrome P450 has been suggested as the reason for the biphasic kinetics (Backes and Eyer 1989). When oxygen concentration is low, the reduced P450 is able to donate its electron to an alternate electron acceptor, for example halothane, producing a carbon radical that may be released to initiate lipid peroxidation (Van Dyke et al. 1988; White 1994).

In the third step of the reaction cycle the one-electron reduced hemoprotein binds oxygen. In contrast to mammalian cytochrome P450, where the oxyferrous complex is decomposed rather quickly to yield ferric P450 and H_2O_2 (Oprian and Coon 1982), in CYP101 the oxyferrous state is fairly stable, autoxidizing only very slowly to ferric CYP101 and $O_2^{\bullet-}$ (Sligar et al. 1974). The ternary complex formed by binding of oxygen to the one-electron reduced hemoprotein is spectroscopically characterized by absorption bands at 418–420 nm and 558 nm in the absolute spectrum and maxima at 440 nm and 590 nm in the difference spectrum to the oxidized P450 (Ishimura et al. 1971; Estabrook et al. 1971; Peterson et al. 1972; Bonfils et al. 1979; Larroque and Van Lier 1980; Tuckey and Kamin 1982). As shown in Fig. 6, from this complex a superoxide anion radical can be released. The negatively charged dioxygen ligand favors binding of a proton (Rein et al. 1986) or at least hydrogen bonding.

Step 4 of the reaction cycle is the introduction of the second electron. In some situations another microsomal hemoprotein, cytochrome b_5, can facilitate catalysis by providing the second electron (Bonfils et al 1981; Schenkman et al. 1987; for review see Schenkman 1993). From the $(RH)Fe^{3+}(O_2^{2-})$ complex hydrogen peroxide can be split off. Thus, due to the fact that cytochrome P450 is a one-electron acceptor, oxygen activation occurs in two steps (Peterson et al. 1977). The structure of the activated oxygen and the precise mechanism of oxygen cleavage are not yet fully understood, however. It is now generally accepted that the next step in the reaction cycle is not substrate hydroxylation but removal of the terminal oxygen atom from the dioxygen ligand, i.e., the cleavage of the dioxygen bond (Rein and Jung 1993). The insertion of the second electron into the dioxygen complex induces a further destabilization of the dioxygen π-bond, leading to the cleavage of the outer oxygen atom accompanied with release of water (Rein et al. 1986). The remaining oxene-iron complex is thereby converted to the iron-oxo intermediate (step 5). Although this is not a radical process, the iron-oxo complex is a radical, which can be written as a triplet oxygen atom coordinated to ferric iron (White 1994). Two principally different mechanisms have been discussed to occur, heterolysis and homolysis (White and Coon 1980; Rein et al. 1984; White 1991; Rein and Jung 1993). The heterolytic cleavage of the peroxide bond is also called the "oxenoid" pathway, while the homolytic cleavage has also been called the "quasi-Fenton" pathway, because the homolytic peroxide scission resembles the Fenton reaction of ferrous ion with various peroxides. The products of the homolysis are the amidyl radical and a thiyl-ferric hydroxide complex. The fundamental difference between the oxenoid and the quasi-Fenton route is the peroxide-lysis step. In the oxenoid pathway heterolysis of the peroxyacid leaves an oxygen atom coordinated to the iron. The formation of a highly resonance-stabilized perferryl species, as shown in Eq. 2, has been postulated (White and Coon 1980):

$$S^{-}Fe(5+)O^{2-} \quad S^{-}Fe(4+)O^{-} \leftrightarrow S^{-}Fe(3+)O \leftrightarrow S^{\bullet}Fe(3+)O^{-} \qquad (2)$$

In the next step hydrogen is abstracted (step 6), followed immediately by recombination of the two radicals to produce a stable product (step 7) . This process was designated as "oxygen rebound" (Groves and McClusky 1976). Finally, the hydroxylated product dissociates and the cycle can start again (step 8).

Interestingly, in a so-called shunt reaction, the substrate can be hydroxylated immediately by peroxides such as hydrogen peroxide, cumene hydroperoxide, and *tert*-butyl hydroperoxide without the necessity of an interaction with an electron-donating system (step9).

There are contradictory results concerning the existence of compounds I or II, which were shown to be spectral intermediates of peroxidases (see Egawa et al. 1994 and references cited therein) in P450-dependent reactions. Whereas the existence of a compound I was not observed in the reaction of ferrous CYP101 with superoxide anion (Kobayashi et al. 1994), the formation of compound I was observed in the reaction of CYP101 with m-chloroperbenzoic acid (Egawa et al. 1994). Spectral simulation based on Gaussian analyses and visible bands indicate that the H_2O_2 and *tert*-butyl hydroperoxide-induced intermediate spectra of CYP1A2 are close to compound I of horseradish peroxidase (Shimizu et al. 1994). Interestingly, Glu318 and Thr319 mutants of CYP1A2 show spectral intermediates with these hydroperoxides which are close to compound II of horseradish peroxidase (Shimizu et al. 1994).

It has been shown that the heme thiolate plays an essential role in O-O bond cleavage in the homolytic as well as in the heterolytic model (for review see Rein and Jung 1993). More recently, proton attachment, or at least hydrogen bonding to the terminal oxygen atom, was demonstrated to be important for the heterolytic dioxygen cleavage. For CYP101, where the three-dimensional structure has been known since 1985 (Poulos et al. 1985), threonine 252 is proposed to be the most potent candidate for inducing the strong polarization of the O-O bond from the distal side of the hemoprotein. The threonine is conserved in most cytochromes P450 (Poulos et al. 1987). Studies using site-directed mutagenesis of these residues confirmed this idea (Imai et al. 1989; Martinis et al. 1989). Solving the crystal structure of the CYP101 active-site mutant Thr252Ala led to the suggestion that the protons attached to the terminal oxygen atom can be released either via the internal solvent channel-linking residues Thr252 and Glu366 or by a proton delivery network including the side chains of Lys178 and/or Arg186, Asp251, the carbonyl oxygens of Val247 and Asp251 (Raag et al. 1991). In this model the side chain of Thr252 must rotate to accept a proton from Asp251. This suggestion has been substantiated by the elegant studies of Sligar and co-workers (Gerber and Sligar 1992). Replacement of Asp251 by asparagine and determination of the accessible fundamental rates and equilibria of the cycle revealed evidence for a participation of Asp251 and Thr252 in a charge relay system for proton delivery (Fig. 7). In this model, Asp251 serves as proton shuttle between the solvent-accessible Asp182-Lys178, Arg186 network and Thr252. A solvent channel consisting of three water molecules between Thr252 and Glu366 was supposed to be the source of protons necessary for dioxygen cleavage (Raag et al. 1991). In addition, this configuration provides a general acid-catalytic mechanism for the dioxygen cleavage and thus favors the heterolytic pathway model. In the hemoprotein domain of the bacterial CYP102 (P450BM-3) no such internal solvent channel has been observed, although these residues are conserved (Thr268, Glu401). It is believed that the conserved threonine in this protein donates a proton to iron-bound oxygen during catalysis. Threonine becomes protonated by the water molecule located in the groove of the I helix in CYP102, whereas the water molecule, in turn, may be protonated by Glu267, which is solvent-exposed in CYP102 (Ravichandran et al. 1993). CYP108 (P450terp) also has a threonine at the position homologous to Thr252 of CYP101 (Haseman et al. 1994). In the very recently crystalized CYP107A1 (P450eryF), however, an alanine residue is present at the position corresponding to Thr252 of CYP101. It has been proposed that the water molecule 564 serves the same hydrogen-bonding function in CYP107A1 as does Thr252 in CYP101 (Cupp-Vickery and Poulos 1995).

In most microsomal cytochromes P450 a conserved "threonine cluster" composed of three threonines (Thr319, Thr321, Thr322 for CYP1A2) has been found at the putative distal side of the hemoprotein (Ishigooka et al. 1992). Interestingly, substitutions in several mammalian cytochromes P450 at the highly conserved threonine (corresponding to Thr252 of CYP101) also affected substrate selectivity and binding (Imai and Nakamura 1989; Chen and Zhou 1992; Hiroya et al. 1992; Ishigooka et al. 1992; Fukuda et al.

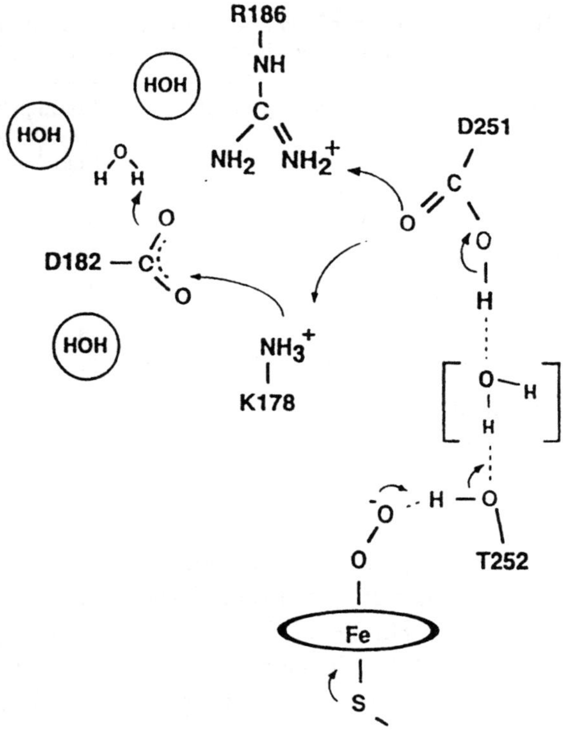

Fig. 7. Charge-relay system of proposed oxygen bond cleavage in CYP101. *Arrows* indicate direction of electron flow; *brackets* around the water bridging Asp251 and Thr252 indicate uncertainty as to the existence of this molecule. (From Gerber and Sligar 1994)

1993). However, while the efficiency of CYP101 for camphor hydroxylation drops to only 5%–6% when Thr252 is replaced with alanine, the corresponding Thr310Ser and Thr310Cys mutants of CYP19 (aromatase) have 48.8 and 2.4% of the wild type activity, respectively (Chen and Zhou 1992). Replacement of the conserved Thr303 of CYP2E1 with serine or valine and of Thr319 of CYP1A2 with alanine (Table 2) also did not lead to a considerable decrease in the substrate conversion compared with the wild-type enzymes (Fukuda et al. 1993; Ishigooka et al. 1992). From these observations, it can be concluded that in microsomal cytochromes P450 the conserved threonine residues may not significantly contribute to the catalytic function. However, replacement of the polar amino acid at position 318 of CYP1A2 significantly influenced the activation of the oxygen molecule in the catalytic function toward 7-ethoxycoumarin, but not toward methanol, implying differences in the hydroxylation mechanism between hydrophilic (methanol) and hydrophobic substrates such as 7-ethoxycoumarin (Hiroya et al. 1992; Ishigooka et al. 1992). Resolution of the three-dimensional structure of a microsomal P450 seems to be necessary to determine the precise mechanism of dioxygen splitting in this group of enzymes. Unfortunately, so far no attempt has been made to check the role of the corresponding conserved threonine and acidic residues in mitochondrial cytochromes P450, which, like CYP101, obtain

Table 2. Turnover numbers toward 7-ethoxycoumarin, consumption rates of oxygen, and formation rates of H_2O_2 by wild-type and mutant CYP1A2 (data from Ishigooka et al. 1992)

CYP1A2	Turnover number (min^{-1})	Consumption Rate of O_2 (min^{-1})	Formation Rate of H_2O_2 (min^{-1})
Wild type	0.33	20	6.3
Glu 318 Ala	0.02	6	3.1
Glu 318 Asp	1.44	27	2.2
Thr 319 Ala	0.90	24	2.5
Val 320 Ser	0.02	11	1.6
Thr 321 Ala	0.11	7	4.3
Thr 322 Ala	0.71	17	5.9

their reducing equivalents from a /2Fe-2S/protein. As shown in Fig. 8, also in mitochondrial cytochromes P450 a threonine residue can be found corresponding to Thr 252 of CYP101. In addition, aspartic acid is found in all but bovine CYP11A1 in the preceding position. Obviously, an acidic amino acid seems not to be essential in this position in mitochondrial cytochromes P450. Site-directed mutagenesis studies with mitochondrial P450 forms as well as with CYP102, CYP108 and CYP107A1, whose three-dimensional structures have recently been solved (Ravichandran et al. 1993; Hasemann et al. 1994; Cupp-Vickery and Poulos 1995), should provide important information about whether a charge relay system as proposed for CYP101 will be applicable for other cytochromes P450.

Variations of the scheme for the reaction mechanism of P450 occur with thromboxane and prostacyclin synthase (Hecker and Ullrich 1989), a plant P450-dependent allene oxide synthase (Song and Brash 1991), a yeast P450 (CYP56, P450DIT2), which is involved in dityrosine formation in spore walls (Briza et al. 1990a,b, 1994), and a nitric oxide reductase (Nakahara et al. 1993).

For prostacyclin and thromboxane A_2 biosynthesis a cage radical mechanism is proposed, according to which the initial radical process is terminated through oxidation of carbon-centered radicals by the iron-sulfur (fifth ligand, cystein) catalytic site, followed by ionic rearrangement to prostacyclin or thromboxane A_2 (Hecker and Ullrich 1989). Mechanistic studies support a homolytic hydroperoxide cleavage as the initial step in the synthesis of allene oxides, a conclusion that is also compatible with the known chemistry of cytochromes P450 (Song et al. 1993).

CYP55A1 acting as nitric oxide reductase, which is found in *Fusarium oxysporum*, was shown to form nitrous oxide (N_2O) from nitric oxide (NO), employing NADH, but not NADPH, as the sole effective electron donor.

CYP	Data Base Entry	
11A1	CPM1_HUMAN.SW	FEDIKANVTEMLAGGVDTTS
11A1	CPM1_BOVIN.SW	LEDVKANITEMLAGGVNTTS
11A1	CPM1_RAT.SW	FKNIQANITEMLAGGVDTTS
11A1	CPM1_PIG.SW	SEDVKANVTEMLAGGVDTTS
11B1	CPN1_HUMAN.SW	PDAIKANSMELTAGSVDTTV
11B1	CPN1_BOVIN.SW	LDTIKANTIDLTAGSVDTTA
11B1	CPN1_RAT.SW	MDAIHANSMELIAGSVDTTA
11B1	CPN1_mouse.SW	LDAIKANSMELTAGSVDTTA
11B2	CPN2_HUMAN.SW	LEAIKANSMELTAGSVDTTA
11B2	CPN2_RAT.SW	LDAIKANSMELTAGSVDTTA
27	P17178.SW	PQETVGTFPELILAGVDTTS
101	2CPD.NRL3d	SDEAKRMCGLLLVGGLDTVG
102	2HPDA.NRL3d	DENIRYQIITFLIAGHETTS
108	S27653.PIR3	DKYINAYYVAIATAGHDTTS
55A1	CP55_FUSOX.SW	KSDAVQIAFLLLVAGNATMV
56	CP56_YEAST.SW	YKQLTDNIVIILVAGHENPQ
5	THAS_HUMAN.SW	VDEIVGQAFIFLIAGYEIIT
5	THAS_MOUSE.SW	VDEIVGQAFLFLIAGHEVIT
	JC2231.PIR3	MGPAAFWLLLFLLKNPEALA
	LU428.PEP	SWGGFKILFPSLMKWIGRAG

Fig. 8. Alignment of various cytochrome P450 forms around the conserved threonine, corresponding to Thr252 of CYP101. *THAS,* Thromboxane-A synthase (human, mouse); *JC2231,* prostaglandin-I synthase (human); *LU428.PEP,* peptide sequence of allene oxide synthase created by the program TRANSLATION from mRNA, data base entry LU428.EM_PL. The multiple sequence alignment was produced by the program TREE (Feng and Doolittle 1987)

The reaction was not inhibited by carbon monoxide at all, suggesting that the free ferrous P450 is not formed during the turnover of this P450 (Nakahara et al. 1993).

The mechanism of dityrosine formation catalyzed by CYP56 includes cross-linking of two molecules of a precursor formed in a preceding step by the sporulation-specific enzyme DIT1 (Briza et al. 1994). The intermediates have not been identified so far.

The molecular basis for the differences in the reaction mechanism of these cytochromes P450 is not yet clear. A comparison of the primary structures of these P450 forms with that of CYP101 shows that the amino acid pair homologous to Asp251/Thr252 of CYP101 has not been conserved (Fig. 8). In prostacyclin (prostaglandin I) synthase (designated in Fig. 8 as JC2231) and thromboxane synthase (CYP5) instead of threonine alanine (prostacyclin synthase), isoleucine (CYP5, human) or valine (CYP5, mouse) are found, while in the preceeding position an acidic amino acid (glutamic acid) is conserved. In CYP56 an asparagine residue has been found in the position corresponding to Thr252 of CYP101.

It is tempting to speculate that this exchange could be responsible for the special function of these proteins, which is not a hydroxylation with preceding oxygen activation, but rather a radical process. However, this hypothesis must be supported by site-directed mutagenesis studies.

Allene oxide synthase (designated in Fig. 8 as LU428) has neither a threonine nor an acidic amino acid in the region close to the conserved residues of the bacterial cytochromes P450, whereas in nitric oxide reductase (designated in Fig. 8 as CYP55) there is a threonine close to Thr252 of CYP101, but the preceding acidic amino acid is missing. Site-directed mutagenesis investigations are necessary in these cases to determine whether the replacements of the conserved threonine and the preceding acidic amino acids are responsible for the divergence of the reaction mechanism of these cytochromes P450.

3 Cytochrome P450/Electron Donor Interactions

The interaction between P450 and the electron donor is a prerequisite for the electron transfer and the following oxygen activation and substrate conversion. Its specificity guarantees a sufficient reaction rate of catalysis and likewise a discrimination between different potential donors and acceptors of electrons to protect the system from shunt reactions.

3.1 Microsomal P450 Systems

Chemical modification of CYP2B4 with fluorescein isothiocyanate (FITC) led to the suggestion that the α-amino group and Lys384 are involved in reductase recognition. Detailed analysis of the gross conformation (P420 content – an enzymatically inactive form of P450 with a maximum of the reduced CO complex at 420 nm, CD measurements in the Soret and the far UV region, hydroperoxide-dependent substrate conversion) and of individual steps of the reaction sequence (product formation, substrate binding affinities, reduction velocity) have shown that only the reduction and the overall activity (25% and 50% at 1 and 2 mol FITC/mol P450, respectively) were disturbed after modification (Bernhardt et al. 1983, 1984). Coincubation of CYP2B4 with FITC in the presence and absence of reductase, moreover, revealed that both residues were protected by the electron donor (about 30% and 50%, respectively) from modification (Bernhardt et al. 1988). Whether the N-terminal methionine is involved in the interaction with reductase also in vivo remains a subject of controversy. While experiments of Bernhardt et al. (1983, 1989a) studying the accessibility of Met1 of CYP2B4 for FITC within microsomes pointed at a location of the N-terminus at the cytoplasmic side of the microsomal membrane, studies of

Vergeres et al. (1991) did not show labeling of microsomal CYP2B1 by FITC, suggesting a location of Met1 of CYP2B1 at the luminal side. Moreover, the N-terminal region of rabbit CYP2E1 was shown to not be essential for interaction with microsomal reductase (Voznessensky et al. 1994). Recently, using a combination of recombinant DNA technology and in vitro techniques, it was demonstrated that 20–30 N-terminal amino acids of two microsomal rabbit cytochromes P450 and of rat CYP2B1 were sufficient for membrane anchoring, placing the N-terminus at the luminal side (Szczesna-Skorupa et al. 1988; Sakaguchi et al. 1987; Monier et al. 1988; DeLemos-Chiarandini et al. 1987). However, extrapolation of the results obtained with the hybrid constructs used in those experiments to the native membrane proteins may not always be possible.

The combination of an in vivo assay with subcellular immunoprecipitation made it possible to show the location of the N-terminus of yeast CYP52A1 and CYP52A2 in the lumen of the endoplasmic reticulum (Sanglard et al. 1993). In contrast, mutagenesis of aromatase favors a model where the N-terminus is located in the cytoplasm (Chen and Zhou 1992). Further studies seem to be necessary to answer this question unambiguously.

Binding of 2-methoxy-5-nitrotropone (MNT), which is much smaller than FITC and uncharged, led to an effect similar to that of FITC binding: a decreased overall activity caused by an impaired electron transfer from reductase to CYP2B4. The label, however, is not bound exclusively to Met1 and to Lys384 as FITC, but reacts with the lysines in positions 49, 100, 139, 144, 251, 384, and 433. The decrease in the overall activity and reduction rate correlates with a relative increase in the modification of lysines 139, 144, 251, and 384, revealing these amino groups to be further candidates for participation in reductase recognition (Bernhardt et al. 1988). Alkylation and acetylation of amino groups of CYP2B1, in contrast, did not cause any alteration of reductase binding (Kunz et al. 1991). Obviously, the modified amino groups are still available for salt-bridge formation, since the substituents at the N-atom have a small volume. From the side of reductase, carboxylic groups were shown to participate in the interaction with CYP2B4 and to be located at or near the binding site for CYP2B4 and CYP2B1 (Bernhardt et al. 1987; Nadler and Strobel 1988), whereby steric constraints seem to play a role in the binding and electron transfer step(s) (Nadler and Strobel 1988, 1991).

Since in liver microsomes many different isoenzymes have to interact with only one type of reductase, it must be expected that the binding sites for reductase on various cytochromes P450 are very similar or identical. In this connection, it is of special interest that Lys384 of CYP2B4 is located in one of the only four regions that are homologous among different P450 forms (Gotoh et al. 1983) and that lysine is conserved in this position at least in all microsomal cytochromes P450 (Nelson and Strobel 1989). Also Lys139 and Lys251 are conserved in several P450 forms, and Lys251 is located in another homologous region (Gotoh et al. 1983). Our chemical modification studies of CYP1A2 revealed, however, that Lys401 (corresponding to Lys384 of CYP2B4) was not accessible to FITC labeling, whereas modification of Lys253 and/or Lys440 and/or Lys453 seemed to interfere with the formation of the protein structure necessary for optimal functional activity as well as with reductase binding (Bernhardt et al. 1992; Bernhardt 1993).

A participation of Tyr243 and Tyr271 of CYP1A2 in reductase binding was concluded from modification experiments with tetranitromethane (Jänig et al. 1987). Interestingly, these residues are close to Lys253. It seems reasonable that salt bridges are responsible for the recognition of reductase and the correct orientation of both proteins to each other, whereas hydrophobic contacts provide the energy necessary for complex formation. A detailed analysis of the function of the above-mentioned lysines and of Lys401 was recently begun by Shimizu and co-workers (Shimizu et al. 1991; Mayuzumi et al. 1993), using site-directed mutagenesis; Lys401 of CYP1A2 does not seem to be involved in the interaction with reductase. However, the KCl-dependent studies suggested that ionic or electrostatic interactions are important. The Arg cluster Arg135-Arg136-Arg137 of CYP1A2 may perhaps participate in the interaction with the reductase, since catalytic activity of the triple mutant toward 7-ethoxycoumarin was less than 10% of the wild-type in the microsomal system (Shimizu et al. 1991). In addition, Lys99 and Lys453 seem to be involved in the interaction with the reductase or in recognition of this protein (Mayuzumi et al. 1993). The contribution of Lys253 and Lys440 to protein-protein interaction unfortunately has not been studied so far. Nevertheless, it has been proposed, based on these experiments, that the interaction of CYP1A2 with the reductase is rather ionic, composed of several ionic interactions between the two proteins, as previously implied for the CYP2B4 system (Bernhardt et al. 1984, 1988).

Experiments of Shen and Strobel (1992, 1993) revealed that in CYP1A1 and CYP2B1 also lysine residues are involved in the interaction with reductase. Lysines homologous to Lys384 (alignment position 466, according to Nelson and Strobel 1988) and Lys251 (alignment position 311) of CYP2B4, together with residues in alignment positions 110, 506, and 511, were shown to participate in these interactions (Strobel and Shen 1994).

Repulsive forces rather than ionic interactions have been proposed by Voznesensky and Schenkman (1992) to play a role in CYP2B4 as well as CYP2E1 and CYP2C2/reductase interaction (Voznesensky and Schenkman 1994). Ionic-strength dependent studies of Bernhardt et al. (1988) demonstrate that, in fact, at lower ionic strength an electrostatic repulsion seems to exist between P450 and reductase in microsomes (Fig. 9), whereas at higher

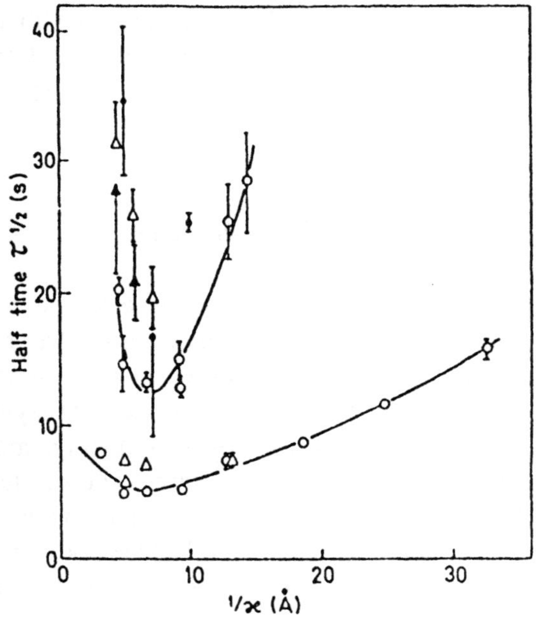

Fig. 9. Dependence of the half-time of the reduction of microsomal cytochrome P450 by microsomal NADPH-cytochrome P450 reductase on the Debye-Hückel length 1/κ at T=293 K. *Lower curve,* aerobic conditions; *upper curve,* anaerobic conditions. The assay solution contained such amounts of potassium phosphate or Tris-HCl buffer, pH 7.4, as were sufficient to bring the ionic strength up to the indicated value. The final P450 concentration was 1.25 *M*. O, Potassium phosphate buffer; •, Tris-HCl buffer; Δ, 0.05 *M* potassium phosphate buffer and respective amounts of KCl; ▲, 0.05 *M* potassium phosphate buffer and respective amounts of NaCl. (From Bernhardt et al. 1988)

ionic strength (222 m*M*, corresponding to 0.1 *M* potassium phosphate buffer) salt bridges between P450 and reductase seem to become damaged (for review see Schenkman 1993 and references cited therein).

In addition to microsomal reductase, some microsomal cytochromes P450 are able to accept the second electron from cytochrome b_5 (for review see Schenkman 1993). Examination of the interaction between these hemoproteins revealed the necessity of carboxyl groups on cytochrome b_5 and lysyl residues on cytochrome P450 for complex formation between CYP2B4 or CYP2C11 and cytochrome b_5 (Tamburini et al. 1985; Tamburini and Schenkman 1986a). In addition, Tyr129 of cytochrome b_5 has been proposed, on the basis of chemical modification studies, to control interaction with CYP2B4 (Hlavica et al. 1994). Charge-pairing has also been observed between NADPH-cytochrome P450 reductase, which is able to reduce cytochrome c and cytochrome b_5 (Nisimoto and Lambeth 1985; Tamburini et al. 1986b).

Cytochrome b_5 has been shown to exert a differential stimulatory action, dependent upon the form of cytochrome P450 and substrate metabolized (Kuwahara and Omura 1980; Morgan and Coon 1984; Jansson et al. 1985; Onoda and Hall 1982; Shet et al. 1993).

Taking together all the data on cytochrome P450/reductase/cytochrome b_5 interaction, a model has been proposed whereby all three proteins, anchored by their respective membrane-binding domain to the endoplasmic reticulum, may interact, forming a ternary complex governed by electrostatic as well as other forces (Schenkman 1993).

Resolution of the three-dimensional structures of a microsomal P450, the NADPH-cytochrome P450 reductase, and the corresponding electron transfer complexes will be necessary to unambiguously evaluate the above-mentioned model.

3.2 Mitochondrial P450 Systems

As in microsomal P450 systems, also in mitochondrial steroid hydroxylases a charge-pair interaction mechanism has been proposed on the basis of chemical modification studies. Incubation of the ferredoxin (called adrenodoxin in the case of adrenal steroid hydroxylase systems) with the water-soluble carbodiimide 1-ethyl-3-(3-dimethylaminopropyl) carbodiimide (EDC) revealed a modification of carboxylic residues in positions 74, 79, and 86, which leads to a reduced ability of adrenodoxin to interact with both adrenodoxin reductase, where one lysine residue was suggested to be involved in adrenodoxin binding (Hamamoto and Ichikawa 1984), and CYP11A1 (Geren et al. 1984; Lambeth et al. 1984). Recently, site-directed mutagenesis studies revealed that Asp76 and Asp79 of adrenodoxin are involved in both CYP11A1 and adrenodoxin reductase binding (Coghlan and Vickery 1991). These results support earlier conclusions that P450 competes with adrenodoxin reductase for binding to a common or overlapping site on the adrenodoxin (Hanukoglu and Jefcoate 1980), as well as data indicating the inability to trap a ternary complex between CYP11A1, adrenodoxin, and adrenodoxin reductase (Lambeth et al. 1984). In the shuttle model derived from these studies adrenodoxin first binds to adrenodoxin reductase. The reduced adrenodoxin then dissociates from adrenodoxin reductase and forms a complex with P450. Oxidized adrenodoxin again dissociates from the P450, and a second cycle begins. This model is supported by the above-mentioned kinetic (Lambeth et al. 1979; Seybert et al. 1979; Hanukoglu and Jefcoat 1980), chemical modification, and cross-linking studies (Lambeth et al. 1984), as well as by site-directed mutagenesis (Coghlan and Vickery 1991, 1992). The cluster model also relies on cross-linking studies (Kido and Kimura 1979; Chashchin et al. 1985; Usanov et al. 1985). It was demonstrated that a cross-linked complex of adrenodoxin and CYP11A1 has catalytic competence with adrenodoxin reductase. From this observation the existence of two different binding sites for the flavo- and the hemoprotein on the surface of adrenodoxin was concluded. Results of Hara and Miyata (1990) support this hypothesis of a ternary complex between adrenodoxin, adrenodoxin reductase, and CYP11A1. Analyzing cross-linked complexes, the authors identified Ile25–Lys98 of adrenodoxin as being involved in adrenodoxin reductase binding, whereas Ile7–Lys22 were shown to be involved in CYP11A1 binding. The cause of the differences seen in cross-linking experiments performed by different groups remains unclear. Recent results from cross-linking and titration experiments suggest the occurrence of a quarternary complex, in which 1 mol of adrenodoxin binds to adrenodoxin reductase and 1 mol to CYP11A1, forming a functional CYP11A1/(adrenodoxin)$_2$/adrenodoxin reductase complex (Hara and Takeshima 1994). Investigation of CYP11A1 activity using expression of a fusion protein, which consists of the construct CYP11A1/adrenodoxin reductase/adrenodoxin, demonstrates that a quarternary complex is at least not essential for cholesterol side-chain cleavage (Harikrishna et al. 1993). The fact that the construct CYP11A1/adrenodoxin/adrenodoxin reductase was less effective in pregnenolone production as compared with the complex CYP11A1/adrenodoxin reductase/adrenodoxin was discussed by the authors in the sense of an increased rotational freedom of the adrenodoxin moiety in the latter one, so that the crucial region of adrenodoxin between residues 72 and 79 can react with both CYP11A1 and the adrenodoxin reductase moiety, supporting the shuttle model (Harikrishna et al. 1993). However, these results are also consistent with a cluster model. Furthermore, studies of Beckert et al. (1994a) revealed that mutation of Y82 of adrenodoxin differentially affects interaction with adrenodoxin reductase and cytochromes P450. Tyr82 of adrenodoxin seems to play an essential role in the interaction with CYP11A1 and CYP11B1, but not with adrenodoxin reductase (as reflected by studying the reduction of the nonphysiological electron acceptor, cytochrome c). Replacement of His56 also led to changes in the reduction of cytochromes c, CYP11A1 and CYP11B1, but this seems to be due to changes of the local protein conformation, rather than to direct involvement in protein-protein interactions (Beckert et al. 1994b, 1995). Furthermore, the C-terminal amino acids exert an effect on interaction with the cytochromes P45011A1 and 11B1, but not with cytochrome c and thus adrenodoxin reductase (Uhlmann et al. 1994). Deletion mutant 4–114, where the N-terminal three serine residues were removed as well as amino acids 115–128, revealed lower K_m-values in CYP11A1-dependent cholesterol side-chain cleavage and in CYP11B1-dependent corticosterone formation. The effect was even more pronounced when a deletion mutant was used where residues 109–128 were removed. While the V_{max} value was not changed in the CYP11A1-dependent reaction, it was increased by a factor of 5 in the CYP11B1-dependent reaction. Furthermore, it was demonstrated that this increase was due to a four- to fivefold increase in the reduction rate. The molecular basis for the drastic increase in the transfer of the first electron by deletion mutant 4–108 is not yet clear, however. Interestingly, in these studies it was clearly demonstrated that there are different requirements or mechanisms of interaction for the recognition not only between adrenodoxin and adrenodoxin reductase, but also between adrenodoxin and CYP11A1, as compared with adrenodoxin and CYP11B1. These data demonstrate that different interaction sites (or mechanisms) are used, at least in part, for the interaction between adrenodoxin and adrenodoxin reductase, on the one hand, and adrenodoxin and cytochromes P450 on the other.

The adrenodoxin-binding sites on CYP11A1 were analyzed using cross-linking, immunological, chemical modification, and site-directed mutagenesis studies (for review see Bernhardt 1993). Binding of pyridoxal 5'-phosphate at a level of 0.82 and 2.85 mol/mol CYP11A1 resulted in 60% and 98% inhibition, respectively, of the electron transfer rate from adrenodoxin to P450. Sequence analysis of modified peptides revealed residues 377 and 381 (residues 338 and 342, respectively, according to the numbering without taking into account the preprotein) as specifically modified (Tsubaki et al. 1989). Modification of CYP11A1 with FITC led to very selective modification of only Lys377 and to 85% inhibition of adrenodoxin binding (Tuls et al. 1989). Thus, Lys377 is suggested to be involved in the interaction with adrenodoxin. Interestingly, in CYP11A1 no lysine has been found in the position homologous to Lys384 of CYP2B4, but Lys377 is very near to Lys384 in the alignment and the region Ala381–Trp400 flanks the homologous region, containing Lys384, from the other side (Nelson and Strobel 1988, 1989; Gotoh and Fujii-Kuriyama 1989). The essential role of lysine residues in adrenodoxin binding has been determined using site-directed mutagenesis. Replacement of Lys377 and Lys381 residues of CYP11A1 by either neutral or negative amino acids revealed greatly increased K_d values for adrenodoxin binding, indicating that these lysines are the key sites in adrenodoxin/CYP11A1 binding (Wada and Waterman 1992).

So far, nothing is known about the interaction of other P450 forms with adrenodoxin. Difficulties in the expression of these enzymes (e.g., CYP11B1, CYP11B2) in *E. coli* or other micro-organisms hamper the study of these systems in vitro.

When everything said above is taken into account, it seems to be possible that two binding sites for adrenodoxin exist on cytochromes P450, one possessing a high affinity for adrenodoxin and reacting with similar groups as adrenodoxin reductase, and another whose affinity is increased after introduction of the first electron and which could probably react with a different binding area of the adrenodoxin molecule, thus giving rise to the formation of a ternary complex under distinct circumstances.

3.3 Bacterial P450 Systems

In CYP101, which is phylogenetically very distant from the mammalian P450s, Lys314, which is homologous to Lys384 of CYP2B4, was shown to be not directly involved in putidaredoxin binding, but to be located at the periphery of the binding region (Bernhardt 1993). This is in accordance with computer modeling and site-directed mutagenesis studies identifying Arg72, Arg112, Lys344, and Arg364 as interacting residues (Stayton et al. 1989, Stayton and Sligar 1990).

Interestingly, modification of CYP101 with 2-methoxy-5-nitrotropone revealed several labeled peptides with lysines in positions 126, 197, 214, 313, 314, 372, and 392, but only Lys197 was protected from modification upon coincubation of CYP101 with putidaredoxin. Whether Lys197 is part of a second binding site for putidaredoxin or is decreased in its accessibility for the chemical label due to conformational changes upon putidaredoxin binding remains to be clarified.

Obviously, the binding sites of various P450s for their electron donors are very similar but not identical to each other. Further support for this assumption comes from heterologous reconstitution studies. It was shown that putidaredoxin and linredoxin bind to CYP2B4 and, moreover, support benzphetamine conversion with the respective bacterial reductases and NADH instead of microsomal reductase. The efficiency of substrate conversion is of the same order of magnitude (cf. Table 8) as that of substrate conversion in a homologous reconstitution mixture with microsomal reductase and dilauroyl phosphatidylcholine (Bernhardt and Gunsalus 1985, 1992). Heterologous reconstitution of CYP2B4 activity was also achieved using adrenodoxin and adrenodoxin reductase as electron donors (Smettan et al. 1985). Very recently, it was demonstrated that CYP2B5 activity toward 4-nitroanisole can be reconstituted with adrenodoxin and adrenodoxin reductase as the electron-supporting system instead of reductase. Surprisingly, the electron flow was about 4.6 times faster from the mitochondrial redox chain to CYP2B5 than from the natural redox partner (Lehnerer et al. 1995).

Thus, an evolutionary conserved binding site for the electron donor seems to exist. Binding is not sufficient for performing catalysis, however. It has been observed in this connection that linredoxin and linredoxin reductase are only weakly able to support camphor hydroxylation by CYP101, although linredoxin binds to CYP101 with only about twofold reduced affinity as compared with putidaredoxin (Bernhardt and Gunsalus 1992). Moreover, NADPH-cytochrome P450 reductase was not able to support CYP101 and CYP111 (P450lin)-dependent substrate conversions (Bernhardt, unpublished results). Obviously, other factors such as redox potentials and correct orientation of proteins to one another play an essential role in determining the ability of a redox couple to perform first and second electron incorporation and subsequent substrate hydroxylation.

4 Electron Transfer in Cytochrome P450 Systems

The mechanism of electron transfer is one of the fundamental problems in life sciences. Electron transfer takes place in photosynthesis and cell respiration, but also in cytochrome P450-dependent substrate conversion.

The P450 mono-oxygenase cycle (Fig. 6) requires one-electron-step reduction of both the ferric and the oxygen-bound forms of the P450 enzyme. The best-studied system with respect to electron transfer mechanism is again the bacterial CYP101 system. In this system, the electrons are transferred by putidaredoxin in a bimolecular complex, where electron transfer is the rate-limiting step at room temperature (Pederson et al. 1977; Hintz and Peterson 1981). The association of putidaredoxin and CYP101 is important in determining the driving force of electron transfer. Direct measurements (Sligar and Gunsalus 1976) and kinetic modeling (Hintz et al. 1982) of the putidaredoxin/CYP101 association reaction show that reduced putidaredoxin binds to CYP101 more tightly than the oxidized protein. It was recently demonstrated that the presence of a C-terminal aromatic residue is required for a relatively high CYP101 affinity of the reduced relative to the oxidized protein (Davies et al. 1990; Davies and Sligar 1992). The "desolvation" of putidaredoxin's C-terminal residue has been discussed as a possible explanation for this behavior (Davies and Sligar 1992). The distance the electron has to be transferred from the prosthetic group of putidaredoxin, the /2Fe-2S/cluster, to the heme of CYP101 is more than 20 Å. X-ray crystallographic data have shown that in CYP101 the heme is deeply embedded into the protein; it is not accessible from the surface and a minimum of ~8 Å away from it (Poulos et al. 1985). Thus, no direct pathway from the prosthetic group of the electron donor to the prosthetic group of CYP101 seems to be possible, but a pathway including protein groups of at least P450cam is necessary. A similar pathway through protein groups has been proposed for the CYP2B4/microsomal reductase electron transfer couple on the basis of chemical modification studies and fluorescence measurements (Bernhardt et al. 1989b).

The role of the protein medium in directly modulating biological electron transfer processes has been the subject of intense theoretical and experimental investigations. There are presently two main models of the electron transfer process in biological systems. One proposes a uniform one-dimensional square tunneling barrier for electron tunneling, and according to Moser et al. (1992) describes a broad range of natural and synthetic electron transfer pathways. The other model describes specific through bond pathways for electron transfer, which account for the unique covalent, hydrogen-bonded, and van der Waals contacts linking donor and acceptor in a protein (the electron is assumed to be transferred via these bonds and contacts). This model gives a consistent description of electron transfer rates not only in

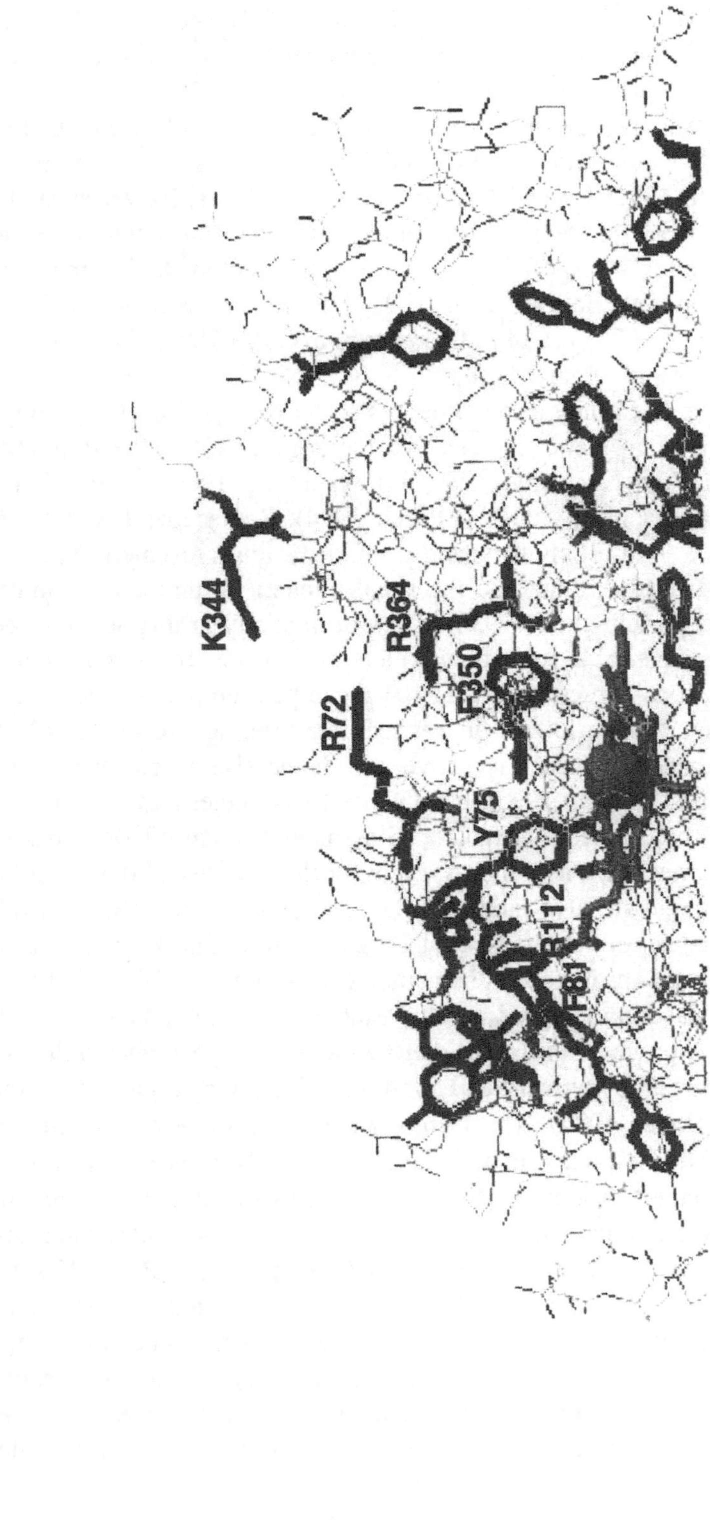

Fig. 10. Three-dimensional structure of CYP101 backbone (upper half of the molecule), according to Poulos et al. (1985). All aromatic side chains and the residues (Arg 72, Arg112, Lys 344, and Arg364) identified by Stayton et al. (1989) and Stayton and Sligar (1990) as being involved in electron donor recognition are drawn in *boldface*. The heme is drawn in *gray*

ruthenated proteins (cytochrome c, myoglobin, cytochrome b_5) (Beratan et al. 1990, 1991, 1992a,b; Onuchic and Beratan 1990), but also in the recently crystallized cytochrome c/cytochrome c peroxidase complex (Pelletier and Kraut 1992; Beratan et al. 1992b).

Whether there is a special role of aromatic residues in electron transfer has been the subject of controversy (Wendoloski et al. 1987; Liang et al. 1988; Everest et al. 1991; He et al. 1991; Inglis et al. 1991; Farver et al. 1993; Willie et al. 1993) and requires further experimental substantiation. A model whereby electron transport is realized by a covalent switching mechanism has been proposed (Baldwin et al. 1991) but does not coincide with recent data on the role of Trp106 of putidaredoxin in CYP101/putidaredoxin interaction (Davies and Sligar 1992).

There have been only a few attempts so far to experimentally study the potential role of aromatic residues of cytochromes P450 in electron transfer (Gotoh and Fujii-Kuriyama 1989; Furuya et al. 1989a,b; Straub et al. 1993a,b; Yasukochi et al. 1994; Porter 1994). Replacement of Try120 of CYP2C2 by a series of aliphatic and aromatic residues revealed a preference for aromatic residues, but also the small aliphatic amino acid, alanine, at position 120 (Straub et al. 1993b). Thus, aromaticity in this position seems not to be indispensable for efficient electron transfer. If one considers the three-dimensional structure of CYP101 it can be seen that several aromatic residues (Fig. 10) are located in between the binding site for the electron donor, putidaredoxin, and the active site, heme (Bernhardt et al. 1994b; Bernhardt and Uhlmann 1995). Phe350 has been suggested to possibly play an essential role in electron transfer (Stayton and Sligar 1990). It has been shown, however, that Phe350 is essential for the integrity of the heme-binding site (Sligar et al. 1991) and contributes to enhance the efficiency of the electron transfer, yet does not seem to be essential for this reaction (Yasukochi et al. 1994). Substitution of the homologous residue in CYP1A2, Phe449, by tyrosine was found to produce an unstable protein (Shimizu et al. 1988). In contrast, it proved possible to replace the homologous phenylalanine in CYP2E1, Phe429, by tyrosine and tryptophan, but not by aspartate, arginine, or leucine (Porter 1994). The tryptophan mutant showed a significantly reduced NADPH oxidation in both the presence and the absence of substrate, suggesting that electron transfer was impaired by this mutation. The author hypothesized that this might be due either to a reduction in electron transfer efficiency or to altered binding of CYP2E1 and reductase (Porter 1994).

Replacement of Tyr75 of CYP101, which is also located between the heme and the putidaredoxin binding site, by serine leed to decreased activity of this P450 by about 30%, due to a decrease in V_{max}. The diminution of the activity has been shown to arise from hindrance of first electron introduction. This alteration was relatively small, however, and thus does not indicate an essential role of Tyr75 in the intracomplex electron transfer between CYP101 and putidaredoxin. Also in mitochondrial adrenodoxin, where involvement of tyrosine 82 in the electron transfer was proposed, based on chemical modification studies (Taniguchi and Kimura 1975, 1976), recent site-directed mutagenesis experiments have provided evidence that this aromatic residue is not involved in the electron transfer, but plays an essential role in binding of CYP11A1 and CYP11B1 (Beckert et al. 1994a,b). However, the replacement of Tyr75 in CYP101 by serine seems to modulate the electron transfer rate (Bernhardt et al. 1994a; Bernhardt and Uhlmann 1995). Calculation of the possible electron transfer pathways in CYP101 using the PATHWAYS II software package (Beratan et al. 1990, 1992a,b) showed this to be probably due to small conformational changes in the intervening sequence, which is involved in electron transfer, induced by the replacement (Bernhardt and Uhlmann 1995). The most probable pathway includes a pathway from the glutamine 360 side chain through the backbone of Gln 360, then through a 3.4-Å through-space jump to the sulfur of Cys357, and in this way directly to the heme iron. Tyr75 is located within about 10 Å of this pathway. Thus, its replacement may cause disturbances of this pathway region (Bernhardt et al. 1994b; Bernhardt and Uhlmann 1995).

Interestingly, when the coupling of every amino acid on the surface of the protein to the active site, heme, is calculated, the strongest coupling is obtained for the area which is surrounded by some of the amino acid residues (Arg72, Lys344) proposed as being involved in the putidaredoxin recognition site (Stayton et al. 1989; Stayton and Sligar 1990). Several groups on the surface of CYP101 are very tightly coupled to the heme. There is no other region so tightly coupled to the heme, demonstrating an excellent agreement of experimental data and theoretical considerations (Bernhardt and Uhlmann 1995). Furthermore, the observation that replacement of Arg112, being part of the recognition site for putidaredoxin, by other amino acids markedly affected the rate of the reduction of CYP101 by putidaredoxin, thus suggesting that this residue could be involved in the electron transfer (Koga et al. 1993; Shimada et al. 1994), is in agreement with this model.

Unfortunately, current models for the calculation of electron transfer pathways do not take into account the dynamics of the protein structure. Thus, it remains open whether different P450/electron donor complexes are possible. Further studies using site-directed mutagenesis of CYP101 and the other cytochromes P450 so far crystallized, as well as investigation of Ru^{+}-labeled cytochromes P450, are necessary to substantiate the pathway model for the P450 family.

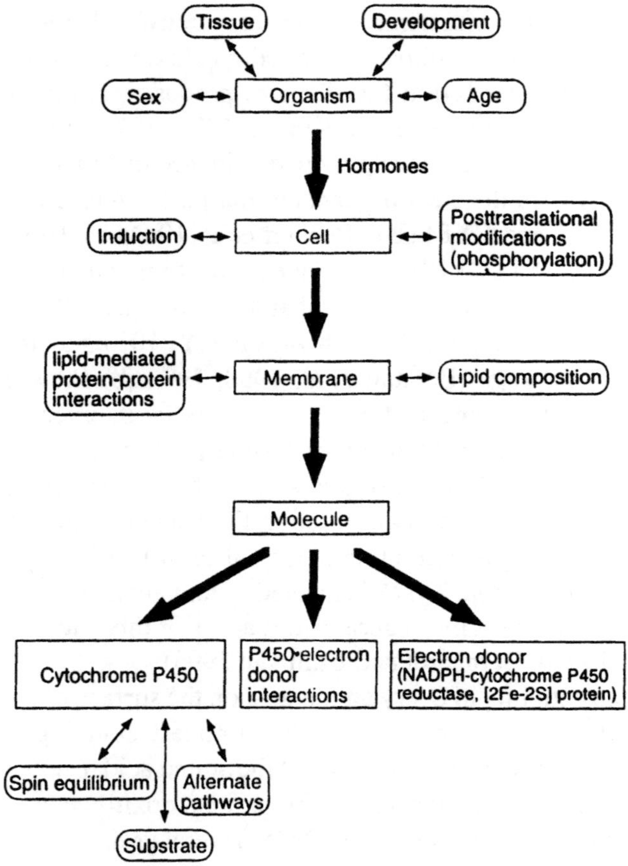

Fig. 11. Levels of regulation of cytochrome P450 systems

5 Regulation of Cytochrome P450 Systems

Regulation of enzyme systems is possible at different levels of integration (Fig. 11). With regard to cytochrome P450 systems, the complexity of the reaction cycle and the organization within membrane systems for most of the cytochromes P450, as well as their organ- and tissue-specific expression, imply various possibilities for the working of regulatory mechanisms (for reviews see Ruckpaul et al. 1989; Ruckpaul 1993). Interaction between these mechanisms then leads to the tuned response of these enzyme systems to endogenous and exogenous signals in terms of acute and long-term reactivities of cytochromes P450. So far, only part of this interesting and complex phenomenon is understood, and further investigation of individual regulation levels as well as combination and integration of the available knowledge is needed.

5.1 Regulation at the Molecular Level

5.1.1 Spin Equilibrium

The spin state of oxidized iron (Fe^{3+}) is generally as follows:

1	1	1	1	1	S=5/2 (high-spin state)
11	11	1			S=1/2 (low-spin state)
11	1	1	1		S=3/2 (intermediate state)

Octahedral d^5 complexes have been shown to exist in two different magnetic and thermodynamic stable forms only, namely the low- or the high-spin state. The temperature dependence of the spin eqilibrium shows a very low energy barrier between the two spin states. Thus, changing the temperature should lead to changes in the spin state of P450. In fact, such changes were observed independently, based on temperature-induced difference spectra (see also Sect. 2.3) with soluble CYP101 (Sligar 1976) and with solubilized liver microsomal P450 from phenobarbital-induced rabbits (Rein et al. 1976a, 1977). The temperature-induced difference spectra of cytochromes P450 exhibit maxima and minima typical for high-spin and low-spin bands of the Soret and visible region (Sligar 1976; Rein et al. 1976b, 1977). Analysis of these spectral changes revealed that at increasing temperatures the concentration of the high-spin conformer increased, and vice versa.

In addition, a correlation between type-I and type-II substrate-induced spectral and spin-state changes has been suggested (Mitani and Horie 1969; Whysner et al. 1970). The substrate-induced spectral changes of the Soret and the Q-bands indicated a spin-state change in the heme complex. The Soret bands of cytochrome P450 are well separated, with a high-spin band at 387 nm and a low-spin band at 418 nm. Type-I substrates induce a high-spin shift concomitant with an increase of the absorption band at 387 nm and a decrease of the band at 418 nm, whereas with so-called inverse type-I substrates the high-spin band decreases concomitant with an increase of the low-spin band. This difference can be used to measure substrate binding to P450. Besides optical characterization, the electron paramagnetic resonance method is well suited to spin-state determination. The appearance of low-spin (g=2) and high-spin (g=6–8) signals in the EPR spetrum of hemoproteins can be measured at low temperatures (< 77 K) (Rein and Ristau 1964). Further substrate-induced changes of the P450 structure can obviously be neglected, in agreement with the observation that only small changes occur in the circular dichroism (CD) spectrum upon substrate binding (Rein et al. 1976b).

Scheme 1

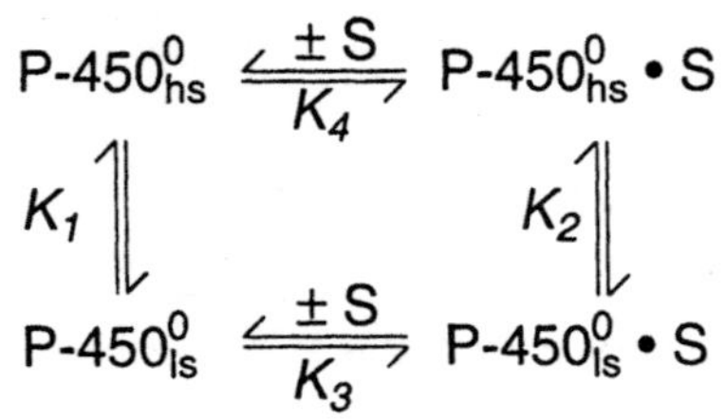

Fig. 12. Thermodynamically closed four-state model of P450 spin and substrate-binding equilibria. (From Ruckpaul et al. 1989)

Both spin equilibria and substrate-binding equilibria can be described in a thermodynamic four-state model (Fig. 12). The difference in the substrate affinity between the two spin states is most pronounced in CYP101, which is characterized by a substrate-induced high-spin shift for camphor of more than 90% (Sligar 1976). The substrate affinities to both spin-state conformers differ by a factor of about 200. In the microsomal CYP2B4 from rabbit liver the affinities of benzphetamine binding to the high- and low-spin conformer vary only by a factor of 5 (Ristau et al. 1978).

Concomitant with a high-spin shift upon substrate binding, the redox potential of the P450 is shifted. With CYP101 the redox potential is –303 mV in the absence of camphor and shifts to –173 mV when camphor is present (Sligar 1976). Similar changes in the redox potential were observed for mitochondrial cytochromes P450 such as CYP11A1 and CYP11B1 upon substrate binding. For microsomal cytochromes P450 the redox potential changes upon substrate binding are much less pronounced. Obviously, species- and isoenzyme-specific differences in the redox potential exist (Ruckpaul et al. 1989 and references cited therein).

A linear dependence of the redox potential E_m on the logarithm of the high-spin fraction alpha was shown using CYP2B4 in the presence of different tertiary amines (Rein et al. 1989). The linear relationship E_m/-log alpha clearly demonstrates the regulation of a cytochrome P450 redox equilibrium via the spin state of the heme iron.

On the other hand, an increase of the k_{app} of cytochrome P450 reduction with a substrate-induced increase in the high-spin fraction alpha has been observed (reviewed by Ruckpaul et al. 1989). For a series of benzphetamine derivatives under saturation conditions and liver microsomal P450 from phenobarbital-induced rat liver, a linear correlation of the apparent rate constant for P450 reduction from the fraction of the high-spin conformer (k_{app}/alpha) has been demonstrated. In this system the low-spin conformer is not reduced (Blanck et al. 1983). The correlation k_{app}/alpha has been supported by studies using reconstituted CYP2B4 and CYP101 (Schwarze et al. 1985; Fisher and Sligar 1985). However, with CYP2B4 also an alternative reduction of P450 via a relatively unfavorable ferrous low-spin state was observed (Schwarze et al. 1985).

Moreover, it must be pointed out that the observed redox/spin-state correlations cannot be generalized to all P450 forms and substrates. Other factors such as specific conformational changes and induced changes of P450-membrane interactions upon substrate binding may be more important for either reduction rate or substrate-induced high-spin shift. It is obvious from the described relationships, however, that effectors changing the spin state or the redox potential of a cytochrome P450 also regulate P450 activity.

5.1.2 Control Function of the Substrate

As described in the previous section, the substrate can regulate P450 activity by changing the fraction of high-spin P450, and in this way the reduction rate, which in many P450 systems is the rate-limiting step of catalysis. Furthermore, it has been demonstrated that the capacity of the substrate to induce an enzymatically active conformation of the P450 strongly depends on structural parameters of the substrate. Using CYP2B4 and a series of benzphetamine derivatives, a correlation of activity with the geometry of the substrate was shown (Schwarze et al. 1985).

In the study of a series of parasubstituted toluene derivatives, enzymatic parameters such as the apparent dissociation constant for the enzyme/substrate complex (K_d), the interaction energy (G_{int}), and the catalytic constant (k_{cat}) have been correlated with several physicochemical constants of the individual substrates such as the Hansch constant for hydrophobicity, the Hammett value, and the molar volume. It was shown that the apparent dissociation constant can be predicted by a linear combination of the molar volume of the substrates and their hydrophobicity (Hansch constant). In addition, a deuterium effect of 2.6 indicated that hydrogen abstraction was partially rate limiting with these substrates (White and McCarthy 1986). Based on other experiments using a series of ten alkylbenzenes, it was concluded that the electron-accepting potential of the hydrocarbons is an important structural feature for P450-dependent catalysis (Lewis et al. 1986). Although these experiments show good correlations with a distinct P450 form, one should be careful in extrapolating these data to other forms and substrates. Interestingly, simultaneous binding of two different molecules to the same P450-active site has been shown for CYP3A4 with phenanthrene and 7,8-benzoflavone, leading to an increase of the V_{max} of phenanthrene metabolism. This is the first evidence that activation, and autoactivation, must be taken into account, at least with this P450 form, when possible drug interactions are considered (Shou et al. 1994).

5.1.3 Regulation by Alternate Pathways

Since the earliest studies of liver microsomes, it has been known that hydrogen peroxide is formed during drug metabolism (Gillette et al. 1957). Subsequent studies by several groups revealed that cytochromes P450 catalyze not only mono-oxygenase (Eq. 3) but also oxidase (Eq. 4a-c) and peroxidase (Eq. 5) reactions:

$$RH + O_2 + NADPH + H^+ \longrightarrow ROH + H_2O + NADP^+ \quad (3)$$

$$2O_2 + NADPH \longrightarrow 2O^{2-} + NADP^+ + H^+ \quad (4a)$$

$$O_2 + NADPH + H^+ \longrightarrow H_2O_2 + NADP^+ \quad (4b)$$

$$O_2 + 2NADPH + 2H^+ \longrightarrow 2H_2O + 2NADP^+ \quad (4c)$$

$$RH + XOOH \longrightarrow ROH + XOH \quad (5)$$

As shown in Fig. 6, H_2O_2 can be formed at different steps of the reaction cycle. However, it was concluded from stoichiometrical measurements that protonation and subsequent release of a two-electron-reduced oxygen species does not represent a significant pathway for the hydrogen peroxide formation (Kuthan and Ullrich 1982). The formation of water from NADPH was suggested by measuring the stoichiometry of NADPH, O_2, and peroxide in microsomes (Zhukov and Archakov 1982). Formation of water was also shown in reconstituted microsomal P450 systems, which, in addition, released H_2O_2 (Gorsky et al. 1984). It was further demonstrated that the ratio P450/oxygen is critical for the amount of water released. While at a fourfold molar excess of CYP1A2 over oxygen water is formed, at a twofold molar excess H_2O_2 is released (Oprian et al. 1983). The release of reduced states of oxygen (1,2, or 4 electron-reduced), as shown in Fig. 6, was suggested to represent an uncoupling of the mono-oxygenase reaction (Gorsky et al. 1984). Again, this uncoupling seems to be dependent upon the isoenzyme and substrate used (see next section).

The possibility that P450 can function as a peroxidase was first proposed by Hrycay and O'Brien (1972). Whether this so-called shunt mechanism (Fig. 6, step 9) is of physiological importance remains unclear at present. Thus, through modulation of the relationship of different pathways, regulation of P450 hydroxylase activity may occur.

5.2 Regulation at the Membrane Level

Since nearly all of the cytochromes P450 are bound to membranes, they can also be regulated at that level. Exclusions are most of the bacterial cytochro-

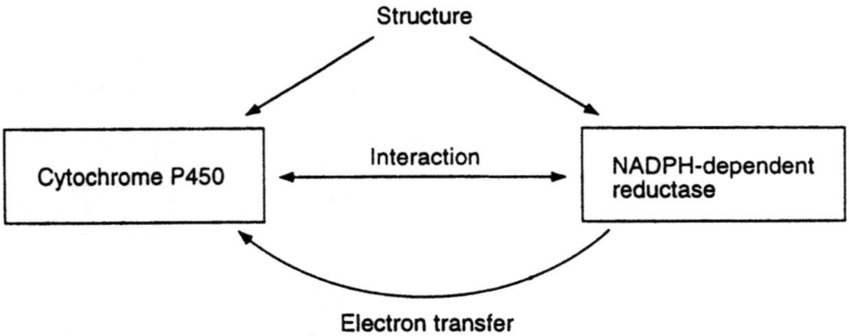

Fig. 13. Lipid regulation of cytochrome P450 systems

mes P450 such as CYP101, CYP102, CYP108, and CYP111, which are soluble.

First reconstitution experiments with microsomal cytochromes P450 revealed that, besides NADPH-dependent reductase, a heat-stable factor, later identified as lipids, is necessary for the functional activity (Lu et al. 1969a,b; Strobel et al. 1970; Lu and Levin 1972). Similar results were obtained when the reconstitution of mitochondrial CYP11A1-dependent activity was attempted (Omura et al. 1966; Lambeth et al. 1980a,b). The composition of the membrane, e.g., protein-lipid as well as protein-protein interactions, thus exerts a functional control on cytochromes P450 (for reviews see Ruckpaul and Bernhardt 1984; Ruckpaul et al. 1989; Blanck and Ruckpaul 1993).

Lipids appear to function in at least three different ways (Fig. 13): (a) Lipids stabilize and/or induce a functionally active conformation of P450 and the corresponding electron transfer system. Such specific lipid binding stabilizing protein conformation has in general been suggested to occur also in other membrane proteins (Abney and Owicki 1985). (b) Lipids modulate the electron transfer. (c) Lipids mediate interactions between P450 and the respective electron donor systems.

The majority of cytochromes P450 is located in the endoplasmic reticulum. This organelle is composed of about 60%–70% protein by weight and 30%–40% phospholipid. The protein moeity contains about 10% P450, which is increased to about 20% upon phenobarbital induction. Thus, the P450/lipid molar ratio amounts to about 1:200 – 1:120 (Depierre and Ernster 1977). The stoichiometry of reductase and P450 was determined to be 1:15 independent of induction (Shephard et al. 1983). The lipid composition of the microsomes is summarized in Table 3 and consists mainly of phosphatidylcholine (PC), phosphatidylethanolamine (PE), and, to a minor extent, the negatively charged phosphatidylserine (PS) and phosphatidylinositol (PI). The specificity of P450/lipid interaction has been widely studied using reconstitution and various techniques to produce liposomes (Ingelman-

Table 3. Phospholipid composition (percentage of total phosphorus) of liver microsomes from untreated rats and of bovine adrenocortical microsomes and mitochondria (from Blanck and Ruckpaul 1993)

Lipid	Liver microsomes	Adrenocortical Microsomes	Mitochondria
CL	n.d.[a]	1.9 ± 1.8	12.5 ± 1.1
PC	50.7 ± 9.9	50.9 ± 3.4	39.7 ± 2.6
PE	23.0 ± 2.3	26.9 ± 1.5	34.3 ± 2.5
PI	12.5 ± 0.5	11.9 ± 1.4	6.0 ± 0.2
PS	7.0 ± 1.7	6.9 ± 2.1	3.5 ± 1.3
SM	6.0 ± 2.6	n.d.	n.d.
Other		1.5 ± 1.7	4.0 ± 2.6

[a]n.d., No data given in the references.

Sundberg and Glaumann 1977; Taniguchi et al. 1979; Bösterling et al. 1979; French et al. 1980; Schwarz et al. 1984).

Dilauroyl-PC (DLPC)-reconstituted systems were shown to have activities similar to those of microsomes (White and Coon 1980). Investigations by Chiang and Coon (1979), as well as Archakov (1982), demonstrate conformational changes of P450 upon interaction with lipids. A high-spin shift was observed upon DLPC binding (Chiang and Coon 1979). Changes of the surroundings of a tyrosine residue of CYP2B4 were detected in the presence of phospholipids and detergents but not other amphiphilic or hydrophobic compounds using derivative spectroscopy (Ruckpaul et al. 1980). Incorporation of P450 into lipid vesicles was shown to stabilize the native conformation of P450 (Uvarov et al. 1980). These data indicate a P450/lipid interaction at the molecular level, leading to activation of the P450.

DLPC was also shown to favor the binding of reductase and substrate to P450 (French et al. 1980). Furthermore, acidic head groups of the phospholipid were observed to favor the formation of the functionally active complex (Ingelman-Sundberg et al. 1983; Blanck et al. 1984). The V_{max} value for microsomal P450-dependent substrate conversion is directly dependent on the concentration of a 1:1 functional complex between P450 and reductase. Thus, phospholipid exerts a clear regulatory effect upon P450/reductase and P450/substrate interaction.

Furthermore, an interaction of P450 with cytochrome b_5 within the microsomal membrane is able to modulate P450 activity (for reviews see Ruckpaul et al. 1989; Schenkman 1993). This interaction should also be dependent on phospholipids, although detailed studies on this topic have not been performed yet. Proteins, on the other hand, were shown to cause a destabilization of the lipid matrix (Akhrem et al. 1982). A decrease of the order of hydrocarbon chains was observed upon P450 and reductase incorporation into different liposomes (Kunz et al. 1985).

In mitochondrial systems, which consist of three proteins, the mitochondrial inner membrane is also able to modulate protein structure and protein-protein interaction. As can be seen from Fig. 5, mitochondrial cytochromes P450 are embedded in the inner mitochondrial membrane, whereas the ferredoxin reductase (called adrenodoxin reductase in the case of the adrenal mitochondrial steroid hydroxylases) is associated only with the membrane. The immediate electron donor to P450, ferredoxin, is soluble and located in the matrix space. Specific P450-lipid interactions have been demonstrated (Lambeth et al. 1980b; Lambeth 1981). As compared with the lipid composition of the endoplasmic reticulum, mitochondria contain sixfold increased amounts of cardiolipin (CL) and slightly decreased levels of PC, PS, and PI (Table 3). Interestingly, an asymmetric distribution with increased levels of negatively charged lipids at the matrix side of the inner mitochondrial membrane, where the steroid hydroxylases are located, has been found (Daum 1985). It has further been shown that lipids can stimulate binding of the substrate, cholesterol, to CYP11A1, and this in the order CL>PS>PC>PE>PI (Pember et al. 1983). This stimulation proceeds via an allosteric effect in a 1:1 (CL) or 2:1 (lipids with other head groups) molar ratio by a decrease in the K_m for cholesterol. Besides the head group, the fatty acids of the lipids can exert a regulatory effect on mitochondrial P450 systems. With CYP11A1, the activity correlates with the degree of unsaturation (Lambeth et al. 1980b; Igarashi and Kimura 1986) as is the case for microsomal cytochromes P450 (Wade et al. 1978; Ruckpaul and Bernhardt 1984). In contrast, CYP11B1 exhibits the highest activity when reconstituted with saturated lipids such as dipalmitoyl phosphatidylcholine, while it is inhibited by lipids containing unsaturated fatty acids. CL inhibits this P450 form via an increase in K_m and a decrease in V_{max} (Seybert 1990). Adrenodoxin favors binding of cholesterol to CYP11A1 and vice versa (Lambeth et al. 1980a), an effect which is used for measuring adrenodoxin binding to CYP11A1 by following the cholesterol-induced high-spin shift. The mechanism and the physiological importance of the differences in CL regulation of CYP11A1 and CYP11B1 are not clear yet and require further investigation, as do the influence of membrane lipid composition on the electron transfer and protein-protein interactions in mitochondrial P450 systems.

5.3 Regulation at the Cellular Level

Cellular regulation of cytochromes P450 is extremely complex and can occur at two different levels, enzyme induction and post-translational modification of enzymes. Since both of these topics have been very extensively studied, a thorough discussion of all aspects by far exceeds the scope of this review, and the reader is referred to recent reviews by Waterman and Estabrook (1983), Ruckpaul et al. (1989), Bresnick (1993), Sogawa and Fujii-Kuriyama (1993), and Jansson (1993).

The phenomenon of P450 induction was observed long before P450 itself was discovered. In 1945, Lacassange et al. found that the number of tumors induced by injection of dibenzofluorene into the skin of mice was drastically reduced when the carcinogen was applied in combination with methylcholanthrene. At that time the observation was attributed to competition of the two structurally related polycyclic hydrocarbons. The induction of drug-metabolizing enzymes was noted in the early 1950s by Miller et al. (1952), while they were performing studies on the effect of carcinogens upon azo dye metabolism. Meanwhile, specific inducers for various P450 families and subfamilies have been identified (Table 4). The mechanism of induction was discovered in studies by the groups of Padmanaban and Waterman (Bhat and Padmanaban 1978, 1979; DuBois and Waterman 1979). It was demonstrated that the inductor functions by enhancing the rate of mRNA synthesis of a special P450. This result was subsequently confirmed by several laboratories following phenobarbital, 3-methylcholanthrene, or β-naphthoflavone treatment of rats or rabbits (for review, see Waterman and Estabrook 1983 and references therein). Of special interest was the observation of the time dependence of phenobarbital-dependent P450 induction. Whereas the peak

Table 4. Inducible subfamilies of cytochrome P450 (from Bresnick 1993)

Family	Subfamily	Other nomenclature	Inducer
P450 1	CYP1A1	P450c; P_1-450; LM6	Polycyclic hydrocarbons, dioxins
	CYP1A2	P450d; P_3-450; LM4	Polycyclic hydrocarbons, dioxins, isosafrole
P450 2	CYP2B1	P450b; LM2	Barbiturates,chlorinated hydrocarbons and aromatics, trans-stilbene oxide, acetylaminofluorene
	CYP2B2	P450e	Barbiturates and above
	CYP2E1	P450j, LM3a	Ethanol, ioniazid, acetone, imidazole
P450 3	CYP3A1	P450p	Pregnenolone-16α-carbonitrile
P450 4	CYP4A1	P452	Clofibrate, phthalates

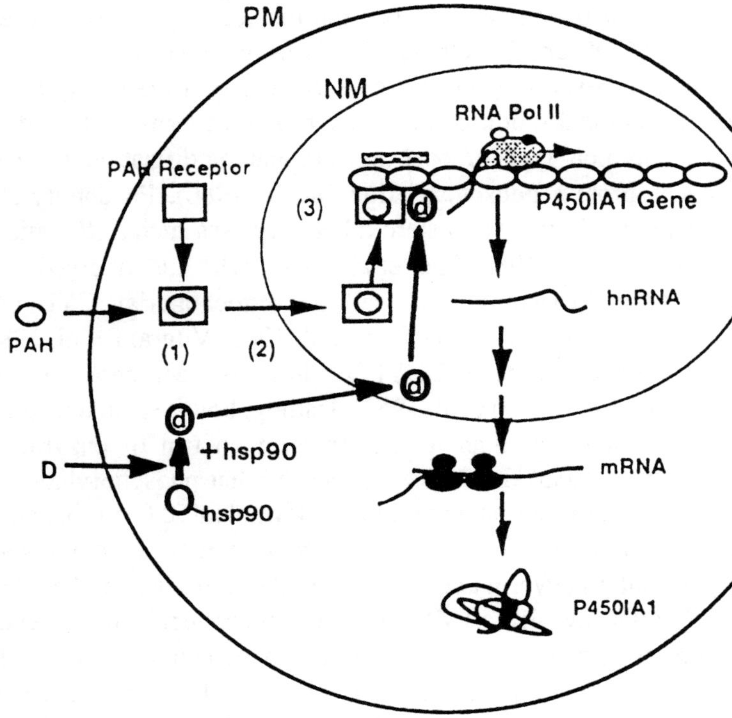

Fig. 14. Model for the induction of CYP1A1 by polycyclic hydrocarbons and dioxins. *PM*, plasma membrane; *NM*, nuclear membrane; *PAH*, polycyclic aromatic hydrocarbon; *D*, dioxins; *0*, Ah receptor; *Pol*, polymerase; *hnRNA*, heterogeneous nuclear RNA. (From Bresnick 1993)

for mRNA and apoprotein induction was observed after 16 h of phenobarbital treatment, the peak for holoprotein detection was found only after 22 h (DuBois and Waterman 1979), indicating that heme synthesis or incorporation are limiting for P450 synthesis.

Nowadays, the best-understood model for P450 induction is the Ah-receptor-mediated induction of CYP1A1-mediated aryl hydrocarbon hydroxylase (Fig. 14). In this model the inducer, 3-methylcholanthrene or 2,3,7,8-tetrachlorodibenzo-p-dioxin (TCDD), which is 30 000 times more effective than 3-methylcholanthrene (Poland and Glover 1974), first binds to a receptor (Okey 1990; Whitlock 1990; Landers and Bunce 1991). The inducer-receptor complex is then transported into the nucleus, where it leads to transcriptional activation of the CYP1A1 gene in conjunction with binding of the heat-shock protein hsp 90. The questions remains whether a normal body ligand for the Ah-receptor exists and what physiological importance this endogenous ligand could have (Beresford 1993). It seems hard to believe, that a mechanism of induction for xenobiotica was created in this P450 subfamily which arose more than one million years ago, long before chemicals such as TCDD and 3-methylcholanthrene were known.

After biosynthesis of cytochromes P450, some of them may be changed by post-translational modification to exert a short-term control of activity. The most common type of post-translational modification in eukaryotic organisms is protein phosphorylation (Krebs 1986). Phosphorylation has been demonstrated for several steroid hydroxylases such as the microsomal CYP7 (Sanghvi et al. 1981; Shet et al. 1994), which is involved in cholesterol 7-alpha-hydroxylation, as well as mitochondrial CYP11A1 and CYP11B1 (Caron et al. 1975; Defaye et al. 1982; Vilgrain et al. 1984). The drug-metabolizing microsomal CYP2B4 has also been shown to be phosphorylated in vitro (Pyerin et al. 1983), leading, however, to denaturation of the hemoprotein rather than to a short-term control of enzyme activity (Taniguchi et al. 1985). It has been proposed that phosphorylation plays a key role in post-translational regulation of several P450 forms in gene family 2 and could be important for short-term regulation in the case of reversibility of phosphorylation (Pyerin and Taniguchi 1989). Studies on the ethanol-inducible CYP2E1 revealed cAMP-dependent phosphorylation in hepatocytes with concomitant denaturation, loss of enzymatic activity, and rapid degradation (Eliasson et al. 1990). CYP3A1 is also phosphorylated with subsequent denaturation, heme loss, and drop in catalytic activity (Mkrtchian et al. 1994). A phosphorylation-dependent regulatory mechanism is obviously responsible for control of cytochrome P450 turnover in different P450 families and could be a general mechanism for post-translational regulation of various isoforms of P450.

It is interesting that neither modification of the N-terminus nor proteolysis has been found in P450 systems as being of regulatory importance so far. Although proteolytic cleavage of the C-terminal part of the mitochondrial electron carrier, adrenodoxin, has been observed (Tanaka et al. 1973) during isolation from bovine adrenals, which was shown to lead to a more active protein with respect to its electron transfer function (Cupp and Vickery 1989; Uhlmann et al. 1994), so far no physiological importance for a truncated adrenodoxin has been reported (Uhlmann et al. 1994).

5.4 Regulation at the Level of the Organism

Some of the cytochromes P450 are expressed in age-, tissue- and sex-dependent manners (Kato 1974; Kato and Kamataki 1982; Keeney and Waterman 1994). These differences are governed by sex or other hormones (for reviews see Kato and Yamazoe 1993; Miller 1988; Ryan and Levin 1993; Birnbaum 1993; Simpson et al. 1993; Waterman and Simpson 1990).

Table 5. Changes in cytochrome P450 level in hypophysectomized rats and growth hormone- or triiodothyronine (T_3)-treated rats (from Kato and Yamazoe 1993)

Cytochrome P450 species	Hypophysectomized (H)		H+ growth hormone-treated male rats		H+T_3-treated male rats
	Male rats	Female rats	Intermittent injection	Continuous infusion	
1A2	↑ ↑	↑ ↑	≠	≠	↓ ↓ ↓
2A1	↑	≠	↓	↑	↓
2B1	↑ ↑ ↑	↑ ↑ ↑	↓ ↓	↓ ↓ ↓	↓ ↓ ↓
2B2	↑ ↑ ↑	↑ ↑ ↑	↓	↓ ↓ ↓	↓ ↓ ↓
2C7	↓	↓	↑	↑ ↑	–
2C11	↓ ↓	↑ ↑ ↑	↑ ↑	↓ ↓ ↓	≠
2C12	≠(ND),	↓ ↓ ↓	≠(ND),	↑ ↑ ↑	≠(ND),
2C13	↑	↑ ↑ ↑	≠	↓ ↓ ↓	–
2C22	↓ ↓	≠(ND),	↑	≠(ND),	–
2E1	↑	↑	↓ ↓	↓ ↓	–
3A2	↑ ↑	↑ ↑ ↑	↓ ↓	↓ ↓ ↓	↓ ↓ ↓

↑, Less than twofold; ↑ ↑, less than fivefold; ↑ ↑ ↑, more than fivefold; ↓, less than 25%; ↓ ↓, less than 60%; ↓ ↓ ↓, more than 60%; ≠, no significant change; ≠(ND), remained nondetectable; –, no data available.

[a] Data shown are derived from changes in specfic protein contents, except for CYP2C7 and CYP2C22, in which the changes of mRNA levels are shown.

Sex-dependent Expression of Cytochromes P450. Sexual dimorphism in hepatic metabolism of drugs and steroids in the rat has been well established (Kato and Kamataki 1982). It was found that the magnitude of observed differences which become apparent at puberty is dependent upon the substrate and metabolic pathway studied. In general, rates of metabolism toward the transformation of many drugs are lower in females than in males, although certain reactions can be catalyzed more efficiently in females (Gustafsson and Ingelman-Sundberg 1975; Gustafsson et al. 1983). Substantial evidence has accumulated showing that sexual dimorphism is "imprinted" in the neonate during the first weeks of life (Gustafsson et al. 1983). Testicular androgens predetermine a male pattern of metabolism. It was demonstrated that hormonal programming of sex-specific expression by androgens or estrogens was not realized by a direct action of the hormones, but was mediated through the hypothalamic-pituitary axis (Gustafsson et al. 1983). Thus, besides gonadal hormones, growth hormone also plays an essential role in determining sex-specific expression of microsomal cytochromes P450. Furthermore, thyroid hormone is known to affect rat microsomal P450 activities (Kato 1974). Recent results demonstrate that this hormone depresses directly some forms of cytochrome P450 without changing the secretion of gonadal hormones or growth hormone (Murayama et al. 1991). Changes in the levels of different cytochrome P450 species in hypophysectomized and growth hormone- or thyroid hormone-treated rats are summarized in Table 5. Changes of the drug-metabolizing enzymes were also observed in diabetic rats (for review see Kato and Yamazoe 1993). The mechanism of insulin action on cytochrome P450 expression and activity is not yet completely clear, however.

Age-dependent Expression of Cytochromes P450. Age-related changes in the metabolism of drugs and xenobiotics are of great practical relevance, since elderly people (> 65 years) are the heaviest consumers of drugs and comprise approximately 20% of the population in developed countries. These age-related differences in metabolism are attributable, at least in part, to changes in the P450 and lipid composition of hepatic microsomes. Thus, CYP2C7 was shown not to be detectable in newborn rats, but rises between 1 and 4 weeks of age to a maximum of 12 weeks (Gonzalez et al. 1986). Similar as well as different age profiles have been observed for other P450 forms (for review see Ryan and Levin 1993). The conclusion can be made that age-related changes are substrate specific and dependent upon species, strain, sex, and tissue.

Tissue-specific Regulation of Cytochromes P450. Tissue-specific regulation of cytochrome P450 expression has been most extensively studied with steroid hydroxylases. It is well known that steroid hydroxylase expression is regulated by ACTH via cAMP-dependent signaling (Simpson et al. 1987; Waterman and Simpson 1990). A scheme of regulation of steroid hydroxylase expression is presented in Fig. 15. Interestingly, this regulation is realized by an all-or-nothing mechanism. While CYP11A1, catalyzing the initial step of steroid hormone synthesis, i.e., the conversion of cholesterol to pregnenolone, is expressed in all steroidogenic tissues (adrenal, testes, ovary; see Fig. 16), CYP11B1, CYP11B2, and CYP21 are present only in the adrenal (Waterman and Simpson 1990). CYP11B1 catalyzes the 11β-hydroxylation of 11-deoxycortisol and 11-deoxycorticosterone to give cortisol and corticosterone, respectively, and CYP11B2, being 93% identical in its nucleotide and protein sequence to CYP11B1 (Kawamoto et al. 1990; Mornet et al. 1989), is involved in the conversion of corticosterone to aldosterone. CYP21 participates in the formation of 11-deoxycorticosterone from progesterone and of 11-deoxycortisol from 17-hydroxyprogesterone (Fig. 16). Besides tissue specificity, differences in zonal distribution of some cytochromes P450 can also be observed. In adrenals, CYP 11A1 was shown to be located in the zona fasciculata as well as in the zona glomerulosa. CYP11B1 has been traced to the zona fasciculata only (Ho and Vinson 1993; Mitani et al. 1994; Erdmann et al. 1995). In contrast, CYP11B2 was found only in the zona glomerulosa (Ho and Vinson 1993; Mitani et al. 1994; Sander et al. 1994). In agreement with differences in the location are differences in the

Fig. 15. Regulation of mitochondrial steroid hydroxylases

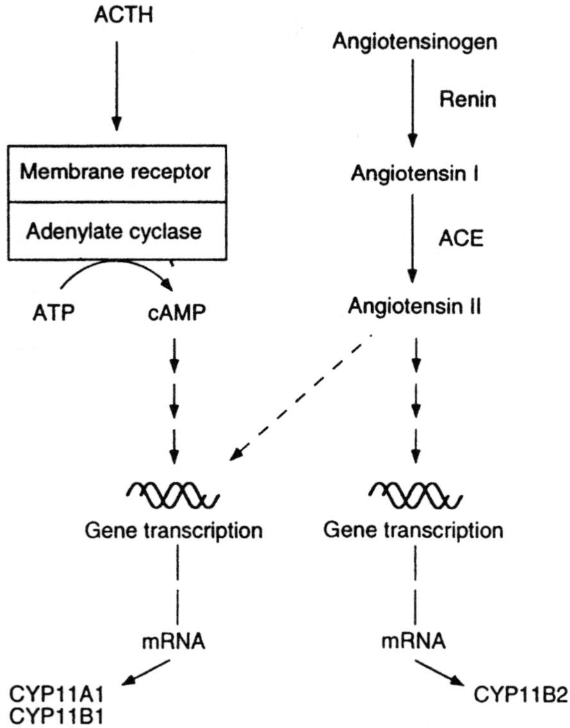

regulation of the genes of the CYP11B family. While CYP11B1 is regulated via ACTH (Fig. 15), CYP11B2 responds to the renin-angiotensin system (Shibata et al. 1991; Curnow et al. 1991; Bird et al. 1993; White et al. 1993). Also Ad4BP, a zinc-finger DNA-binding protein functioning as a transcription factor in the regulation of steroidogenic P450 genes in a cAMP-dependent manner, was shown to be expressed to a varying extent in different tissues (Morohashi et al. 1994).

Aromatase, catalyzing the aromatization of the A ring of androgens to form estrogens, is expressed in a number of cells and tissues including ovary, testis, placenta, adipose tissue, and brain (for review see Simpson et al. 1993). It was clearly demonstrated that tissue-specific regulation of human aromatase expression is achieved using alternate promotors (Simpson et al. 1993; Bulun et al. 1994). The expression of CYP19 in human ovary utilizes a promotor which is proximal to the translation start site. This promotor, however, is not utilized in placenta; instead, a promotor at least 40 kb upstream from the translation start site is utilized. It appears that this type of regulation was first shown to be effective in a P450 system. Whether some other cytochromes P450 will use a similar mechanism for tissue-specific regulation remains to be elucidated.

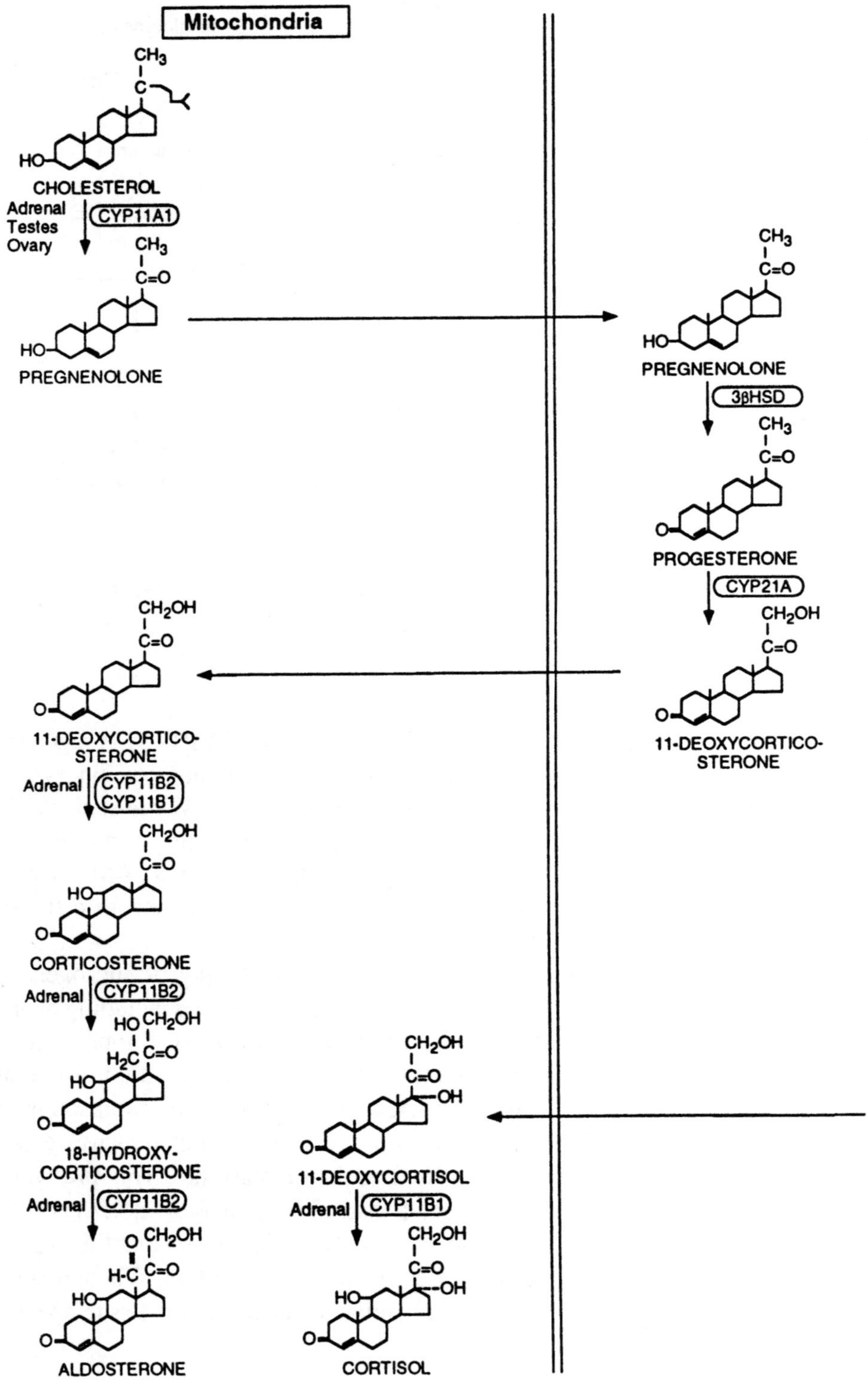
Mitochondria
CHOLESTEROL
Adrenal
Testes
Ovary
CYP11A1
PREGNENOLONE
PREGNENOLONE
3βHSD
PROGESTERONE
CYP21A
11-DEOXYCORTICO-STERONE
11-DEOXYCORTICO-STERONE
Adrenal
CYP11B2
CYP11B1
CORTICOSTERONE
Adrenal
CYP11B2
18-HYDROXY-CORTICOSTERONE
Adrenal
CYP11B2
ALDOSTERONE
11-DEOXYCORTISOL
Adrenal
CYP11B1
CORTISOL

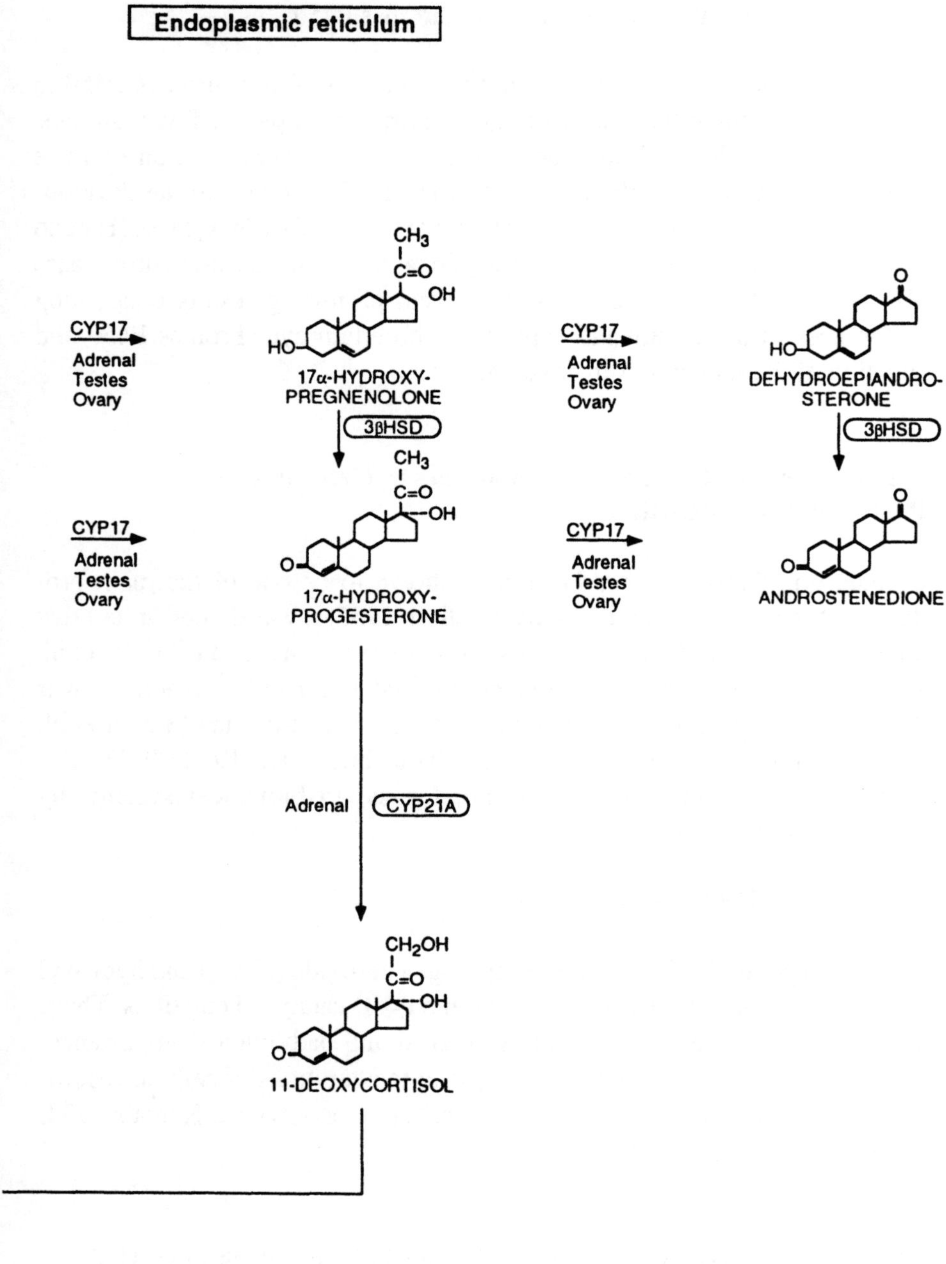

Fig. 16. Biosynthesis of steroid hormones. *3βHSD*, 3β-hydroxy-Δ^5-steroid dehydrogenase-Δ^5-isomerase

5.5 Developmental Regulation of Cytochromes P450

Although little is known so far about the function of cytochromes P450 in development, this subject seems to be an intriguing topic for future studies, since first experimental approaches such as in situ hybridization of mice fetuses with different P450 probes (Keeney 1995), studies on the developmental expression of transcriptional factors for steroid hydroxylases (Hatano et al. 1994), and knockout of a transcriptional factor for steroid hydroxylases (Ikeda et al. 1994; Luo et al. 1995) reveal promising results concerning differences in developmental expression of certain cytochromes P450 and function of these enzymes in development.

6 Formation of Reactive Oxygen Species in Cytochrome P450-Catalyzed Reactions

Cytochrome P450 enzymes have been shown to be one of the main producers of reactive oxygen in the liver cell. In 1957, it was demonstrated that during NADPH oxidation in microsomes H_2O_2 is generated (Gillette et al. 1957). Later, "uncoupling" of cytochrome P450-dependent reactions was described for different drugs and chemicals such as n-hexane (Staudt et al. 1974), hexobarbital, and benzphetamine (Hildebrandt and Roots 1975). Reactive oxygen species are, in general, formed in biological systems by reduction of dioxygen:

$$O_2 \rightarrow O_2^{\bullet-} \rightarrow H_2O_2 \rightarrow HO^{\bullet} \rightarrow H_2O$$

Superoxide anion ($O_2^{\bullet-}$), as well as hydrogen peroxide (H_2O_2) and hydroxyl radical ($OH^{\bullet}$) can be formed in cytochrome P450-catalyzed reactions. These reactive oxygen species are of physiological and pathological importance; due to their damaging effect, they may lead to alterations of cellular macromolecules such as DNA, proteins, and lipids (for reviews see Kappus 1993; Karuzina and Archakov 1994).

6.1 Reactive Oxygen Generation on the Level of the Cytochrome P450

6.1.1 Possibilities of Reactive Oxygen Generation

As described in Sect. 2.3 and depicted in Fig. 6, hydrogen peroxide can be generated in two ways: protonation and decay of the peroxy complex (the direct way) (Nordblom and Coon 1977; Kuthan and Ullrich 1982; Zhukov and Archakov 1982), and dismutation of preformed $O_2^{\bullet-}$ released from the oxy ferrous complex (the indirect way) (Estabrook et al. 1979; Kuthan et al. 1978). Direct formation through peroxy complex breakdown was observed during the oxidation of tertiary methyl-substituted amines (Ristau et al. 1989) and during camphor hydroxylation using the bacterial CYP101 system containing a mutant CYP101 (Imai et al. 1989). The measurement of the stoichiometry of $O_2^{\bullet-}$ and H_2O_2 generation makes it possible to decide between direct and indirect H_2O_2 formation in the P450 reaction cycle. In the case of indirect H_2O_2 formation, the $O_2^{\bullet-}$ to H_2O_2 ratio will be 2, due to the stoichiometry of $O_2^{\bullet-}$ dismutation, whereas a smaller ratio will be obtained if H_2O_2 is partly formed directly (Zhukov and Archakov 1982; Gorsky et al. 1984). H_2O_2 and $O_2^{\bullet-}$ can react with reduced transition metals such as Fe^{2+} in the so-called Fenton or Haber-Weiss reaction to form the most potent oxidant, the hydroxyl radical:

$$Fe^{3+} + O_2^{\bullet-} \rightarrow Fe^{2+} + O_2 \qquad (6)$$
$$H_2O_2 + Fe^{2+} \rightarrow HO^{\bullet} + OH^{-} + Fe^{3+} \qquad (7)$$

That P450 is the main site of reactive oxygen generation in P450-dependent reactions has been shown by different lines of evidence such as CO-sensitivity of H_2O_2 formation and inhibition of this reaction by well-known P450 inhibitors (Karuzina and Archakov 1994). Furthermore, human liver microsomes produced superoxide and hydrogen peroxide at rates about 20%–30% of that with rat liver microsomes, which was shown to be due, in part, to the lower content of total P450 in the human liver microsomes, being only one-third the content of rat liver microsomes (Rashba-Step and Cederbaum 1994).

6.1.2 *Factors Influencing the Generation of Reactive Oxygen Species*

Early studies on the H_2O_2 formation caused by P450 revealed that this process is affected by the substrate as well as by the isoenzyme used. Especially high reactive oxygen generation has been observed in ethanol-induced liver microsomes and with isolated CYP2E1 (Ekström et al. 1986; Ekström and Ingelman-Sundberg 1986; Persson et al. 1990; Albano et al. 1991; Rashba-Step et al. 1993). In HepG2 cells stably expressing human CYP2E1, it was demonstrated that the rate of microsomal H_2O_2 production was two times higher than in the control cells without CYP2E1. The rate of H_2O_2 production is half the rate of $O_2^{\bullet-}$ production, suggesting that this H_2O_2 is derived largely from indirect H_2O_2 formation by superoxide radical dismutation (Dai et al. 1993). Since the rates of reactive oxygen formation

Table 6. Stoichiometry of cytochrome P450-catalyzed reactions in reconstituted systems

Substrate	H_2O_2 formed (nmol/min/nmol P450)	$NADPH/O_2/product/H_2O_2$
None	21[a]	1.0 /1.0/0 /1.0 [a]
Benzphetamine	42[a]	1.0/1.0/0.4/0.6 [a]
Cyclohexane	17[a]	1.0/0.9/0.8/0.3 [a]
N,N-dimethylaniline	25[a]	1.0/1.1/0.6/0.5 [a]
ethanol	90[b,c]	1.0/0.8/0.2/0.2 [b]
1-pentanol	134[b,c]	1.0/0.8/0.3/0.3 [b]

[a] Data from Nordblom and Coon (1977).
[b] Data from Gorsky et al. (1984).
[c] nmol H_2O_2 formed/reaction mixture.

and lipid peroxidation were seven to ten times higher on a per nmol P450 basis with this recombinant cell line compared with human liver microsomes, it was suggested that CYP2E1 is especially reactive in the production of reactive oxygen species and in catalysis of lipid peroxidation. So far, it is not clear why this form shows such a high level of uncoupling; but this process could play an important role in pathophysiological changes due to heavy alcohol consumption.

Since many cytochromes P450 are inducible, the amount of reactive oxygen species can increase by the effect of an inducer, e.g., halogenated aromatic compounds, including TCDD, lindane, or phenobarbital (Al-Bayati and Stohs 1987; Junqueira et al. 1986; Saito 1990; Scholz et al. 1990). In addition, an influence of the cytochrome P450 substrate on reactive oxygen generation has been evidenced (Table 6) (Nordblom and Coon 1977; Gorsky et al. 1984). It can be seen from Table 6 that in the case of CYP2E1-dependent ethanol oxidation the sum of product plus H_2O_2 formed is less than 1.0. This was shown to be due to the formation of water by an overall four-electron transfer to molecular oxygen in this reconstitution system (Gorsky et al. 1984). Using a series of benzphetamine analogs, it was further demonstrated that the high-spin shift caused by substrate binding correlates well with the ratio of $HCHO_{formed}/(HCHO \text{ plus } H_2O)_{formed}$ but not with the ratio of H_2O_2 formed /NADPH consumed (Zhukov et al. 1989). When hepatic microsomes were studied, free radical production was shown to be maximal in the absence of substrates subject to oxidation by P450 enzymes (Bondy and Naderi 1994).

Stimulation of H_2O_2 production can also be achieved by the inhibitory action of certain compounds on oxygenase reactions. It was demostrated that low concentrations of quercetin, a naturally occuring flavonoid, inhibit the oxidation of various substrates, e.g., ethoxyresorufin, p-nitroanisole, and benzo[a]pyrene, by increasing the amount of H_2O_2 generation without influencing the rate of NADPH utilization (Sousa and Marletta 1985). Also stobadine, a novel drug of pyridoindole structure proven to have a cardioprotective effect, led to an increase in H_2O_2 production when added to rat liver microsomes (Stefek 1993).

Finally, the degree of uncoupling can be modulated by changing some of the active site residues. Due to the fact that the three-dimensional structure of CYP101 has been known since 1985 (Poulos et al. 1985), most of the crucial experiments in the study of this problem have been performed using CYP101. As illustrated in Fig. 7, activation of molecular oxygen seems to be realized, at least in CYP101, via a charge relay system. As can be seen from Table 7, CYP101 shows essentially no uncoupling. All the reducing equivalents used produce the hydroxylated product. When Thr252 of CYP101 was replaced by alanine, glycine, or cysteine, however, a decrease in the formation of 5-exo-hydroxycamphor was obtained concomitant with an increase in H_2O_2 production (Table 7) (Imai et al. 1989; Martinis et al. 1989; Shimada et al. 1994). The ferrous Thr252Ala and Thr252Val mutants (Imai et al. 1989) were converted very rapidly to oxy-P450 and slowly decomposed to the ferric form. Since these rates were too slow to account for the overall rates of molecular oxygen consumption, the formation of H_2O_2 was suggested to arise directly by the decomposition of a peroxy complex, and not through dismutation of superoxide anion (Imai et al. 1989). Resolution of the crystal structure of mutant Thr252Ala to 2.1 Å revealed that the opening between residues 252 and 248 is wider in the mutant, so that an extra solvent may move in (Raag et al. 1991). From model studies with heme systems it is well known that a polar environment can stabilize an oxy-heme complex but that a protic environment destabilizes this complex (Brinigar et al. 1974), presumably by providing protons to dioxygen. It has been discussed (Poulos and Raag 1992) that reduction of the hypothetical $(FeO)^{3+}$ species, although not very well understood, might be promoted in a protic environment (i.e., excess solvent). In this model, a Thr252-Ala366 solvent channel exists, and the Thr252 side chain to peptide hydrogen bond prevents ready access of protons to the active site, so that this residue is crucial for a controlled delivery of protons to dioxygen within the CYP101 active site. This hypothesis is supported by the observation that the Thr252Ser mutant exibits essentially the same mono-oxygenase activity as the wild-type enzyme (Table 7). It is interesting to note in this connection that despite participation of Asp251 in the proton conduit (Gerber and Sligar 1992, 1994), conversion of this residue to alanine does not result in H_2O_2 production (Shimada et al. 1991). A conclusive explanation of this behavior can be given only after refinement of the Asp251Ala crystal structure.

Table 7. Relationship of products formed per oxygen consumed by wild-type and mutant CYP101 (data taken from Shimada et al. 1994)

CYP101	Product formed per 5-exo-OH-camphor [%]	O_2 consumed H_2O_2 [%]
Thr 252 (Wild type)	97	3
Thr 252 Ala	5	89
Thr 252 Gly	3	88
Thr 252 Ser	85	15
Thr 252 Cys	7	86
Thr 252 Asn	57	12
Thr 252 Pro	10	77

On the other hand, when molecules smaller than d-camphor have been used as substrates for CYP101, less coupling has generally been observed: nearly 100% for d-camphor and 2% for most of the other substrates reported (Atkins and Sligar 1988, 1989). Using a Phe96Tyr mutant, a sensitive probe for determination of the active-site polarity has been created (Atkins and Sligar 1990). The sensitivity of Tyr96 to environmental polarity has been used to demonstrate that the product/enzyme complex formed with 5-*exo*-hydroxy-camphor may be associated with increased water access to the heme. It has been supposed that unoccupied space in the active site could allow greater substrate mobility or water access, which could promote production of reactive oxygen species (Atkins and Sligar 1987; Wade 1990). This model can be tested by re-engineering the protein, such that the substrate is packed more tightly in the active site, allowing no additional water molecule to come in. In fact, when CYP101-catalyzed ethylbenzene hydroxylation, which has a coupling (stoichiometry of reducing equivalents to product) of only 5%, is studied, it can be demonstrated that this coupling can increase to 13% when steric bulk in the form of large hydrophobic residues is added to the top of the buried binding pocket (Thr101Met/Thr185Phe/Val247Met), whereas similar substitutions on the other side of the cleft result in reduced product formation (Loida and Sligar 1993a). Detailed studies of the relation between NADH utilization, product formation, and hydrogen peroxide, as well as water formation (Loida and Sligar 1993b), clearly demonstrate that active-site hydration is important in the uncoupling to form H_2O_2. While the liberation of H_2O_2 does not correlate with the location of the mutation in the pocket, this location, as well as the size of the amino acid side chains in the substrate-binding site, is of decisive relevance for the water production. The partitioning between hydroxylation and oxidase activities varies over 65% due to the location of nonpolar substituents engineered into the active site. Thus, the side chains that form the internal surface of the active site – although structurally distinct from the charge relay and not directly participating in dioxygen activation – through their role in determining substrate orientation and dynamics and active-site hydration, have a dramatic effect on the fate of reaction cycle intermediates and the coupling of reducing equivalents to product. An influence of the substrate mobility and protein flexibility has also been shown on the position and width of the CO stretching signals in CYP101 (Jung et al. 1992).

There are, unfortunately, almost no comparable experiments on microsomal or mitochondrial cytochromes P450. In a study of CYP1A2 mutants (Ishigooka et al. 1992), where the conserved threonine cluster composed of three threonines (Thr319, Thr321, Thr322) had been mutated, it was found that the Thr319Ala mutant incorporated oxygen into the substrate much more efficiently than the wild type. In contrast, the Glu318Ala mutant had a much lower efficiency of incorporated oxygen atom to the substrate versus produced H_2O_2 compared with the wild type CYP1A2 (Table 2). This indicates that in microsomal P450 forms polar amino acids at position 318 prior to the conserved threonine could contribute significantly to the activation of the oxygen molecule and to the regulation of the access of water to the active site. A participation of the conserved threonine in determining the substrate specificity has been evidenced in some microsomal P450 forms (Furuya et al. 1989a,b; Imai and Nakamura 1988, 1989; Fukuda et al. 1993). Unfortunately, the effect of mutation with respect to reactive oxygen formation was not investigated in these studies.

6.2 Formation of Reactive Oxygen Species at the Level of the Electron-Supporting System

As depicted in Fig. 5, P450 is able to accept reducing equivalents under physiological conditions from different electron donors, via a FAD- and FMN-containing reductase in microsomal P450 systems and via a FAD-containing reductase and a /2Fe-2S/protein in most bacterial and in mitochondrial P450 systems. These electron donor systems are also a possible source for the generation of reactive oxygen species.

6.2.1 Generation of Reactive Oxygen Species at the Level of Microsomal Reductase

In the early 1960s, a flavin-containing reductase was found in the microsomal fraction (Philips and Langdon 1962; Williams and Kamin 1962). Although the observation that the reductase is concomitantly induced with drug hydroxylation led to the suggestion that it functions under physiological conditions as an electron donor to microsomal drug-metabolizing enzymes (Ernster and Orrhenius 1965; Jick and Shuster 1966), direct evidence for this proposal came only in the late 1960s, when the laurate-ω-hydroxylase activity was reconstituted from a detergent-solubilized preparation of cytochrome P450, reductase, and a heat-stable component (Lu and Coon 1968; Lu et al. 1969a) later identified as phosphatidylcholine (Strobel et al. 1970; reviewed by Backes 1993). The microsomal reductase was shown to have a molecular weight of approximately 78 kD and to contain one molecule each of FAD and FMN per molecule of protein (Van der Hoeven and Coon 1974; Yasukochi and Masters 1976; Dignam and Strobel 1977). Interestingly, recent studies on the dissection of NADPH-cytochrome P450 reductase into distinct functional domains support the hypothesis that FMN/FAD-containing proteins have evolved as a fusion of two ancestral genes (Smith et al. 1994). This raises the possibility that other structurally related proteins such as nitric oxide synthase and sulfite reductase could have evolved by a similar procedure. Since each flavin is able to accept up to two electrons, a reductase molecule may contain between zero and four electrons. The reduction state of the reductase can be followed by monitoring changes of the absorption spectrum. The oxidized enzyme has absorption maxima at 275, 380, and 455 nm, whereas the addition of excess NADPH under anaerobic conditions caused a decrease at 455 nm, with a small increase of the absorption between 510 and 700 nm (Masters and Kamin 1965). When the reduced reductase is exposed to air, it does not become completely oxidized but forms an air-stable semiquinone and the superoxide anion (Masters and Kamin 1965; Iyanagi and Mason 1973). This reaction proceeds especially at high O_2 and low enzyme concentrations and in the absence of cytochromes P450 (Yasukochi et al. 1979). Following initial controversy, it is now generally accepted that the air-stable semiquinone contains a single electron per mole of enzyme (Iyanagi et al. 1978; Vermilion and Coon 1978; Yasukochi et al. 1979). Titration of reductase with reducing equivalents indicated that the flavins were not equivalent and that one of the flavins (FAD) acts by accepting the electrons from NADPH, whereas the other (FMN) acts as an electron carrier donating electrons to cytochromes P450 (reviewed by Backes 1993). The order of electron transfer into the reductase molecule is controlled by the reduction potentials for the electron transfer to the various flavin species. These reduction potentials have been determined by Iyanagi et al. (1974).

		E'o (V)
FAD	$FADH^{\bullet}$	-0.110
$FADH^{\bullet}$	$FADH_2$	-0.270
FMN	$FMNH^{\bullet}$	-0.290
$FMNH^{\bullet}$	$FMNH_2$	-0.365

The nine potential reductase subforms are depicted in Fig. 17. It has to be taken into account, however, that NADPH is an obligate two-electron donor and therefore cannot transfer electrons to a reductase molecule containing three electrons. A number of investigators have suggested that the reductase cycles between the one- and three-electron reduced state (Iyanagi et al. 1978, 1981; Vermilion et al. 1981). Others have described cycling between the two- and four-electron-containing forms (Oprian and Coon 1982; Backes and Reker-Backes 1988). In both models cytochrome P450 is reduced by two one-electron transfer steps, as shown by Peterson et al. (1977), with the proteins forming a 1:1 complex (Miwa et al. 1979; Miwa and Lu 1984). According to Backes (1993), it seems quite possible that under different conditions one or the other mechanism may predominate.

Whether production of the superoxide anion by reoxidation of the semiquinone is of greater toxicological importance remains open. It seems conceivable that under conditions where the amount of cytochromes P450 is lowered or the interaction between P450 and reductase is disturbed, this process could play a role.

The ability of reductase to catalyze one-electron quinone reduction, leading to the formation of biologically active semiquinones (Powis et al. 1987; Spencer and Rifkind 1990), is another source of reactive oxygen species at

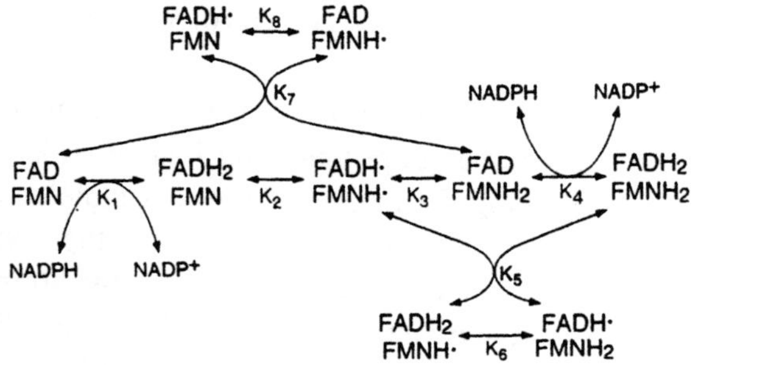

Fig. 17. Potential reduction states of NADPH-cytochrome P450 reductase. (From Backes and Reker-Backes 1988)

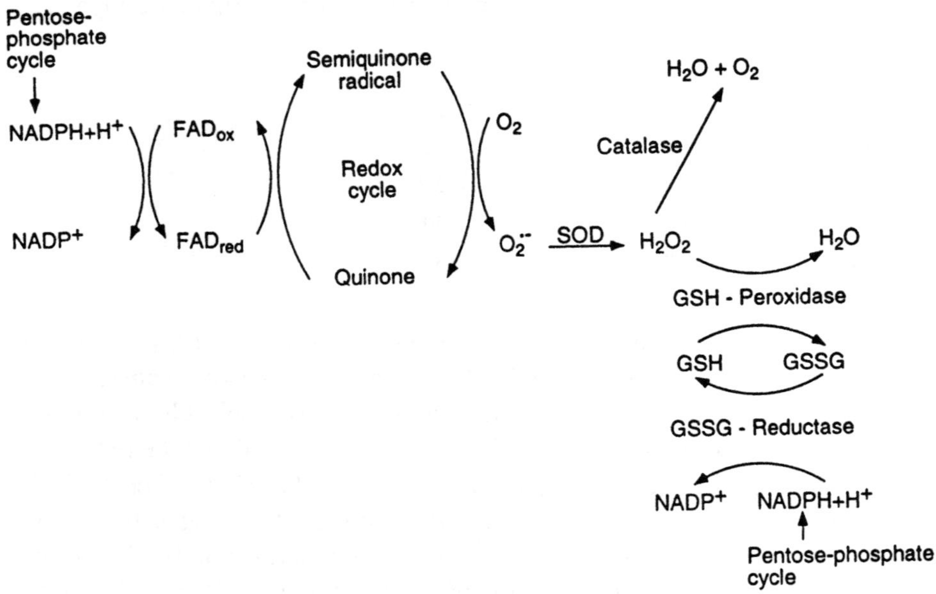

Fig. 18. Redox cycling of a quinone compound by NADPH-cytochrome P450 reductase

the level of reductase. As shown in Fig. 18, these reactive oxygen species can be formed from the respective semiquinones by redox cycling (Kappus 1985, 1986, 1993). The best-studied example of a drug undergoing redox cycling is paracetamol (acetaminophen), which is metabolized to the benzoquinoneimine. This product then undergoes redox cycling to give $O_2^{\bullet-}$ (Van de Straat 1987; Van de Straat et al. 1988). Other compounds undergoing redox cycling of quinone intermediates, and thus leading to reactive oxygen species, are t-butyl-hydroxyanisol (Kahl et al. 1989; Kahl 1991; Cummings et al. 1990), naphthalene, and benzene (Kappus 1993).

6.2.2 Generation of Reactive Oxygen Species at the Level of the Microsomal Mono-Oxygenase System

As indicated above, relatively high amounts of oxygen at low levels of reductase stimulate production of superoxide anions by microsomal reductase following formation of the air-stable semiquinone, especially in the absence of cytochrome P450. On the other hand, the relationship between the amounts of P450/reductase, the ionic strength of the medium, and the presence of cytochrome b_5 was shown to affect the amount of reactive oxygen species formed in P450 systems (Zhukov et al. 1989). These results

Table 8. CYP2B4-dependent benzphetamine demethylase activity in homologously and heterologously reconstituted systems (from Bernhardt and Gunsalus 1992)

	Redox donor µM	V_{max} min^{-1}	Product yield HCHO/NAD(P)H
Microsomal		41[a]	0.48
PdR, Pd	6.0: 42–105	42	0.37
LdR, Ld	0.7: 7– 34	52	0.30

100 m*M* potassium phosphate buffer, 10% glycerol, pH 7, 298 K; 1 m*M* benzphetamine added to initiate reaction.

[a]With dilauroyl phosphatidylcholine; data from Koop and Coon (1979).

indicate that the interaction between P450 and the electron donors (reductase or cytochrome b_5) obviously plays an important role in determining the amount of reactive oxygen species. It has been shown in this connection that heterologous reconstitution of CYP2B4 activity with bacterial electron donor systems leads to comparable efficiency of benzphetamine-N-demethylation when compared with a "native" reconstitution system, consisting of CYP2B4, microsomal reductase, and dilauroyl phosphatidylcholine. However, the relative amount of NADPH used for product formation was lower than in the case of the reconstituted microsomal system (Table 8), indicating that P450/electron donor interaction is crucial for reactive oxygen formation. This conclusion is supported by the observation that replacement of Arg112 of CYP101, which is involved in the interaction with the electron donor putidaredoxin, by lysine, cysteine, and tyrosine led to an increase in H_2O_2 production, which was 30% in the case of Arg112Tyr (Shimada et al. 1994).

Thus, the "tightness" of the electron donor coupling to P450 will influence the formation of reactive oxygen species in microsomal P450 systems at the level of the P450 and, to a minor extent, at the level of reductase. Since this interaction is affected not only by the relative amounts of P450/electron donor but also by the lipid composition, pathophysiological as well as nutritional aspects will modulate the amount of reactive oxygen species formed.

6.3 Generation of Reactive Oxygen Species in Mitochondrial P450 Systems

Mitochondrial P450 systems are involved in several key hydroxylation steps leading to physiologically active metabolites such as glucocorticoids, mineralocorticoids, sex hormones, bile acids, and hydroxyvitamins D3. As de-

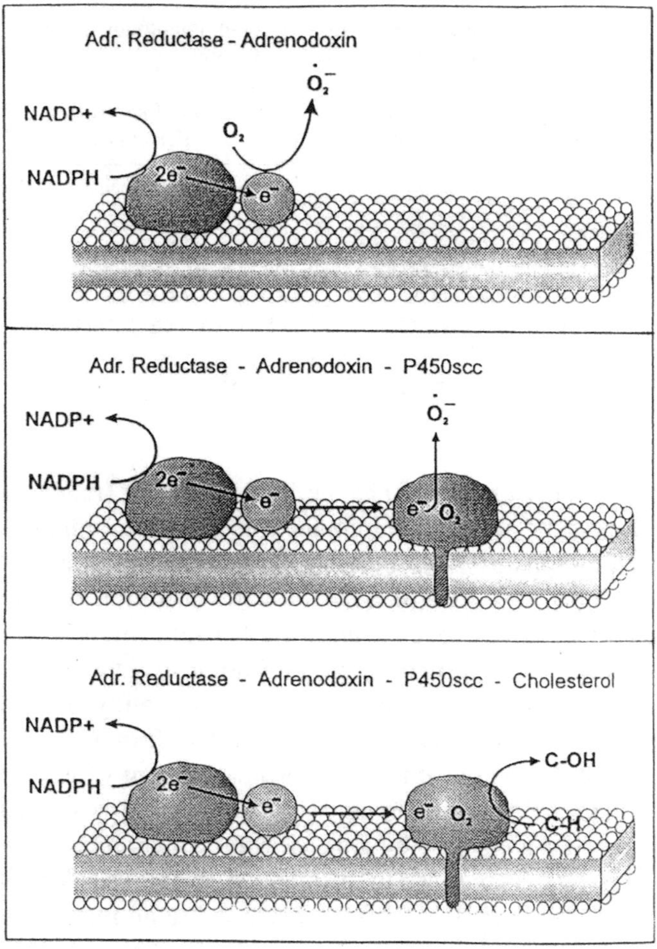

Fig. 19. Possible pathways of electron flows and reactive oxygen species generation (leakage) in mitochondrial cytochrome P450 systems. (From Hanukoglu et al. 1993)

picted in Fig. 5, mitochondrial cytochromes P450 obtain their reducing equivalents from NADPH via a FAD-containing mitochondrial reductase and a ferredoxin of the /2Fe-2S/ type. Their general protein composition thus resembles the composition of most bacterial cytochrome P450 systems, e.g., the CYP101 and CYP108 systems. Ferredoxin and the ferredoxin reductase were first discovered by Omura et al. (1966) as proteins supporting the 11β-hydroxylase cytochrome P450 (CYP11B1). Due to the property of /2Fe-2S/ proteins as one-electron shuttles, P450 is reduced also in these systems by two one-electron transfer steps. Whether the three proteins form a ternary complex or whether the ferredoxin functions as a mobile shuttle between the ferredoxin reductase and the steroid hydroxylases has been the subject of controversy (reviewed by Lambeth 1990, see also Sect. 3.2).

Reactive oxygen species can be formed in mitochondrial P450 systems, as in bacterial and microsomal systems, at the level of the P450 (see Sect. 6.1), but also at the level of the electron-supporting system. The elegant studies of Hanukoglu et al. (1993) demostrate that mitochondrial P450 systems can leak electrons, producing O_2-derived free radicals (Fig. 19). It was measured that 7.8 μM electrons/min arise when the system is reconstituted without P450, whereas only 3.5 μM electrons are produced in the presence of CYP11A1, and even < 0.5 μM electrons/min arise when cholesterol is present as substrate. When CYP11B1 instead of CYP11A1 was used, a relatively higher amount of leakage was observed in the presence of substrate (Rapoport et al. 1994). The tighter coupling during cholesterol and deoxycorticosterone conversion may be explained by the observations that the substrate enhances binding of adrenodoxin and induces a high-spin shift of the heme iron, that it increases the redox potential of P450, and that it stabilizes the oxyferro complex of P450 (Hanukoglu and Jefcoat 1980; Lambeth and Pember 1983). In addition, the substrate can occupy the active site, thus excluding water and decreasing the polarity of the microenvironment of the FeO_2 complex, and consequently inhibiting the release of $O_2^{\bullet-}$, as was shown for CYP101 (Poulos et al. 1985; Loida and Sligar 1993a,b).

The present findings raise the question of whether the mitochondrial P450 systems form reactive oxygen species under physiological conditions and what cellular mechanisms operate to control this effect. It was shown in cultured adrenocortical cells that CYP11B1 undergoes rapid inactivation in the presence of so-called pseudosubstrates, whereas no such drastic inactivation has been observed with CYP11A1. The inactivation was suggested to result from generation of reactive oxygen species (Hornsby 1986). Thus, it seems possible that the availability of substrates and pseudosubstrates could be an important factor in reactive oxygen generation in mitochondria. So far, no correlation has been made between the availability of these substrates and toxicological effects. Furthermore, it is not known whether pathophysiological conditions arise from a dysregulation of NADPH synthesis and cholesterol availability.

6.4 Self-Inactivation of Cytochrome P450 by Reactive Oxygen Species

As discussed for CYP11B1 in cultured adrenocortical cells, many enzymes may undergo self-inactivation by reactive oxygen species. Among these are superoxide dismutases, chloroperoxidase, glutathione peroxidase, NADH oxidase, and xanthine oxidase. Prostaglandin synthase and thromboxane synthase also undergo self-inactivation, but the mechanism is not yet clear (see Karuzina and Archakov 1994 and references cited therein). Inactivation caused by reactive oxygen species (Guengerich 1978; Loosemore et al. 1980; Karuzina and Archakov 1985) has also been described for microsomal cytochromes P450. Furthermore, oxidative damage of microsomal proteins can be mediated by reactive oxygen species formed by P450. Interestingly, ascorbate was shown to specifically inhibit free-metal-independent P450-mediated protein oxidation and thereby to prevent subsequent proteolytic degradation, whereas other scavengers of reactive oxygen species, including superoxide dismutase, catalase, and glutathione, were ineffective. Interaction of ascorbate with the P450 $Fe^{3+}O_2^-$ complex has been suggested as a mechanism of protection (Mukhopadyay and Chatterjee 1994). As discussed before, reactive intermediates in oxygenase reaction can be formed at the stage of the oxy or peroxy complex (Fig. 6). Since neither H_2O_2 nor $O_2^{\bullet-}$ is a strong oxidant by itself, $OH^{\cdot}$ seems to be the most likely candidate for the ultimate oxidant (Halliwell and Gutteridge 1984, 1986), although a direct damaging effect of $O_2^{\bullet-}$ has also been demonstrated (Fridovich 1986). Inactivation by reactive oxygen species has been shown for CYP2B4, whereas CYP1A2 and CYP101 were not inactivated (Karuzina and Archakov 1994). This observation implies that the possibility of self-inactivation of some P450 forms by active oxygen species should be taken into consideration in situations where an increased uptake of xenobiotics, e.g., special drugs, occurs. Diminution of the amount of P450 can then lead to a further increase in reactive oxygen production due to electron leakage (see Sect. 6.1.2), unless this process is not counterregulated by a decrease in the supply of reducing equivalents.

6.5 Participation of Cytochrome P450 in Lipid Peroxidation

Direct initiation of lipid peroxidation by cytochrome P450 using CYP2B4 has been suggested (Ekström and Ingelman-Sundberg 1986), although this finding has been subject to controversy (for review see Kappus 1993). Recent studies suggest that CYP1A1 catalyzes lipid peroxidation by a peroxide-induced reaction, whereas CYP1A2 functions mainly in NADPH-induced lipid peroxidation via generation of a reactive oxygen species (Ohmori et al. 1993). Furthermore, reactive oxygen species generated during the metabolism of cyclosporin A and inducing microsomal lipid peroxidation were shown to be a possible explanation for the toxicity of this immunosuppressive drug (Serino et al. 1993) In addition, lipid peroxidation can be induced by metabolites of P450-dependent reactions. Thus, conversion of carbon tetrachloride, leading to the CCl_3 radical, obviously starts lipid peroxidation (for review see Kappus 1993). It was recently demonstrated that rats treated with ethanol had higher CYP2E1-dependent hepatic microsomal activities and CCl_4-induced lipid peroxidation than controls. Simultaneous chlormetriazole treatment inhibited CYP2E1 expression in rat liver and abolished CCl_4-induced lipid peroxidation (Hu et al. 1994). Studies to prevent $CCl_3^{\cdot}$-induced lipid peroxidation and liver injury were performed using cimetidine as the protective agent. It was shown that the protective effect of cimetidine can be attributed to a reduction in cytochrome P450 (Mera et al. 1994). Another possibility for lipid peroxidation seems to be catalysis induced by destruction products of the heme, presumably by release of iron ions, which initiate lipid peroxidation through a Fenton reaction (see Sect. 6.1.1). The latter observation may lead to the conclusion that conditions where cytochromes P450 undergo self-inactivation may induce lipid peroxidation. Phosphorylation of P450 was demonstrated to increase lipid peroxidation by about 30%. As a mechanism, it was hypothesized that phosphorylation of P450 alters the interaction with reductase, which may enhance production of free radical species, initiating lipid peroxidation (Mkrchian and Andersson 1990). Another possible explanation, however, is the observed phosphorylation-dependent denaturation of P450, leading to a release of iron ions and allowing the Fenton reaction to proceed. However, it seems that our understanding of these processes is still very incomplete and requires further intensive investigation.

7 Cytochrome P450 and Nitric Oxide

It was recently discovered that NO synthase, catalyzing the conversion of L-arginine via N^G-hydroxy-L-arginine to citrulline and nitric oxide (NO), is a P450-type hemoprotein, possessing also a FAD- and FMN-containing reductase domain (Bredt et al. 1991; Stuehr and Ikeda-Saito 1992; McMillan et al. 1992; White and Marletta 1992). Current knowledge about the enzymatic pathway catalyzed by NO synthase and its physiological implications has been reviewed elsewhere (Moncada et al. 1991; Lowenstein and Snyder 1992; White and Marletta 1993; Masters 1994). For this reason, this section will cover only possible interactions between cytochrome P450 and NO synthase or NO.

At present, several levels of interference between NO synthase and P450 have been described, which are not yet understood in detail. First, it was shown that NO synthase may participate in cellular electron transfer processes and that a variety of electron acceptors may interfere with NO formation, due to the broad substrate specificity of the reductase domain of NO synthase (Klatt et al. 1992). Second, it was demonstrated that cytochromes P450 of the CYP3A family are able to catalyze the second step in NO synthase-dependent NO production, the oxidation of N^G-hydroxy-L-arginine to citrulline and NO (Renaud et al. 1993), thus presumably leading to a competition between the "normal" substrates of this P450 family and the NO synthase intermediates. Furthermore, NADPH-dependent microsomal metabolism of N-nitrosamines was shown to result in both oxidative dealkylation and denitrosation of the molecule. The latter reaction leads to the formation of nitric oxide, either by a P450-dependent one-electron reduction of the nitrosamine or by liberation of NO via an oxidative mechanism mediated by a P450-dependent one-electron abstraction (Appel et al. 1991). In addition, cytochrome P450 can be inhibited by NO in the same manner as by CO (Khatsenko et al. 1993). Inhibition of CYP1A1 and CYP1A2 was also demonstrated using V79 Chinese hamster cells genetically engineered for stable expression of these P450 forms and NO donors (Stadler et al. 1994). Stimulation of endogenous NO synthesis in rat hepatocytes also led to a decrease in CYP1A1-dependent benzo[a]pyrene turnover. The competitive inhibitor of NO synthesis, N^G-monomethyl-L-arginine, was able to restore CYP1A1 activity (Stadler et al. 1994). Moreover, immunostimulants, which are now known to cause NO overproduction, led to a decrease in both hepatic microsomal P450 activity and CO binding (Khatsenko et al. 1993). On the other hand, binding of NO to both ferric and ferrous NO synthase was recently demonstrated, thus giving rise to product inhibition during the catalytic process of this enzyme (Wang et al. 1994). It is obvious that such a type of interaction with P450 can take place only in tissues where both proteins are expressed, such as brain.

Finally, it was very recently found that endogenous nitric oxide production may be linked to 1,25-dihydroxyvitamin D3 synthesis in HD-11 cells in vitro, indicating that macrophage NO-generating capacity could be functionally linked to endogenous synthesis of the active vitamin D metabolite (Adams et al. 1994). The physiological importance of this finding has to be further substantiated.

Taken together, these results suggest that there is not only a similarity between the structures and the reaction mechanisms of P450 and NO synthase, but also a variety of possibilities for interference which contribute to the complexity of regulation of both enzymes and reactive oxygen production.

8 Concluding Remarks and Outlook

Cytochromes P450 are multicatalysts possessing high complexity and displaying a broad field of activity. Although more than 300 different P450 forms have already been described, new forms are constantly being discovered, opening new research fields. While in the 1960s the function of cytochromes P450 was discovered and their hemoprotein nature identified, the 1970s were characterized by the purification of different P450 forms and by the identification of the components of mono-oxygenase systems. In the 1980s the hallmarks of P450 research were the resolution of primary structures and of the first crystal structure, as well as breakthroughs in understanding the molecular biology of these systems. The past few years have brought new crystal structures of soluble cytochromes P450, deeper insights into the mechanism of gene expression of various P450 forms, and the expansion of interest to other fields of biology such as plant, insect, and fish tissues. These new directions will expand the magnitude of contributions to P450 research and open new fields of application. There are still many questions waiting to be answered during the coming years. There is no three-dimensional structure of a membrane-bound P450 available, and the mechanism of electron transfer in P450 systems is still poorly understood, as is the folding of the protein components of P450 systems. In addition, the importance of polymorphisms and microheterogeneities in this family and the regulation of gene expression need to be further investigated. Moreover, the role of cytochromes P450 in developmental regulation and in pathological processes has to be clarified.

During the past few years, the importance of cytochromes P450 in pathological processes has seemed to increase. An involvement of CYP2D6 in the pathogenesis of Parkinson's disease has been postulated, although this suggestion has been a matter of controversy (Tanner 1991; Sturman and Williams 1991; Smith et al. 1992). Furthermore, cytochromes P450 have been shown to play an essential role in some forms of hepatitis (Leeder et al. 1992; Seelig et al. 1993; Lecoeur et al. 1994). The role of reactive oxygen species produced by cytochrome P450 systems in pathological processes has hardly been considered so far. It has been demonstrated, however, that free radicals are involved in the pathophysiology of chronic pancreatitis, and a participation of P450 in this process has been proposed (Niederau et al. 1991). Moreover, it has been shown that cytochromes P450 mediate hydroxyl radical formation during reoxygenation of the kidney leading to lethal cell injury (Paller and Jacob 1994). Although it is conceivable that natural mutants of cytochromes P450 might exist which show uncoupling of the reaction cycle and an increased production of reactive oxygen species, no such mutants have been described so far. This is due to the fact that in most such studies the mutants were tested in an in vitro assay only for their enzymatic activity, but not with respect to the stoichiometry of substrates used and reaction products formed. Thus, the naturally occurring mutants of cytochromes P450 with point mutations described so far were shown or supposed to influence heme binding (White et al. 1991; Wu and Chung 1991; Geley et al. 1995), protein stability (Faletto et al. 1992), and membrane incorporation (Tusie-Luna et al. 1991). In other cases decreases in the activity have been observed, but the molecular basis of these effects is not yet understood (Higashi et al. 1991; Partanen and Campbell 1991; Imai et al. 1993; Curnow et al. 1993; Nakashima et al. 1994). In this connection it would be of great interest to investigate whether reactive oxygen species can be produced by such mutants and can be correlated to pathophysiological changes.

Acknowledgements. The author wishes to acknowledge the helpful discussions with her colleagues, especially Prof. Dr. K. Ruckpaul and Dr. C. Jung. I am grateful to Dr. H. Schrauber for performing the alignments, to P. Haink for the typing, and to K. Müller for the graphical work. The support by grants from the *Deutsche Forschungsgemeinschaft* Be 1343/1-2, Be 1343/2-4, and Be 1343/6-1 is also appreciated.

References

Abney IR, Owicki IC (1985) Theories of protein-lipid and protein-protein interactions in membranes. In: Watts A, De Pont JJHHM (eds) Progress in protein-lipid interactions, vol I. Elsevier, Amsterdam, pp 1–60

Adams JS, Ren SY, Arbelle JE, Clemens TL, Shany S (1994) A role for nitric oxide in the regulated expression of the 25-hydroxy-vitamin D-1-hydroxylation reaction in the chick myelomonocytic cell line HD-11. Endocrinology 134: 499–502

Akhrem AA, Adrianov VT, Bokut SB, Luka ZA, Kissel MA, Skornyakova TG, Kisselev PA (1982) Thermotropic behaviour of phospholipid vesicles reconstituted with rat liver microsomal cytochrome P450. Biochim Biophys Acta 692: 287–295

Al-Bayati ZAF, Stohs SJ (1987) The role of iron in 2,3,7,8-tetrachlorodibenzo-p-dioxin-induced lipid peroxidation by rat liver microsomes. Toxicol Lett 38: 115–121

Albano E, Tomasi A, Persson JO, Terelius Y, Goria-Gatti L, Ingelman-Sundberg M, Dianzani MU (1991) Role of ethanol-inducible cytochrome P450 (P450IIE1) in catalysing the free radical activation of aliphatic alcohols. Biochem Pharmacol 41: 1895–1902

Appel KE, Gorsdorf S, Scheper T, Spiegelhalder B, Wiessler M, Schoepke M, Engelhom C, Kramer R (1991) Metabolic denitrosation of N-nitrosamines: mechanism and biological consequences. IARC Sci Publ 105: 351–357

Archakov AI (1982) Stability, conformational rigidity and life-time of microsomal redox enzymes in soluble and membrane-bound state. In: Hietanen E, Laitinen M, Hänninen O (eds) Cytochrome P450. Biochemistry, biophysics and environmental implications. Elsevier/North Holland, Amsterdam, pp 487–495

Atkins WM, Sligar SG (1987) Metablic switching in cytochrome P450cam: deuterium isotope effects on regiospecificity and the monooxygenase/oxidase reaction. J Am Chem Soc 109: 3754–3760

Atkins WM, Sligar SG (1988) Deuterium isotope effects in norcamphor metabolism by cytochrome P450cam: kinetic evidence for the two-electron reduction of a high-valent iron-oxo intermediate. Biochemistry 27: 1610–1616

Atkins WM, Sligar SG (1989) Molecular recognition in cytochrome P450: alteration of regioselective alkane hydroxylation via protein engineering. J Am Chem Soc 111: 2715-2717

Atkins WM, Sligar SG (1990) Tyrosine-96 as a natural spectroscopic probe of the cytochrome $P450_{cam}$ active site. Biochemistry 29: 1271–1275

Axelrod J (1954) An enzyme for the deamination of sympathomimetic amines: properties and distribution. J Pharmacol Exp Ther 110: 2

Axelrod J (1955) The enzymatic demethylation of ephedrine. J Pharmacol Exp Ther 114: 430-438
Axelrod J (1982) The discovery of the microsomal drug-metabolizing enzymes. Trends Pharmacol Sci 3: 383–386
Backes WL (1993) NADPH-cytochrome P450 reductase: function. In: Schenkman JB, Greim H (eds) Cytochrome P450. Springer, Berlin Heidelberg New York (Handbook of experimental pharmacology, vol 105), pp 15–34
Backes WL, Eyer CS (1989) Cytochrome P450 LM2 reduction: substrate effects on the rate of reductase-LM2 association. J Biol Chem 264: 6252–6259
Backes WL, Reker-Backes CE (1988) The effect of NADPH concentration on the reduction of cytochrome P450 LM2. J Biol Chem 263: 247–253
Backes WL, Sligar SG, Schenkman JB (1980) Cytochrome P450 reduction exhibits burst kinetics. Biochem Biophys Res Commun 97: 860–867
Backes WL, Sligar SG, Schenkman JB (1982) Kinetics of hepatic cytochrome P450 reduction: correlation with spin state of the ferric heme. Biochemistry 21: 1324–1330
Backes WL, Tamburini PP, Jansson I, Gibson GG, Sligar SG, Schenkman JB (1985) Kinetics of cytochrome P450 reduction: evidence for faster reduction of the high-spin ferric state. Biochemistry 24: 5130–5136
Baldwin JE, Morris GM, Richards WG (1991) Electron transport in cytochromes P450 by covalent switching. Proc R Soc Lond [Biol] 245: 43–51
Beckert V, Dettmer R, Bernhardt R (1994a) Mutations of tyrosine 82 in bovine adrenodoxin that affect binding to cytochrome P45011A1 and P45011B1 but not electron transfer. J Biol Chem 269: 2568–2573
Beckert V, Dettmer R, Bernhardt R (1994b) Structural requirements for ferredoxin associated electron transfer evaluated by analysis of Tyr-82 and His-56 bovine adrenodoxin mutants. In: Lechner MC (ed) Cytochrome P450. 8th international conference. Libbey Eurotext, Paris, pp 869–872
Beckert V, Schrauber H, Bernhardt R, Van Dijk AA, Kakoschke C, Wray V (1995) Mutational effects on the spectroscopic properties and biological activities of oxidized bovine adrenodoxin, and their structural implications. Eur J Biochem 231: 226–235
Beratan DN, Onuchic JN, Betts JN, Bowler BE, Gray HB (1990) Electron-tunneling pathways in rutheneted proteins. J Am Chem Soc 112: 7915–7921
Beratan DN, Betts BE, Onuchic JN (1991) Protein elecron transfer rates set by the bridging secondary and tertiary structure. Science 252: 1285–1288
Beratan DN, Betts BE, Onuchic JN (1992a) Tunneling pathway and redox state-dependent electronic couplings at nearly fixed distance in electron transfer proteins. J Phys Chem 7: 2852–2855
Beratan DN, Onuchic JN, Winkler JR, Gray HB (1992b) Electron-tunneling pathways in proteins. Science 258: 1740–1741
Beresford AP (1993) CYP1A1: Friend or foe? Drug Metab Rev 25: 503–517
Bernhardt R (1993) Chemical probes of cytochrome P450 structure. In: Schenkman JB, Greim H (eds) Cytochrome P450. Springer, Berlin Heidelberg New York, pp 547–560 (Handbook of experimental pharmacology, vol 105)
Bernhardt R, Gunsalus IC (1985) Heterologous reconstitution of cytochrome P450 LM2 activity with bacterial electron transfer systems. In: Vereczky L, Magyar K (eds) Cytochrome P450, biochemistry, biophysics and induction. Akademia Kiado, Budapest, pp 159–162
Bernhardt R, Gunsalus IC (1992) Reconstitution of cytochrome P4502B4 (LM2) activity with camphor and linalool monooxygenase electron donors. Biochem Biophys Res Commun 187: 310–317
Bernhardt R, UhlmannH (1995) Role of tyrosine 75 in cytochrome P450cam and analysis of possible electron transfer pathways. Eur J Biochem (submitted)
Bernhardt R, Ngoc Dao NT, Stiel H, Schwarze W, Friedrich J, Jänig GR, Ruckpaul K (1983) Modification of cytochrome P450 with fluorescein isothiocyanate. Biochim Biophys Acta 745: 140–148

Bernhardt R, Makower A, Jänig GR, Ruckpaul K (1984) Selective chemical modification of a functionally linked lysine in cytochrome P450 LM2. Biochim Biophys Acta 785: 186–190

Bernhardt R, Pommerening K, Ruckpaul K (1987) Modification of carboxyl groups on NADPH-cytochrome P450 reductase involved in binding of cytochromes c and P450 LM2. Biochem Int 14: 823–832

Bernhardt R, Kraft R, Otto A, Ruckpaul K (1988) Electrostatic interactions between cytochrome P450 LM2 and NADPH-cytochrome P450 reductase. Biomed Biochim Acta 47: 581–592

Bernhardt R, Kraft R, Ruckpaul (1989a) A simple determination of the sideness of the NH_2-terminus in the membrane-bound cytochrome P450 LM2. Biochem Int 17: 1143–1150

Bernhardt R, Stiel H, Ruckpaul K (1989b) Distance between lysine 384 and heme of cytochrome P450 LM2 (P450IIB4) studied by fluorescence energy transfer measurements. Biochem Biophys Res Commun 163: 1282–1289

Bernhardt R, Kraft R, Alterman M, Otto A, Schrauber H, Gunsalus IC, Ruckpaul K (1992) Common mechanism of interaction between cytochrome P450 and electron donors in different monooxygenase systems. In: Archakow AI, Bachmanova GI (eds) Cytochrome P450. Biochemistry and biophysics. Proceedings of the 7th international meeting, Cytochrome P450: structure and function, biotechnological and ecological aspects. Moscow, 28 July-2 Aug 1991, INCO-TNC. Joint Stock Company, Moscow, Russia, pp 204–209

Bernhardt R, Beckert V, Uhlmann H, Sligar SG (1994a) Studies on electron transfer pathways in cytochrome P450 systems. In: Lechner MC (ed) Cytochrome P450. 8th international conference. Libbey Eurotext, Paris, pp 387–394

Bernhardt R, Kraft R, Uhlmann H, Beckert V (1994b) Investigation of protein-protein interactions in mitochondrial steroid hydroxylase systems using site-directed mutagenesis. J Protein Chem 13: 482–483

Bhat KS, Padmanaban G (1978) Cytochrome P450 synthesis in vivo and in a cell-free system from rat liver. FEBS Lett 89: 337–340

Bhat KS, Padmanaban G (1979) Studies on the biosynthesis of cytochrome P450 in rat liver – a probe with phenobarbital. Arch Biochem Biophys 198: 110–116

Bird IM, Neil A, Hanley R, Word A, Mathis JM, McCarthy JL, Mason JI, Rainey WE (1993) Human NCI-H295 adrenocortical carcinoma cells: a model for angiotensin-II-responsive aldeosterone secretion. Endocrinology 133: 1555–1561

Birnbaum LS (1993) Changes in cytochrome P450 in senescence. In: Schenkman JB, Greim H (eds) Cytochrome P450. Springer, Berlin Heidelberg New York (Handbook of experimental pharmacology, vol 105), pp 477–492

Blanck J, Ruckpaul K (1993) Lipid-protein interactions. In: Schenkman JB, Greim H (eds) Cytochrome P450. Springer, Berlin Heidelberg New York (Handbook of experimental pharmacology, vol 105), pp 581–597

Blanck J, Rein H, Sommer M, Ristau O, Smettan G, Ruckpaul K (1983) Correlations between spin equilibrium shift, reduction rate and N-demethylation activiy in liver microsomal cytochrome P450 and a series of benzphetamine analogues as substrates. Biochem Pharmacol 32: 1683–1688

Blanck J, Smettan G, Ristau O, Ingelman-Sundberg M, Ruckpaul K (1984) Mechanism of rate control of the NADPH-dependent reduction of cytochrome P450 by lipids in reconstituted phospholipid vesicles. Eur J Biochem 144: 509–513

Blanck J, Jänig G-R, Schwarz D, Ruckpaul K (1989) Role of lipid in the electron transfer between NADPH-cytochrome P450 reductase and cytochrome P450 from mammalian liver cells. Xenobiotica 19: 1231–1246

Boddupalli SS, Oster T, Estabrook RW, Peterson JA (1992) Reconstitution of the fatty acid hydroxylation function of cytochrome P450 BM-3 utilizing its individual recombinant hemo- and flavoprotein domains. J Biol Chem 267: 10375–10380

Bondy SC, Naderi S (1994) Contribution of hepatic cytochrome P450 systems to the generation of reactive oxygen species. Biochem Pharmacol 48: 155–159

Bonfils C, Debey P, Maurel P (1979) Highly purified microsomal cytochrome P450: the oxyferro intermediate stabilized at low temperature. Biochem Biophys Res Commun 88: 1301–1307

Bonfils C, Balny C, Maurel P (1981) Direct evidence for electron transfer from ferrous cytochrome to the oxy ferrous cytochrome P450 LM2. J Biol Chem 256: 9457–9465

Bösterling B, Stier A, Hildebrandt AG, Dawson JH, Trudell JR (1979) Reconstitution of cytochrome P450 and cytochrome P450 reductase into phosphatidylcholine-phosphatidylethanolamine bilayers: characterization of structure and metabolic activity. Mol Pharmacol 16: 332–342

Bredt DS, Hwang PM, Glatt CE, Lowenstein C, Reed RR, Snyder SH (1991) Cloned and expressed nitric oxide synthase structurally resembles cytochrome P450 reductase. Nature 351: 714–718

Bresnick E (1993) Induction of cytochromes P450 1 and P450 2 by xenobiotics. In: Schenkman JB, Greim H (eds) Cytochrome P450. Springer, Berlin Heidelberg New York, pp 503–524 (Handbook of experimental pharmacology, vol 105)

Brinigar WS, Chang CK, Geibel J, Traylor TG (1974) Solvent effects on reversible formation and oxidative stability of heme-oxygen complexes. J Am Chem Soc 96: 5597–5599

Briza P, Breitenbach M, Ellinger A, Segall J (1990a) Isolation of two developmentally regulated genes involved in spore wall maturation in *Saccharomyces cerevisiae*. Genes Dev 4: 1775–1789

Briza P, Ellinger A, Winkler G, Breitenbach M (1990b) Characterization of a dl-dityrosine-containing macromolecule from yeast ascospore walls. J Biol Chem 265: 15118–15123

Briza P, Eckerstorfer M, Breitenbach M (1994) The sporulation-specific enzymes encoded by the DIT1 and DIT2 genes catalyze a two-step reaction leading to a soluble ll-dityrosine-containing precursor of the yeast spore wall. Proc Natl Acad Sci USA 91: 4524–4528

Brodie BB (1956) Pathways of drug metabolism. J Pharm Pharmacol 8: 1–17

Brown RR, Miller JA, Miller EC (1954) The metabolism of methylated aminoazo dyes. IV. Dietary factors enhancing demethylation in vitro. J Biol Chem 209: 211–222

Bulun SE, Rosenthal IM, Brodie AM, Inkster SE, Zeller WP, DiGeorge AM, Frasier SD, Kilgore MW, Simpson ER (1994) Use of tissue-specific promoters in the regulation of aromatase cytochrome P450 gene expression in human testicular and ovarian sex cord tumors, as well as in normal fetal and adult gonads [corrected and republished with original paging, article originally printed in J Clin Endocrinol Metab 1993 Dec, 77: 1616–1621]. J Clin Endocrinol Metab 78: 1616–1621

Caron MG, Goldstein S, Savard K, March JM (1975) Protein kinase stimulation of a reconstituted cholesterol side chain cleavage enzyme system in the bovine corpus luteum. J Biol Chem 250: 5137–5143

Chashchin VL, Turko IV, Akhrem AA, Usanov SA (1985) Cross-linking studies of adreno-cortical cytochrome P450scc. Evidence for a covalent complex with adrenodoxin and localization of the adrenodoxin-binding domain. Biochim Biophys Acta 828: 313–324

Chen S, Zhou D (1992) Functional domains of aromatase cytochrome P450 inferred from comparative analyses of amino acid sequences and substantiated by site-directed mutagenesis experiments. J Biol Chem 267: 22587–22594

Chiang Y-L, Coon MJ (1979) Comparative study of two highly purified forms of liver microsomal cytochrome P450: circular dichroism and other properties. Arch Biochem Biophys 195: 178–187

Coghlan VM, Vickery LE (1991) Site-specific mutations in human ferredoxin that affect binding to ferridoxin reductase and cytochrome P450scc. J Biol Chem 266: 18606–18612

Coghlan VM, Vickery LE (1992) Electrostatic interactions stabilizing ferredoxin electron transfer complexes. Disruption by "conservative" mutations. J Biol Chem 267: 8932–8935

Conney AH (1967) Pharmacological implications of microsomal enzyme induction. Pharmacol Rev 19: 317–366

Conney AH, Miller EC, Miller JA (1956) The metabolism of methylated aminoazo dyes. V. Evidence for induction of enzyme synthesis in the rat by 3-methylcholanthrene.Cancer Res 16: 450–459

Conney AH, Davison C, Gastel R, Burns JJ (1960) Adaptive increases in drug-metabolizing enzymes induced by phenobarbital and other drugs. J Pharmacol Exp Ther 130: 1–8

Cooper DY, Levine S, Narasimhulu S, Rosenthal O, Estabrook RW (1965) Photochemical action spectrum of the terminal oxidase of mixed function oxidase systems. Science 147: 400–402

Cooper JR, Brodie BB (1954) Enzyme systems involved in the biotransformation of barbiturates. J Pharmacol Exp Ther 110: 12

Cummings SW, Curtis BB, Peterson JA, Prough RA (1990) The effect of the tert-butylquinone metabolite of butylated hydroxyanisole on cytochrome P450 monooxygenase activity. Xenobiotica 20: 915–924

Cupp JR, Vickery LE (1989) Adrenodoxin with a COOH-terminal deletion (des 116–128) exhibits enhanced activity. J Biol Chem 264: 1602–1607

Cupp-Vickery JR, Poulos TL (1995) Structure of cytochrome P450eryF involved in erythromycin biosynthesis. Struct Biol 2: 144–153

Curnow KM, Tusie-Luna M-T, Pascoe L, Natarajan R, Gu J-L, Nadler JL, White PC (1991) The product of the CYP11B2 gene is required for aldosterone biosynthesis in the human adrenal cortex. Mol Endocrinol 5: 1513–1522

Curnow KM, Slutsker L, Vitek J, Cole T, Speiser PW, New MI, White PC, Pascoe L (1993) Mutations in the CYP11B1 gene causing congenital adrenal hyperplasia and hypertension cluster in exons 6, 7, and 8. Proc Natl Acad Sci USA 90: 4552–4556

Dai Y, Rashba-Step J, Cederbaum AI (1993) Stable expression of human cytochrome P4502E1 in HepG2 cells: characterization of catalytic activities and production of reactive oxygen intermediates. Biochemistry 32: 6928–6937

Daum G (1985) Lipids of mitochondria. Biochim Biophys Acta 822: 1–42

Davies MD, Sligar SG (1992) Genetic variants in the putidaredoxin-cytochrome P450cam eletron-transfer complex: identification of the residue responsible for redox-state-dependent conformers. Biochemistry 31: 11383–11389

Davies MD, Qin L, Beck JL, Suslick KS, Koga H, Horiuchi T, Sligar SG (1990) Putidaredoxin reduction of cytochrome P450cam: dependence of electron transfer on the identity of putidaredoxin's C-terminal amino acid. J Am Chem Soc 112: 7396–7398

Defaye G, Monnier N, Guidicelli C, Chambaz EM (1982) Phosphorylation of purified mitochondrial cytochromes P450 (cholesterol desmolase and 11β-hydroxylase) from bovine adrenal cortex. Mol Cell Endocrinol 27: 157–168

DeLemos-Chiarandini C, Frey AB, Sabatini DD, Kreibich G (1987) Determination of membrane topology of the phenobarbital-inducible rat liver cytochrome P450 isoenzyme PB-4 using site-specific antibodies. J Cell Biol 104: 209–219

Depierre JW, Ernster L (1977) Enzyme topology of intracellular membranes. Annu Rev Biochem 46: 201–262

Dignam JD, Strobel HW (1977) NADPH-cytochrome P450 reductase from rat liver: purification by affinity chromatography and characterization. Biochemistry 16: 1116–1123

DuBois RN, Waterman MR (1979) Effect of phenobarbital administration to rats on the level of the in vitro synthesis of cytochrome P450 directed by total rat liver RNA. Biochem Biophys Res Commun 90: 150–157

Egawa T, Shimada H, Ishimura Y (1994) Evidence for compound-I formation in the reaction of cytochrome P450cam with m-chloroperbenzoic acid. Biochem Biophys Res Commun 201: 1464–1469

Ekström G, Cronholm T, Ingelman-Sundberg M (1986) Hydroxyl-radical production and ethanol oxidation by liver microsomes isolated from ethanol-treated rats. Biochem J 233: 755–761

Ekström G, Ingelman-Sundberg M (1986) Mechanism of lipid peroxidation dependent upon cytochrome P450 LM_2. Eur J Biochem 158: 195–201

Eliasson E, Johansson I, Ingelman-Sundberg M (1990) Substrate-, hormone-, and cAMP-regulated cytochrome P450 degradation. Proc Natl Acad Sci USA 87: 3225–3229

Erdmann B, Gerst H, Lenz D, Bähr V, Bernhardt R (1995) Zone-specific localization of cytochrome P45011B1 in the human adrenal cortex by PCR-derived riboprobes. Histochemistry (in press)

Ernster L, Orrenius S (1965) Substrate-induced synthesis of the hydroxylating enzyme system of liver microsomes. Fed Proc 24: 1190–1199

Estabrook RW, Cooper DY, Rosenthal O (1963) The light-reversible carbon monoxide inhibition of the steroid C21-hydroxylase system of the adrenal cortex. Biochem Z 338: 741–755

Estabrook RW, Hildebrandt AG, Baron J, Netter KJ, Leibman KC (1971) New spectral intermediate associated with cytochrome P450 function in liver microsomes. Biochem Biophys Res Commun 42: 132–139

Estabrook RW, Kawano S, Weringloer J, Kuthan H, Tsuji H, Graf H, Ullrich V (1979) Oxycytochrome P450: its breakdown to superoxide for the formation of hydrogen peroxide. Acta Biol Med Germ 38: 423–434

Everest AM, Wallin SA, Stemp EDA, Nocek JM, Mauk AG, Hoffman BM (1991) Aromatic hole superexchange through position 82 of cytochrome c is not required for intracomplex electron transfer to zinc cytochrome c peroxidase. J. Am. Chem. Soc. 113: 4337–4338

Faletto MB, Linko P, Goldstein JA (1992) A single amino acid mutation (Ser180—Cys) determines the polymorphism in cytochrome P450 g (P4502C13) by altering protein stability. J Biol Chem 267: 2032–2037

Farver O, Skov LK, Pascher T, Karlsson BG, Nordling M, Lundberg LG, Vänngard T, Pecht I (1993) Intramolecular electron transfer in single-site-mutated azurins. Biochemistry 32: 7317–7322

Feng DF, Doolittle RF (1987) Progressive sequence alignment as a prerequisite to correct phylogenetic trees. J Mol Evol 25: 351–360

Fisher MT, Sligar SG (1985) Control of heme protein redox potential and reduction rate: linear free energy relation between potential and ferric spin state equilibrium. J Am Chem Soc 107: 5018–5019

Fisher MT, Sligar SG (1987) Temperature jump relaxation kinetics of the $P450_{cam}$ spin equilibrium. Biochemistry 26: 4797–4803

French JS, Guengerich FP, Coon MJ (1980) Interactions of cytochrome P450, NADPH-cytochrome P450 reductase, phospholipid, and substrate in the reconstituted liver microsomal enzyme system. J Biol Chem 255: 4112–4119

Fridovich I (1986) Biological effects of the superoxide radical. Arch Biochem Biophys 247: 1–11

Fujii-Kuriyama Y, Mizukami Y, Kawajiri K, Sogawa K, Muramatsu M (1982) Primary structure of a cytochrome P450: coding sequence of phenobarbital-inducible cytochrome P450 cDNA from rat liver. Proc Natl Acad Sci USA 79: 2793–2797

Fukuda T, Imai Y, Komori M, Nakamura M, Kusunose E, Satouchi K, Kusunose M (1993) Replacement of Thr-303 of P450 2E1 with serine modifies the regioselectivity of its fatty acid hydroxylase activity. J Biochem 113: 7–12

Furuya H, Shimizu T, Hatano M, Fujii-Kuriyama Y (1989a) Mutations at the distal and proximal sites of cytochrome $P450_d$ changed regio-specificity of acetanilide hydroxylations. Biochem Biophys Res Commun 160: 669–676

Furuya H, Shimizu T, Hirano K, Hatano M, Fujii-Kuriyama Y (1989b) Site-directed mutagenesis of rat liver cytochrome $P450_d$: catalytic activities toward benzphetamine and 7-ethoxycoumarin. Biochemistry 28: 6848–6857

Garfinkel (1958) Studies on pig liver microsomes. I. Enzymic and pigment composition of different microsomal fractions. Arch Biochem Biophys 77: 493–509

Geley S, Jöhrer K, Peter M, Denner K, Bernhardt R, Sippell WG, Kofler R (1995) Amino acid substitution R384P in aldosterone synthase causes CMO-I deficiency. J Clin Endocrinol Metab 80: 424–429

Gerber NC, Sligar SG (1992) Catalytic mechanism of cytochrome P450: evidence for a distal charge relay. J Am Chem Soc 114: 8742–8743

Gerber NC, Sligar SG (1994) A role for Asp-251 in cytochrome P450cam oxygen activation. J Biol Chem 269: 4260–4266
Geren LM, O'Brien P, Stonehuerner J, Millett F (1984) Identification of specific carboxylate groups on adrenodoxin that are involved in the interaction with adrenodoxin reductase. J Biol Chem 259: 2155–2160
Gillette JR, Brodie BB, LaDu BN (1957) The oxidation of drugs by liver microsomes: on the role of TPNH and oxygen. J Pharmacol Exp Ther 119: 532–540
Gonzalez FJ (1990) Molecular genetics of the P450 superfamily. Pharmacol Ther 45: 1–38
Gonzalez FJ, Gelboin HV (1994) Role of human cytochromes P450 in the metabolic activation of chemical carcinogens and toxins. Drug Metab Rev 26: 165-183
Gonzalez FJ, Kimura S, Son BJ, Pastewka J, Gelboin HV, Hardwick JP (1986) Sequence of two related P450 mRNAs transcriptionally increased during rat development. J Biol Chem 261: 10667–10672
Gorsky LD, Koop DR, Coon MJ (1984) On the stoichiometry of the oxidase and monooxygenase reactions catalyzed by liver microsomal cytochrome P450. J Biol Chem 259: 6812–6817
Gotoh O, Fujii-Kuriyama Y (1989) Evolution, structure, and gene regulation of cytochrome P450. In: Ruckpaul K, Rein H (eds) Frontiers in biotransformation, vol 1. Akademie, Berlin, pp 195–243
Gotoh O, Tagashira Y, Iizuka T, Fujii-Kuriyama Y (1983) Structural characterization of cytochrome P450. Possible location of the heme binding cysteine in determined amino-acid sequence. J Biochem (Tokyo) 93: 807–817
Groves JT, McClusky GA (1976) Aliphatic hydroxylation via oxygen rebound. Oxygen transfer catalyzed by iron. J Am Chem Soc 98: 859–861
Guengerich FP (1978) Destruction of heme and hemoproteins mediated by liver microsomal reduced nicotinamide adenine dinucleotide phosphate-cytochrome P450 reductase. Biochemistry 17: 3633–3639
Guengerich FP, Gillam EMJ, Ohmori S, Sandhu P, Brain WR, Sari M-A, Iwasaki M (1993) Expression of human cytochrome P450 enzymes in yeast and bacteria and relevance to studies on catalytic specificity. Technology 82: 21–37
Gunsalus IC, Wagner GC (1978) Bacterial P450cam methylene monooxygenase components: cytochrome m, putidaredoxin, and putidaredoxin reductase. Methods Enzymol 52: 166–188
Gustafsson JA, Ingelman-Sundberg M (1975) Regulation and substrate specificity of a steroid sulfate-specific hydroxylase system in female rat liver microsomes. J Biol Chem 250: 3451-3458
Gustafsson JA, Mode A, Norstedt G, Skett P (1983) Sex-steroid induced changes in hepatic enzymes. Annu Rev Physiol 45: 51–60
Halliwell B, Gutteridge JMC (1984) Oxygen toxicity, oxygen radicals, transition metals and disease. Biochem J 219: 1–14
Halliwell B, Gutteridge JMC (1986) Oxygen free radicals and iron in relation to biology and medicine. Some problems and concepts. Arch Biochem Biophys 246: 501–514
Hamamoto I, Ichikawa Y (1984) Modification of a lysine residue of adrenodoxin reductase, essential for complex formation with adrenodoxin. Biochim Biophys Acta 786: 32–41
Haniu M, Armes LG, Yasunobu KT, Shastry BA, Gunsalus IC (1982) Amino acid sequence of the *Pseudomonas putida* cytochrome P450. J Biol Chem 257: 12664–12671
Hanukoglu I, Jefcoate CR (1980) Mitochondrial cytochrome P450scc. Mechanism of electron transport by adrenodoxin. J Biol Chem 255: 3057–3061
Hanukoglu I, Rapoport R, Weiner L, Sklan D (1993) Electron leakage from the mitochondrial NADPH-adrenodoxin reductase-adrenodoxin-P450scc (cholesterol side chain cleavage) system. Arch Biochem Biophys 305: 489–498
Hara T, Miyata T (1990) Structure-activity relationship of mitochondrial steroid hydroxylase covalent complexes. In: Ingelman-Sundberg M, Gustafsson J-A, Orrenius S (eds) Abstracts of the 8th international symposium on microsomes and drug oxidations, Stockholm, 25–29 July 1990, Karolinska Institute, p 129

Hara T, Takeshima M (1994) Conclusive evidence of a quaternary cluster model for cholesterol side chain cleavage reaction catalyzed by cytochrome P450scc. In: Lechner MC (ed) Cytochrome P450. 8th international conference. Libbey Eurotext, Paris, pp 417–420
Harikrishna JA, Black SM, Szklarz GD, Miller WL (1993) Construction and function of fusion enzymes of the human cytochrome P450scc system. DNA Cell Biol 12: 371–379
Hasemann CA, Ravichandran KG, Peterson JA, Deisenhofer J (1994) Crystal structure and refinement of cytochrome P450terp at 2.3 Å resolution. J Mol Biol 236: 1169–1185
Hatano O, Takayama K, Imai T, Waterman MR, Takakusu A, Omura T, Morohashi K (1994) Sex-dependent expression of a transcription factor, Ad4BP, regulating steroidogenic P450 genes in the gonads during prenatal and postnatal rat development. Development 120: 2787–2797
Hayaishi D, Nozaki M (1969) Nature and mechanisms of oxygenases. Science 164: 389–396
He S, Modi S, Bendall DS, Gray JC (1991) The surface-exposed-tyrosine residue Tyr83 of pea plastocyanin is involved in both binding and electron transfer reactions with cytochrome f. EMBO J 10: 4011–4016
Hecker M, Ullrich H (1989) On the mechanism of prostacyclin and thromboxane A2 biosynthesis. J Biol Chem 264: 141–150
Higashi Y, Hiromasa T, Tanae A, Miki T, Nakura J, Kondo T, Ohura T, Ogawa E, Nakayama K, Fujii-Kuriyama Y (1991) Effects of individual mutations in the P450(C21) pseudogene on the P450(C21) activity and their distribution in the patient genomes of congenital steroid 21-hydroxylase deficiency. J Biochem (Tokyo) 109: 638–644
Hildebrandt AG, Roots I (1975) Reduced nicotinamide adenine dinucleoitde phosphate (NADPH)-dependent formation and breakdown of hydrogen peroxide during mixed-function oxidation reactions in liver microsomes. Arch Biochem Biophys 171: 385–397
Hintz MJ, Peterson JA (1981) The kinetics of reduction of cytochrome P-450cam by reduced putidaredoxin. J Biol Chem 256: 6721–6728
Hintz MJ, Mock DM, Peterson LL, Tuttle K, Peterson JA (1982) Equilibrium and kinetic studies on the interaction of cytochrome P-450cam and putidaredoxin. J Biol Chem 257: 14324–14332
Hiroya K, Ishigooka M, Shimizu T, Hatano M (1992) Role of Glu318 and Thr319 in the catalytic function of cytochrome $P450_d$ (P4501A2): effects of mutations on the methanol hydroxylation. FASEB J 6: 749–751
Hlavica P, Kellermann J, Golly I, Lehnerer M (1994) Chemical modification of Tyr34 and Tyr129 in rabbit liver microsomal cytochrome b_5 affects interaction with cytochrome P450 2B4. Eur J Biochem 224: 1039–1046
Ho MM, Vinson GP (1993) 11β-Hydroxylase gene expression in the rat adrenal cortex. J Endocrinol 139: 301–306
Hornsby PJ (1986) Cytochrome P450/pseudosubstrate interactions and the role of antioxidants in the adrenal cortex. Endocr Res 12: 469–494
Hrycay EG, O'Brien PJ (1972) Cytochrome P450 as a microsomal peroxidase in steroid hydroperoxide reduction. Arch Biochem Biophys 153: 480–494
Hu Y, Mishin V, Johansson I, Von-Bahr C, Cross A, Ronis MJ, Badger TM, Ingelman-Sundberg M (1994) Chlormethiazole as an efficient inhibitor of cytochrome P450 2E1 expression in rat liver. J Pharmacol Exp Ther 269: 1286–1291
Igarashi Y, Kimura T (1986) Importance of the unsaturated fatty acyl group of phospholipids in their stimulatory role on rat adrenal mitochondrial steroidogenesis. Biochemistry 25: 6461–6466
Ikeda Y, Shen WH, Ingraham HA, Parker KL (1994) Developmental expression of mouse steroidogenic factor-1, an essential regulator of the steroid hydroxylases. Mol Endocrinol 8: 654–662
Imai M, Shimada H, WatanabeY, Matsushima-Hibiga Y, Makino R, Koga H, Horiuchi T, Ishimura Y (1989) Uncoupling of cytochrome P450cam monooxygenase reaction by a single mutation, threonine-252 to alanine or valine. A possible role of the hydroxy amino acid in oxygen activation. Proc Natl Acad Sci USA 86: 7823–7827

Imai Y, Nakamura M (1988) The importance of threonine-301 from cytochromes P450 (laurate (ω-1)-hydroxylase and testosterone 16-hydroxylase) in substrate binding as demonstrated by site-directed mutagenesis. FEBS Lett 234: 313–315

Imai Y, Nakamura M (1989) Point mutations at threonine-301 modify substrate specificity of rabbit liver microsomal cytochromes P450 (laurate (ω-1)-hydroxylase and testosterone 16-hydroxylase). Biochem Biophys Res Commun 158: 717–722

Imai T, Globerman H, Gertner JM, Kagawa N, Waterman MR (1993) Expression and purification of functional human 17 alpha-hydroxylase/17,20-lyase (P450c17) in *Escherichia coli*. Use of this system for study of a novel form of combined 17 alpha-hydroxylase/17,20-lyase deficiency. J Biol Chem 268: 19681–19689

Ingelman-Sundberg M, Glaumann H (1977) Reconstitution of the liver microsomal hydroxylase system into liposomes. FEBS Lett 78: 72–76

Ingelman-Sundberg M, Blanck J, Smettan G, Ruckpaul K (1983) Reduction of cytochrome P450 LM2 by NADPH in reconstituted phospholipid vesicles is dependent on membrane charge. Eur J Biochem 134: 157–162

Inglis SC, Guillemette JG, Johnson JA, Smith M (1991) Analysis of the invariant Phe82 residue of yeast iso-1-cytochrome *c* by site-directed mutagenesis using a phagemid yeast shuttle vector. Protein Eng 4: 569–574

Ishigooka M, Shimizu T, Hiroya K, Hatano M (1992) Role of Glu318 at the putative distal site in the catalytic function of cytochrome $P450_d$. Biochemistry 31: 1528–1531

Ishimura Y, Ullrich V, Peterson JA (1971) Oxygenated cytochrome P450 and its possible role in enzymic hydroxylation. Biochem Biophys Res Commun 42: 140–146

Iyanagi T, Mason HS (1973) Some properties of hepatic reduced nicotinamide adenine dinucleotide phosphate-cytochrome c reductase. Biochemistry 12: 2297–2308

Iyanagi T, Makino N, Mason HS (1974) Redox properties of the reduced nicotinamide adenine dinucleotide phosphate-cytochrome P450 and reduced nicotinamide adenine dinucleotide-cytochrome b_5 reductase. Biochemistry 13: 1701–1710

Iyanagi T, Anan FK, Imai Y, Mason HS (1978) Studies on the microsomal mixed function oxidase system: redox properties of detergent-solubilized NADPH-cytochrome P450 reductase. Biochemistry 17: 2224–2230

Iyanagi T, Makino R, Anan FK (1981) Studies on the microsomal mixed-function oxidase system: mechanism of actionof hepatic NADPH-cytochrome P450 reductase. Biochemistry 20: 1722–1730

Jänig GR, Kraft R, Blanck J, Ristau O, Rabe H, Ruckpaul K (1987) Chemical modification of cytochrome P450 LM4. Identification of functionally linked tyrosine residues. Biochim Biophys Acta 916: 512–523

Jansson I (1993) Posttranslational modification of cytochrome P450. In: Schenkman JB, Greim H (eds) Cytochrome P450. Springer, Berlin Heidelberg New York (Handbook of experimental pharmacology, vol 105), pp 561–580

Jansson I, Tamburini PP, Favreau LV, Schenkman JB (1985) The interaction of cytochrome b_5 with four cytochrome P450 enzymes from the untreated rat. Drug Metab Dispos 13: 453–458

Jick H, Shuster L (1966) The turnover of microsomal reduced nicotinamide adenine dinucleotide phosphate-cytochrome c reductase in the livers of mice treated with phenobarbital. J Biol Chem 241: 5366–5369

Jung C, Hui Bon Hoa G, Schröder K-L, Simon M, Doucet JP (1992) Substrate analogue induced changes of the CO-stretching mode in the cytochrome P450cam-carbon monoxide complex. Biochemistry 31: 12855–12862

Junqueira VBC, Simizu K, Videla LA, Barros SB (1986) Dose-dependent study of the effects of acute lindane administration on rat liver superoxide anion production, antioxidant enzyme activities and lipid peroxidation. Toxicology 41: 193–204

Kahl R (1991) Protective and adverse biological actions of phenolic antioxidants. In: Sies H (ed) Oxidative stress: oxidants and antioxidants. Academic, London, pp 245–273

Kahl R, Weinke S, Kappus H (1989) Production of reactive oxygen species due to metabolic activation of butylated hydroxyanisole. Toxicology 59: 179–94

Kappus H (1985) Lipid peroxidation: mechanisms, analysis, enzymology and biological relevance. In: Sies H (ed) Oxidative stress. Academic, London, pp 273–310

Kappus H (1986) Overview of enzymes involved in bioreduction of drugs and in redox cycling. Biochem Pharmacol 35: 1–6

Kappus H (1993) Metabolic reactions: role of cytochrome P450 in the formation of reactive oxygen species. In: Schenkman JB, Greim H (eds) Cytochrome P450. Springer, Berlin Heidelberg New York (Handbook of experimental pharmacology, vol 105), pp 145–154

Karuzina I, Archakov AA (1985) Inactivation of cytochrome P450 in hydroxylase reactions. Biokhimiya 50: 1805–1810

Karuzina I, Archakov AA (1994) The oxidative inactivation of cytochrome P450 in monooxy-genase reactions. Free Radic Biol Med 16: 73–97

Katagiri M, Ganguli BN, Gunsalus IC (1968) A soluble cytochrome P450 functional in methylene hydroxylation. J Biol Chem 243: 3543–3546

Kato R (1974) Sex-related differences in drug metabolism. Drug Metab Rev 3: 1–32

Kato R, Kamataki T (1982) Cytochrome P450 as a determinant of sex difference of drug metabolism in the rat. Xenobiotica 12: 787–800

Kato R, Yamazoe Y (1993) Hormonal regulation of cytochrome P450 in rat liver. In: Schenkman JB, Greim H (eds) Cytochrome P450. Springer, Berlin Heidelberg New York (Handbook of experimental pharmacology, vol 105), pp 447–459

Kawamoto T, Mitsuuchi Y, Ohnishi T, Ichikawa Y, Yokoyama Y, Sumimoto H, Toda K, Miyahara K, Kuribayashi I, Nakao K, Hosofa K, Yamamoto Y, Imura H, Shizuta Y (1990) Cloning and expression of a cDNA for human cytochrome P450aldo as related to primary aldosteronism. Biochem Biophys Res Commun 173: 309–316

Keeney DS (1995) Perspectives in steroid hydroxylase gene expression: novel sites of expression during embryonic development. Endocrine Res 21: 103–107

Khatsenko OG, Gross SS, Rifkind AB, Vane JR (1993) Nitric oxide is a mediator of the decrease in cytochrome P450-dependent metabolism caused by immunostimulants. Proc Natl Acad Sci USA 90: 11147–11151

Kido T, Kimura T (1979) The formation of binary and ternary complexes of cytochrome P450scc with adrenodoxin and adrenodoxin reductase-adrenodoxin complex. J Biol Chem 254: 11806–11815

Klatt P, Heinzel B, John M, Kastner M, Bohme E, Mayer B (1992) Ca^{2+}/calmodulin-dependent cytochrome c reductase activity of brain nitric oxide synthase. J Biol Chem 267: 11374–11378

Klingenberg M (1958) Pigments of rat liver microsomes. Arch Biochem Biophys 75: 376–386

Kobayashi K, Iwamoto T, Honda K (1994) Spectral intermediate in the reaction of ferrous cytochrome $P450_{cam}$ with superoxide anion. Biochem Biophys Res Commun 201: 1348–1355

Koga H, Sagara Y, Yaoi T, Tsujimura M, Nakamura K, Sekimizu K, Makino R, Shimada H, Ishimura Y, Yura K, Co M, Ikeguchi M, Horiuchi T (1993) Essential role of the ARG 112 residue of cytochrome P450cam for electron transfer from reduced putidaredoxin. FEBS Lett 331: 109–113

Koop DR, Coon MJ (1979) Purification and properties of P-4503b, a constitutive form of cytochrome P-450, from rabbit liver microsomes. Biochem Biophys Res Commun 91: 1075–1081

Krebs EG (1986) The enzymology of control by phosphorylation. In: Boyer PD, Krebs EG (eds) The enzymes, 3rd edn, vol XVII/A. Academic, New York, pp 3–20

Kunz BC, Rehorek M, Hauser H, Winterhalter KH, Richter C (1985) Decreased lipid order induced by microsomal cytochrome P450 and NADPH-cytochrome P450 reductase in model membranes: fluorescence and electron spin resonance studies. Biochemistry 24: 2889- 2895

Kunz BC, Vergeres G, Winterhalter K, Richter C (1991) Chemical modification of rat liver microsomal cytochrome P450: study of enzymic properties and membrane topology. Biochim Biophys Acta 1063: 226–234

Kuthan H, Ullrich V (1982) Oxidase and oxygenase function of the microsomal cytochrome P450 monooxygenase system. Eur J Biochem 126: 583–588

Kuthan H, Tsuji H, Graf H, Ullrich V, Werringloer J, Estabrook RW (1978) Generation of superoxide anion as a source of hydrogen peroxide in a reconstituted monooxygenase system. FEBS Lett 91: 343–345

Kuwahara S, Omura T (1980) Different requirement for cytochrome b_5 in NADPH-microsomal cytochrome P450. Biochem Biophys Res Commun 96: 1562–1568

La Du BN, Trousof N, Brodie BB (1953) Enzymatic dealkylation of aminopyrine and other alkylamines in vitro. Fed Proc 12: 339

La Du BN, Gaudette L, Trousof N, Brodie BB (1955) Enzymatic dealkylation of aminopyrine (Pyramidon) and other alkylamines.J Biol Chem 214: 741–752

Lacassagne A, Buu-Hoi NP, Rudali G (1945) Inhibition of the carcinogenic action produced by a weakly carcinogenic hydrocarbon on a highly active carcinogeneic hydrocarbon. Br J Exp Pathol 26: 5–12

Lambeth JD (1981) Cytochrome P450scc – cardiolipin as an effector of activity of a mitochondrial cytochrome P450. J Biol Chem 256: 4757–4762

Lambeth JD (1990) Enzymology of mitochondrial side-chain cleavage by cytochrome P450scc. In: Ruckpaul K, Rein H (eds) Frontiers in biotransformation, vol 3. Akademie, Berlin, pp 58–100

Lambeth JD, Geren LM, Millet F (1984) Adrenodoxin interaction with adrenodoxin reductase and cytochrome P450scc. Cross-linking of protein complexes and effects of adrenodoxin modification by EDC. J Biol Chem 259: 10025–10029

Lambeth JD, Kriengsiri S (1985) Cytochrome P450scc-adrenodoxin interactions: ionic effects on binding and regulation of cytochrome P450 reduction of bound steroid substrates. J Biol Chem 260: 8810–8816

Lambeth JD, Pember SO (1983) Cytochrome P450scc-adrenodoxin complex. Reduction properties of the substrate-associated cytochrome and relation of the reduction states of heme and iron-sulfur centers to association of the proteins. J Biol Chem 258: 5596–5602

Lambeth JD, Seybert DW, Kamin H (1979) Ionic effects on adrenal steroidogenic electron transport. The role of adrenodoxin as an electron shuttle. J Biol Chem 254: 7255–7264

Lambeth JD, Seybert DW, Kamin H (1980a) Phospholipid vesicle reconstituted cytochrome P450scc – mutually facilitated binding of cholesterol and adrenodoxin. J Biol Chem 255: 138–143

Lambeth JD, Kamin H, Seybert DW (1980b) Phosphatidylcholine vesicle reconstituted cytochrome P450scc. Role of the membrane in control of activity and spin state of the cytochrome. J Biol Chem 255: 8282–8288

Landers JP, Bunce NJ (1991) The Ah receptor and the mechanism of dioxin toxicity. Biochem J 276: 273–287

Larroque C, Van Lier JE (1980) The subzero temperature stabilized oxyferro complex of purified cytochrome P450scc. FEBS Lett 115: 175–177

Lecoeur S, Bonierbale E, Challine D, Gautier JC, Valadon P, Dansette PM, Catinot R, Ballet F, Mansuy D, Beaune PH (1994) Specificity of in vitro covalent binding of tienilic acid metabolites to human liver microsomes in relationship to the type of hepatotoxicity: comparison with two directly hepatotoxic drugs. Chem Res Toxicol 7: 434–442

Leeder JS, Riley RJ, Cook VA, Spielberg SP (1992) Human anti-cytochrome P450 antibodies in aromatic anticonvulsant-induced hypersensitivity reactions. J Pharmacol Exp Ther 263: 360–367

Lehnerer M, Schulze J, Petzold A, Bernhardt R, Hlavica P (1995) Rabbit liver cytochrome P450 2B5: high-level expression of the full-length protein in *Escherichia coli*, purification, and reconstitution of catalytic activity. Biochim Biophys Acta (submitted)

Lewis DFV, Tamburini PP, Gibson GG (1986) The interaction of a homologous series of hydrocarbons with hepatic cytochrome P450. Molecular orbital derived electronic and structural parameters influencing the haemoprotein spin state. Chem Biol Interact 58: 289–300

Liang N, Mauk AG, Pielak GJ, Johnson JA, Smith M, Hoffman BM (1988) Regulation of interprotein electron transfer by residue 82 of yeast cytochrome c. Science 240: 311–313
Loida PJ, Sligar SG (1993a) Engineering cytochrome $P450_{cam}$ to increase the stereospecificity and coupling of aliphatic hydroxylation. Protein Eng 6: 207–212
Loida PJ, Sligar SG (1993b) Molecular recognition in cytochrome P450: mechanism for the control of uncoupling reactions. Biochemistry 32: 11530–11538
Loosemore M, Light DR, Wash CH (1980) Studies on the autoinactivation behaviour of pure reconstituted phenobarbital-induced cytochrome P450 isozyme from rat liver. J. Biol Chem 255: 9017–9020
Lowenstein CJ, Snyder SH (1992) Nitric oxide, a novel biologic messenger. Cell 70: 705–707
Lu AYH, Coon MJ (1968) Role of hemoprotein P450 in fatty acid -hydroxylation in soluble enzyme system from liver microsomes. J Biol Chem 243: 1331–1332
Lu AYH, Levin W (1972) Partial purification of cytochromes P450 and P448 from rat liver microsomes. Biochem Biophys Res Commun 46: 1334–1339
Lu AYH, Junk KW, Coon MJ (1969a) Resolution of the cytochrome P450-containing ω-hydroxylation system of the liver microsomes into three components. J Biol Chem 244: 3714–3721
Lu AYH, Strobel HW, Coon MJ (1969b)Hydroxylation of benzphetamine and other drugs by a solubilized form of cytochrome P450 from liver microsomes: lipid requirement for drug demethylation. Biochem Biophys Res Commun 36: 545–551
Luo X, Ikeda Y, Lala DS, Baity LA, Maede C, Parker KL (1995) A cell-specific nuclear receptor plays essential roles in adrenal and gonadal development. Endocrine Res 21: 517–524
Martinis SA, Atkins WM, Stayton PS, Sligar SG (1989) A conserved residue of cytochrome P450 is involved in heme-oxygen stability and activation. J Am Chem Soc 111: 9252–9253
Mason HS (1957) Mechanisms of oxygen metabolism. Science 125: 1185–1188
Mason HS, Fowlks WL, Peterson E (1955) Oxygen transfer and electron transport by the phenolase complex. J Am Chem Soc 77: 2914–2915
Masters BSS (1994) Nitric oxide syntheses: why so complex? Annu Rev Nutr 14: 131–145
Masters BSS, Kamin H (1965) Studies on the mechanism of microsomal triphosphopyridine nucleotide-cytochrome c reductase. J Biol Chem 240: 921–931
Mayuzumi H, Sambongi C, Hiroya K, Shimizu T, Tateishi T, Hatano M (1993) Effect of mutations of ionic amino acids of cytochrome P450 1A2 on catalytic activities toward 7-ethoxycoumarin and methanol. Biochemistry 32: 5622–5628
McMillian K, Bredt DS, Hirsch J, Snyder SH, Mastes BSS (1992) Cloned, expressed rat cerebellar nitric oxide synthase contains stoichiometric amounts of heme which binds carbon monoxide. Proc Natl Acad Sci USA 89: 1141–1145
Mera E, Muriel P, Castillo C, Mourelle M (1994) Cimetidine prevents and partially reverses CCl_4-induced liver cirrhosis. J Appl Toxicol 14: 87–90
Miller EC, Miller JA, Brown RR (1952) On the inhibitory action of certain polycyclic hydrocarbons on azo dye carcinogenesis. Cancer Res 12: 282–283
Miller WL (1988) Molecular biology of seroid hormone synthesis. Endocr Rev 9: 295–318
Mitani F, Horie S (1969) Studies on P450. VI. The spin state of P450 solubilized from bovine adrenocortical mitochondria. J Biochem (Tokyo) 66: 139–149
Mitani F, Suzuki H, Hata J-I, Ogishima T, Shimada H, Ishimura Y (1994) A novel cell layer without corticosteroid-synthesizing enzymes in rat adrenal cortex: histochemical detection and possible physiological role. Endocrinology 135: 431–438
Miwa GT, Lu AYH (1984) The association of cytochrome P450 and NADPH-cytochrome P450 reductase in phospholipid membranes. Arch Biochem Biophys 234: 161–166
Miwa GT, West SB, Huang MT, Lu AYH (1979) Studies on the association of cytochrome P450 and NADPH-cytochrome c reductase during catalysis in a reconstituted hydroxylating system. J Biol Chem 254: 5695–5700

Mkrtchian SL, Andersson KK (1990) A possible role of cAMP-dependent phosphorylation of hepatic microsomal cytochrome P450: a mechanism to increase lipid peroxidation in response to hormone. Biochem Biophys Res Commun 166: 787–793

Mkrtchian S, Eliasson E, Halpert JR, Ingelman-Sundberg M (1994) Substrate-regulated, cAMP-dependent phosphorylation of cytochrome P450 3A1. In: Lechner MC (ed) Cytochrome P450. 8th international conference. Libbey Eurotext, Paris, pp 829–832

Moncada S, Palmer RMJ, Higgs EA (1991) Nitric oxide: physiology, pathophysiology, and pharmacology. Pharmacol Exp Ther 43: 109–142

Monier S, Van Luc P, Kreibich G, Sabatini DD, Adesnik M (1988) Signals for the incorporation and orientation of cytochrome P450 in the endoplasmic reticulum membrane. J Cell Biol 107: 457–470

Morgan ET, Coon MJ (1984) Effect of cytochrome b_5 on cytochrome catalyzed reactions. Drug Metab Dispos 12: 358–364

Mornet E, Dupont J, Vitek A, White PC (1989) Characterization of two genes encoding human steroid 11β-hydroxylase (P45011β). J Biol Chem 264: 20961–20967

Morohashi K, Iida H, Nomura M, Hatano O, Honda S, Tsukiyama T, Niwa O, Hara T, Takakusu A, Shibata Y (1994) Functional difference between Ad4BP and ELP, and their distributions in steroidogenic tissues. Mol Endocrinol 8: 643–653

Moser CC, Keske JM, Warncke K, Farid RS, Dutton PL (1992) Nature of biological electron transfer. Nature 355: 796–802

Mukhopadhyay CK, Chatterjee IB (1994) NADPH-initiated cytochrome P450-mediated free metal ion-independent oxidative damage of microsomal proteins. J Biol Chem 269: 13390–13397

Müller GC, Miller JA (1953) The metabolism of methylated amino azo dyes. II. Oxidative demethylation by rat liver homogenates. J Biol Chem 202: 579–587

Murayama N, Shimada M, Yamazoe Y, Kato R (1991) Difference in the susceptibility of two phenobarbital-inducible forms, P450IIB1 and P450IIB2, to thyroid hormone-induced and growth hormone-induced suppression in rat liver: phenobarbital-inducible P450IIB2 suppression by thyroid hormone acting directly, but not through the pituitary system. Mol Pharmacol 39: 811–817

Nadler SG, Strobel HW (1988) Role of electrostatic interactions in the reaction of NADPH-cytochrome P450 reductase with cytochromes P450. Arch Biochem Biophys 261: 418–429

Nadler SG, Strobel HW (1991) Identification and characterization of an NADPH-cytochrome P450 reductase-derived peptide involved in binding to cytochrome P450. Arch Biochem Biophys 290: 277–284

Nakahara K, Tanimoto T, Hatano K, Usuda K, Shoun H (1993) Cytochrome P450 55A1 (P450dNIR) acts as nitric oxide reductase employing NADH as the direct electron donor. J Biol Chem 268: 8350–8355

Nakashima N, Sakai Y, Sakai H, Yanase T, Haji M, Umeda F, Koga S, Hoshita T, Nawata H (1994) A point mutation in the bile acid biosynthetic enzyme sterol 27-hydroxylase in a family with cerebrotendinous xanthomatosis. J Lipid Res 35: 663–668

Narhi LO, Fulco AJ (1987) Identification and characterization of two functional domains in cytochrome P450 BM-3, a catalytically self-sufficient monooxygenase induced by barbiturates in *Bacillus megaterium*. J Biol Chem 262: 6683–6690

Nebert DW, Nelson DR, Coon MJ, Estabrook RW, Feyereisen R, Fujii-Kuriyama Y, Gonzalez FJ, Guengerich FP, Gunsalus IC, Johnson EF, Loper JC, Sato R, Waterman MR, Waxman DJ (1991) The P450 superfamily: update on new sequences, gene mapping, and recommended nomenclature. DNA Cell Biol 10: 1–14

Nelson DR, Strobel HW (1988) On the membrane topology of vertebrate cytochrome P450 proteins. J Biol Chem 263: 6038–6050

Nelson DR, Strobel HW (1989) Secondary structure prediction of 52 membrane-bound cytochromes P450 show a strong structural similarity to P450cam. Biochemistry 28: 656–660

Nelson DR, Kamataki T, Waxman DJ, Guengerich FP, Estabrook RW, Feyereisen R, Gonzalez FJ, Coon MJ, Gunsalus IC, Gotoh O, Okuda K, Nebert DW (1993) The P450 superfamily: update on new sequences, gene mapping, accession numbers, early trivial names of enzymes, and nomenclature. DNA Cell Biol 12: 1–51

Niederau C, Schultz HU, Letko G (1991) Involvement of free radicals in the pathophysiology of chronic pancreatitis: potential of treatment with antioxidant and scavenger substances. Klin Wochenschr 69: 1018–1024

Nisimoto Y, Lambeth JD (1985) NADPH-cytochrome P450 reductase-cytochrome b_5 interactions: cross-linking of the phospholipid vesicle-associated proteins by a water-soluble carbodiimide. Arch Biochem Biophys 241: 386–396

Nordblom GD, Coon MJ (1977) H_2O_2 formation and stoichiometry of hydroxylation reactions catalyzed by highly purified liver microsomal cytochrome P450. Arch Biochem Biophys 180: 343–347

Ohmori S, Misaizu T, Nakamura T, Takano N, Kitagawa H, Kitada M (1993) Differential role in lipid peroxidation between rat P450 1A1 and P450 1A2. Biochem Pharmacol 46: 55–60

Okey AB (1990) Enzyme induction in the cytochrome P450 system. Pharmacol Ther 45: 241–298

Omura T, Sato R (1964) The carbon monoxide-binding pigment of liver microsomes. I. Evidence for its hemoprotein nature. J Biol Chem 239: 2370–2378

Omura T, Sanders E, Estabrook RW, Cooper DY, Rosenthal O (1966) Isolation from adrenal cortex of a nonheme iron protein and a flavoprotein functional as a reduced triphosphopyridine nucleotide-cytochrome P450 reductase. Arch Biochem Biophys 117: 660–673

Onoda M, Hall PF (1982) Cytochrome b_5 stimulates purified testicular microsomal cytochrome P450 (C_{21} side-chain cleavage). Biochem Biophys Res Commun 108: 454–460

Onuchic JN, Beratan DN (1990) A predicitive theoretical model for electron tunneling pathways in proteins. J Chem Phys 92: 722–733

Oprian DD, Coon MJ (1982) Reactions of oxygenated P450 LM4. In: Sato R, Kato R (eds) Microsomes, drug oxidation and drug toxicity. Wiley, New York, pp 139–145

Oprian DD, Gorsky LD, Coon MJ (1983) Properties of the oxygenated form of liver microsomal cytochrome P450. J Biol Chem 258: 8684–8691

Ortiz de Montellano PR (1989) Cytochrome P450 catalysis: radical intermediates and dehydrogenation reactions. Pharmacol Sci 10: 354–359

Paller MS, Jacob HS (1994) Cytochrome P450 mediates tissue-damaging hydroxyl radical formation during reoxygenation of the kidney. Proc Natl Acad Sci USA 91: 7002–7006

Partanen J, Campbell RD (1991) Substitution of Ile-172 to Asn in the steroid 21-hydroxylase B (P450c21B) gene in a Finnish patient with the simple virilizing form of congenital adrenal hyperplasia. Hum Genet 87: 716–720

Pederson TC, Austin RH, Gunsalus IC (1977) Redox and ligand dynamics in P450cam-putidaredoxin complexes. In: Ullrich V (ed) Microsomes and drug oxidations. Pergamon, Oxford, pp 275–283

Pelletier H, Kraut J (1992) Crystal structure of a complex between electron transfer partners, cytochrome c peroxidase and cytochrome c. Science 258: 1748–1755

Pember SO, Powell GL, Lambeth JD (1983) Cytochrome P450scc – phospholipid interactions. Evidence for a cardiolipin binding site and thermodynamics of enzyme interactions with cardiolipin, cholesterol and adrenodoxin. J Biol Chem 258: 3198–3206

Persson JO, Terelius Y, Ingelman-Sundberg M (1990) Cytochrome P450-dependent formation of reactive oxygen radicals: isozyme-specific inhibition of P450-mediated reduction of oxygen and carbon tetrachloride. Xenobiotica 20: 887–900

Peterson JA, Ishimura Y, Griffin BW (1972) *Pseudomonas putida* cytochrome P450: characterization of an oxygenated form of the hemoprotein. Arch Biochem Biophys 149: 197–208

Peterson JA, Ebel RE, O'Keefe DH, Matsubara T, Estabrook RW (1976) Temperature dependence of cytochrome P450 reduction. J Biol Chem 251: 4010–4016

Peterson JA, White RE, Yasukochi Y, Coomes ML, O'Keeffe DH, Eble RE, Masters BSS, Ballou DP, Coon MJ (1977) Evidence that purified liver microsomal cytochrome P450 is a one-electron acceptor. J Biol Chem 252: 4431–4434

Peterson JA, Lu J-Y, Geisselsoder J, Graham-Lorence S, Carmona C, Witney F, Lorence MC (1992) Cytochrome P450terp isolation and purification of the protein and cloning and sequencing of its operon. J Biol Chem 267: 14193–14203

Phillips AH, Langdon RG (1962) Hepatic triphosphopyridine nucleotide cytochrome c reductase: isolation, characterization, and kinetic studies. J Biol Chem 237: 2652–2660

Poland A, Glover E (1974) Comparison of 2,3,7,8-tetrachlorodibenzo-p-dioxin, a potent inducer of aryl hydrocarbon hydroxylase with 3-methylcholanthrene. Mol Pharmacol 10: 349–359

Porter TD (1994) Mutagenesis at a highly conserved phenylalanine in cytochrome P450 2E1 affects heme incorporation and catalytic activity. Biochemistry 33: 5942–5946

Poulos TL, Raag R (1992) Cytochrome P450cam: crystallography, oxygen activation, and electron transfer. FASEB J 6: 674–679

Poulos TL, Finzel B, Gunsalus IC, Wagner GC, Kraut J (1985) The 2.6 Å crystal structure of *Pseudomonas putida* cytochrome P-450. J Biol Chem 260: 16122–16130

Poulos TL, Finzel BC, Howard AJ (1987) High-resolution crystal structure of cytochrome P450cam. J Mol Biol 195: 687–700

Powis G, See KL, Santone KS, Melder DC, Hodnett EM (1987) Quinoneimines as substrates for quinone reductase (NAD(P)H: (quinone-acceptor)oxidoreductase) and the effect of dicumarol on their activity. Biochem Pharmacol 36: 2473–2479

Pyerin W, Taniguchi H (1989) Phosphorylation of hepatic phenobarbital-inducible cytochrome P450. EMBO J 8: 3003–3010

Pyerin W, Wolf CR, Kinzel V, Kübler D, Oesch F (1983) Phosphorylation of cytochrome P450-dependent monooxygenase components. Carcinogenesis 4: 573–576

Raag R, Martinis SA, Sligar SG, Poulos TL (1991) Crystal structure of the cytochrome $P450_{cam}$ active-site mutant Thr252Ala.Biochemistry 30: 11420–11429

Rapoport R, Raikhinstein M, Sklan D, Hanukoglu I (1994) Electron pathways in adrenal mitochondrial cytochrome P450 systems: relative rates of leakage and hydroxylation. In: Lechner MC (ed) Cytochrome P450. 8th international conference. Libbey Eurotext, Paris, pp 357–363

Rashba-Step J, Cederbaum AI (1994) Generation of reactive oxygen intermediates by human liver microsomes in the presence of NADPH or NADH. Mol Pharmacol 45: 150–157

Rashba-Step J, Turro NJ, Cederbaum AI (1993) Increased NADPH- and NADH-dependent production of superoxide and hydroxyl radical by microsomes after chronic ethanol treatment. Arch Biochem Biophys 300: 401–408

Rauschenbach R, Isernhagen M, Noeske-Jungblut C, Boidol W, Siewert G (1993) Cloning sequencing and expression of the gene for cytochrome P450meg, the steroid-15β-monooxygenase from *Bacillus megaterium* ATCC 13368. Mol Gen Genet 241: 170–176

Ravichandran KG, Boddupalli SS, Hasemann CA, Peterson JA, Deisenhofer J (1993) Crystal structure of hemoprotein domain of P450BM-3, a prototype for microsomal P450s. Science 261: 731–736

Rein H, Jung C (1993) Metabolic reactions: mechanism of substrate oxygenation. In: Schenkman JB, Greim H (eds) Cytochrome P450. Springer, Berlin Heidelberg New York (Handbook of experimental pharmacology, vol 105), pp 105–122

Rein H, Ristau O (1964) Über den Nachweis von zwei magnetischen Formen bei einigen Methämoglobinkomplexen. Naturwissenschaften 20: 480–481

Rein H, Jänig GR, Winkler W, Ruckpaul K (1976a) Circular dichroism of partially purified cytochrome P450 from rabbit liver microsomes. Acta Biol Med Ger 35: K41-K50

Rein H, Maricic S, Jänig GR, Vuk-Pavlovic S, Benko B, Ristau O, Ruckpaul K (1976b) Haem accessibility in cytochrome P450 from rabbit liver. A proton magnetic relaxation study by stereochemical probes. Biochim Biophys Acta 446: 325–330

Rein H, Ristau O, Friedrich J, Jänig GR, Ruckpaul K (1977) Evidence for the existence of a high-spin low-spin equilibrium in liver microsomal cytochrome P450. FEBS Lett 75: 19–22
Rein H, Jung C, Ristau O, Friedrich J (1984) Biophysical properties of cytochrome P450, analysis of the reaction mechanism – thermodynamic aspects. In: Ruckpaul K, Rein H (eds) Cytochrome P450. Akademie, Berlin, pp 163–249
Rein H, Jung C, Ristau O, Ruckpaul K (1986) Biological activation of oxygen. In: Shilov AE (ed) Fundamental research in homogeneous catalysis, vol 2. Gordon and Breach, London, pp 733–744
Rein H, Ristau O, Blanck J, Ruckpaul (1989) The spin-redox couple as regulator of the catalytic activity of cytochrome P450. In: Schuster I (ed) Cytochrome P450: biochemistry and biophysics. Taylor and Francis, London, p 284
Remmer H (1959) Der beschleunigte Abbau von Pharmaka in den Lebermikrosomen unter dem Einfluß von Luminal. Naunyn Schmiedebergs Arch Exp Pathol Pharmacol 235: 279–290
Renaud JP, Boucher JL, Vadon S, Delaforge M, Mansuy D (1993) Particular ability of liver P450s 3A to catalyze the oxidation of N omega-hydroxyarginine to citrulline and nitrogen oxides and occurrence in No synthases of a sequence very similar to the heme-binding sequence in P450s. Biochem Biophys Res Commun 192: 53–60
Ristau O, Rein H, Jänig GR, Ruckpaul K (1978) Quantitative analysis of the spin equilibrium of cytochrome P450 LM2 fraction from rabbit liver microsomes. Biochim Biophys Acta 536: 226–234
Ristau O, Wagnerova DM, Rein H, Ruckpaul K (1989) Cytochrome P450 enzyme system – kinetics of oxygen consumption and hydrogen peroxide formation. J Inorg Biochem 37: 111–118
Ruckpaul K (1993) Cytochrom P450 abhängige Enzyme – Targetenzyme für die Arzneistoffentwicklung? Pharm Unserer Zeit 22: 296–304
Ruckpaul K, Bernhardt R (1984) Biochemical aspects of the monooxygenase system in the endoplasmic reticulum of mammalian liver. In: Ruckpaul K, Rein H (eds) Cytochrome P450. Akademie, Berlin pp 9–57
Ruckpaul K, Rein H, Ballou DP, Coon MJ (1980) Analysis of interaction among purified components of the liver microsomal cytochrome P450 containing monooxygenase system by second derivative spectroscopy. Biochim Biophys Acta 626: 41–56
Ruckpaul K, Rein H, Blanck J (1989) Regulation mechanisms of the activity of the hepatic endoplasmic cytochrome P450. In: Ruckpaul K, Rein H (eds) Frontiers in biotransformation, vol I. Akademie, Berlin, pp 1–65
Ryan DE, Levin W (1993) Age- and gender-related expression of rat liver cytochrome P450. In: Schenkman JB, Greim H (eds) Cytochrome P450. Springer, Berlin Heidelberg (Handbook of experimental pharmacology, vol 105), pp 461–476
Saito M (1990) Polychlorinated biphenyl-induced lipid peroxidation as measured by thiobarbituric acid-reactive substances in liver subcellular fractions of rats. Biochim Biophys Acta 1040: 301–308
Sakaguchi M, Mihara K, Sato R (1987) A short amino-terminal segment of microsomal cytochrome P450 functions both as an insertion and as a stop-transfer sequence. EMBO J 6: 2425–2431
Sander M, Ganten D, Mellon SH (1994) Role of adrenal renin in the regulation of adrenal steroidogenesis by corticotropin. Proc Natl Acad Sci USA 91: 148–152
Sanghvi A, Grassi E, Warty V, Diven W, Wight C, Lester R (1981) Reversible activation-inactivation of cholesterol 7-hydroxylase possibly due to phosphorylation-dephosphorylation. Biochem Biophys Res Commun 103: 886–892
Sanglard D, Sengstag C, Seghezzi W (1993) Probing the membrane topology of *Candida tropicalis* cytochrome P450. Eur J Biochem 216: 477–485
Schenkman JB (1982) Brief history of cytochrome P450. In: Schenkman JB, Kupfer D (eds) Hepatic cytochrome P450 monooxygenase system. Pergamon, New York, pp 1–5

Schenkman JB (1993) Metabolic reactions: mechanism of substrate oxygenation. In: Schenkman JB, Greim H (eds) Cytochrome P450. Springer, Berlin Heidelberg New York (Handbook of experimental pharmacology, vol 105), pp 527–545
Schenkman JB, Remmer H, Estabrook RW (1967) Spectral studies of drug interaction with hepatic microsomal cytochrome. Mol Pharmacol 8: 178–188
Schenkman JB, Tamburini PP, Jansson I, Epstein PM (1987) Interactions between cytochrome P450 and other components of the microsomal electron transfer system. In: Sato R, Omura T, Imai Y, Fujii-Kuriyama Y (eds) Cytochrome P450: new trends. Yamada Science Foundation, Japan, pp 59–64
Scholz W, Schütze K, Kunz W, Schwarz M (1990) Phenobarbital enhances the formation of reactive oxygen in neoplastic rat liver nodules. Cancer 50: 7015–7022
Schwarz D (1991) Rotational motion and membrane topology of the microsomal cytochrome P450 system as analyzed by saturation transfer EPR. In: Ruckpaul K, Rein H (eds) Frontiers in biotransformation, vol 5. Akademie, Berlin, pp 93–137
Schwarz D, Gast K, Meyer HW, Lachmann U, Coon MJ, Ruckpaul K (1984) Incorporation of the cytochrome P450 monooxygenase system into large unilamellar liposomes using octylglucoside, especially for measurements of protein diffusion in membranes. Biochem Biophys Res Commun 121: 118–125
Schwarze W, Blanck J, Ristau O, Jänig GR, Pommerening K, Rein H, Ruckpaul K (1985) Spin state control of cytochrome P450 reduction and catalytic activity in a reconstituted P450 LM2 system as induced by a series of benzphetamine analogues. Chem Biol Interact 54: 127–141
Seelig R, Renz M, Bunger G, Schroter H, Seelig HP (1993) Anti-LKM-1 antibodies determined by use of recombinant P450 2D6 in ELISA and western blot and their association with anti-HCV and HCV-RNA. Clin Exp Immunol 92: 373–380
Serino F, Grevel J, Napoli KL, Kahan BD, Strobel HW (1993) Generation of oxygen free radicals during the metabolism of cyclosporin A: a cause-effect relationship with metabolism inhibition. Mol Cell Biochem 122: 101–112
Seybert DW (1990) Lipid regulation of bovine cytochrome P450 11β activity. Arch Biochem Biophys 279: 188–194
Seybert DW, Lancaster JR jr, Lambeth JD, Kamin H (1979) Participation of the membrane in the side chain cleavage of cholesterol. Reconstitution of cytochrome P450scc into phospholipid vesicles. J Biol Chem 254: 12088–12098
Shen SJ, Strobel HW (1992) The role of cytochrome P450 lysine residues in the interaction between cytochrome P4501A1 and NADPH cytochrome P450 reductase. Arch Biochem Biophys 294: 83–90
Shen SJ, Strobel HW (1993) Role of lysine and arginine residues of cytochrome P450 in the interaction between cytochrome P4502B1 and NADPH cytochrome P450 reductase. Arch Biochem Biophys 304: 257–265
Shephard EA, Phillips IR, Bayney RM, Pike SF, Rabin BR (1983) Quantification of NADPH: cytochrome P450 reductase in liver microsomes by a specific radioimmunoassay technique. Biochem J 211: 333–340
Shet MS, Fisher CW, Holmans PL, Estabrook RW (1993) Human cytochrome P450 3A4: enzymatic properties of a purified recombinant fusion protein containing NADPH-P450 reductase. Proc Natl Acad Sci USA 90: 11748–11752
Shet MS, Fisher CW, Arlotto MP, Shackleton CH, Holmans PL, Martin-Wixtrom CA, Saeki Y, Estabrook RW (1994) Purification and enzymatic properties of a recombinant fusion protein expressed in *Escherichia coli* containing the domains of bovine P450 17A and rat NADPH-P450 reductase. Arch Biochem Biophys 311: 402–417
Shibata H, Ogishima T, Mitani F, Suzuki H, Murakami , Saruta T (1991) Regulation of aldosterone synthase cytochrome P450 in rat adrenals by angiotensin II and potassium. Endocrinology 128: 2534–2539
Shimada H, Makino R, Imai M, Horiuchi T, Ishimura Y (1991) Mechanism of oxygen activation by cytochrome P450cam. In: Yamamoto S, Nozaki M, Ishimura Y (eds)

International symposium on oxygenases and oxygen activation. Yamada Science Foundation, Japan, pp 133–136
Shimada H, Makino R, Unno M, Horiuchi T, Ishimura Y (1994) Protein and electron transfer mechanism in dioxygen activation by cytochromes P450cam. In: Lechner MC (ed) Cytochrome P450. 8th international conference. Libbey Eurotext, Paris, pp 229–306
Shimizu T, Hirano K, Takahashi M, Hatano M, Fujii-Kuriyama (1988) Site-directed mutagenesis of rat liver cytochrome $P450_d$: axial ligand and heme incorporation. Biochemistry 27: 4138–4141
Shimizu T, Murakami Y, Hatano M (1994) Glu^{318} and Thr^{319} mutations of cytochrome P450 1A2 remarkably enhance homolytic O-O cleavage of alkyl hydroperoxides. J Biol Chem 269: 13296–13304
Shimizu T, Tateishi T, Hatano M, Fujii-Kuriyama Y (1991) Probing the role of lysines and arginines in the catalytic function of cytochrome P450d by site-directed mutagenesis. Interaction with NADPH-cytochrome P450 reductase. J Biol Chem 266: 3372–3375
Shou M, Grogan J, Mancewicz JA, Krausz KW, Gonzalez FJ, Gelboin HV, Korzekwa KR (1994) Activation of CYP3A4: evidence for the simultaneous binding of two substrates in a cytochrome P450-active site. Biochemistry 33: 6450–6455
Simpson ER, Mason JI, John ME, Zuber MX, Rodgers RJ, Waterman MR (1987) Regulation of the biosynthesis of steroidogenic enzymes. J Steroid Biochem 27: 801–805
Simpson ER, Mahendroo MS, Means GD, Kilgore MW, Corbin CJ, Mendelson CR (1993) Tissue-specific regulation of aromatase cytochrome P450 (CYP19) expression. In: Schenkman JB, Greim H (eds) Cytochrome P450. Springer, Berlin Heidelberg New York (Handbook of experimental pharmacology, vol 105), pp 611–625
Sligar SG (1976) Coupling of spin, substrate and redox equilibria in cytochrome P450. Biochemistry 15: 5399–5406
Sligar SG, Gunsalus IC (1976) A thermodynamic model of regulation: modulation of redox equilibria in camphor monooxygenase. Proc Natl Acad Sci USA 73: 1078–1082
Sligar SG, Lipscomb JD, Debrunner PG, Gunsalus IC (1974) Superoxide anion production by the autoxidation of cytochrome P450cam. Biochem Biophys Res Commun 61: 290–296
Sligar SG, Filipovic D, Stayton P (1991) Mutagenesis of cytochromes P450cam and b_5. Methods Enzymol 206: 31–49
Smettan G, Shkumatov VM, Pommerening K, Ruckpaul K (1985) Properties of reconstituted hybrid cytochrome P450 systems. In: Vereczky L, Magyar K (eds) Cytochrome P450, biochemistry biophysics and induction. Akademia Kiado, Budapest, pp 207–210
Smith CAD, Gough AC, Leigh PN (1992) Debrisoquine hydroxylase gene polymorphism and susceptibility to Parkinson's disease. Lancet 339: 1375–1377
Smith GCM, Tew DG, Wolf CR (1994) Dissection of NADPH-cytochrome P450 oxidoreductase into distinct functional domains. Proc Natl Acad Sci USA 91: 8710–8714
Sogawa K, Fujii-Kuriyama Y (1993) Regulation of cytochrome P450 expression. In: Schenkman JB, Greim H (eds) Cytochrome P450. Springer, Berlin Heidelberg New York (Handbook of experimental pharmacology, vol 105), pp 493–501
Song W-C, Brash AR (1991) Purification of an allene oxide synthase and identification of the enzyme as a cytochrome P450. Science 253: 781–784
Song W-C, Baertschi SW, Boeglin WE, Harris TM, Brash AR (1993) Formation of epoxyalcohols by a purified allene oxide synthase. J Biol Chem 268: 6293–6298
Sousa RL, Marletta MA (1985) Inhibition of cytochrome P450 activity in rat liver microsomes by the naturally occurring flavonoid, quercetin. Arch Biochem Biophys 240: 345–357
Spencer CB, Rifkind AB (1990) NAD(P)H: quinone oxidoreductase (DT-diaphorase) in chick embryo liver. Comparison to activity in rat and guinea pig liver and differences in coinduction with 7-ethoxyresorufin deethylase by 2,3,7,8-tetrachlorodibenzo-p-dioxin. Biochem Pharmacol 39: 327–335
Stadler J, Trockfeld J, Schmalix WA, Brill T, Siewert JR, Greim H, Doehmer J (1994) Inhibition of cytochromes P4501A by nitric oxide. Proc Natl Acad Sci USA 91: 3559–3563

Staudt H, Lichtenberger F, Ullrich V (1974) The role of NADH in uncoupled microsomal monooxygenations. Eur J Biochem 46: 99–106

Stayton PS, Sligar SG (1990) The cytochrome P450cam binding surface as defined by site-directed mutagenesis and electrostatic modeling. Biochemistry 29: 7381–7386

Stayton PS, Poulos TL, Sligar SG (1989) Putidaredoxin competitively inhibits cytochrome b_5-cytochrome P450cam association: a proposed molecular model for a cytochrome P450cam electron transfer complex. Biochemistry 28: 8201–8205

Stefek M (1993) In vitro studies on the interaction of the pyridoindole antioxidant stobadine with rat liver microsomal P450. Xenobiotica 23: 983–993

Straub P, Johnson EF, Kemper B (1993a) Hydrophobic side chain requirements for lauric acid and progesterone hydroxylation at amino acid 113 in cytochrome P450 2C2, a potential determinant of substrate specificity. Arch Biochem Biophys 306: 521–527

Straub P, Ramarao MK, Kemper B (1993b) Preference for aromatic substitutions at tryptophan-120, which is highly conserved and a potential mediator of electron transfer in cytochrome P450 2C2. Biochem Biophys Res Commun 197: 433–439

Strobel HW, Shen S (1994) Studies of the interactions of microsomal cytochrome P450 reductase with cytochromes P450. In: Lechner MC (ed) Cytochrome P450. 8th international conference. Libbey Eurotext, Paris, pp 341–348

Strobel HW, Lu AYH, Heidema J, Coon MJ (1970) Phosphatidylcholine requirement in the enzymatic reduction of hemoprotein P450 and in fatty acid, hydrocarbon, and drug hydroxylation. J Biol Chem 245: 4851–4854

Stuehr DJ, Ikeda-Saito M (1992) Spectral characterization of brain and macrophage nitric oxide synthases. Cytochrome P450-like hemoproteins that contain a flavin semiquione radical. J Biol Chem 267: 20547–20550

Sturman SG, Williams AC (1991) Pathogenesis of Parkinson's disease. Curr Sci 4: 323–330

Szczesna-Skorupa E, Browne N, Mead D, Kemper B (1988) Positive charges at the NH_2-terminus convert the membrane-anchor signal peptide of cytochrome P450 to a secretory signal peptide. Proc Natl Acad Sci USA 85: 738–742

Tamburini PP, Schenkman JB (1986a) Mechanism of interaction between cytochromes P450 RLM5 and b_5. Evidence for an electrostatic mechanism involving cytochrome b_5 heme propionate groups. Arch Biochem Biophys 245: 512–522

Tamburini PP, Schenkman JB (1986b) Differences in the mechanism of functional interaction between NADPH-cytochrome P450 reductase and its redox partners. Mol Pharmacol 30: 178: 185

Tamburini PP, Gibson GG, Backes WL, Sligar SG, Schenkman JB (1984) Reduction kinetics of purified rat liver cytochrome P450. Evidence for a sequential reaction mechanism dependent on the hemoprotein spin state. Biochemistry 23: 4526–4533

Tamburini PP, White RW, Schenkman JB (1985) Chemical characterization of protein-protein interactions between cytochrome P450 and cytochrome b_5. J Biol Chem 260: 4007–4015

Tanaka M, Haniu M, Yasunobu KT (1973) The amino acid sequence of bovine adrenodoxin. J Biol Chem 248: 1141–1157

Taniguchi T, Kimura T (1975) Studies on NO_2-Tyr^{82} and NH_2-Tyr^{82} derivatives of adrenodoxin. Effects of chemical modification on electron transferrring activity. Biochemistry 14: 5573–5578

Taniguchi T, Kimura T (1976) Studies on nitrotyrosine-82 and aminotyrosine-82 derivatives of adrenodoxin. Effects of chemical modification on the complex formation with adrenodoxin reductase. Biochemistry 15: 2849–2853

Taniguchi H, Imai Y, Iyanagi T, Sato R (1979) Interaction between NADPH-cytochrome P450 reductase and cytochrome P450 in the membrane of phosphatidylcholine vesicles. Biochim Biophys Acta 550: 341–356

Taniguchi H, Pyerin W, Stier A (1985) Conversion of hepatic microsomal cytochrome P450 to P420 upon phosphorylation by cyclic AMP-dependent protein kinase. Biochem Pharmacol 34: 1835–1837

Tanner CM (1991) Abnormal liver enzyme-mediated metabolism in Parkinson's disease: a second look. Neurology 4: 89–91

Tsubaki M, Iwamoto Y, Hiwatashi A, Ichikawa Y (1989) Inhibition of electron transfer from adrenodoxin to cytochrome P450scc by chemical modification with pyridoxal -5'-phosphate: identification of adrenodoxin-binding site of cytochrome P450scc. Biochemistry 28: 6899–6907

Tuckey RC, Kamin H (1982) The oxyferro complex of adrenal cytochrome P450scc: effect of cholesterol and intermediates on its stability and optical characteristics. J Biol Chem 257: 9309–9314

Tuls J, Geren L, Millett F (1989) Fluorescein isothiocyanate specifically modifies lysine 338 of cytochrome P450scc and inhibits adrenodoxin binding. J Biol Chem 264: 16421–16425

Tusie-Luna MT, Speiser PW, Dumic M, New MI, White PC (1991) A mutation (Pro-30 to Leu) in CYP21 represents a potential nonclassic steroid 21-hydroxylase deficiency allele. Mol Endocrinol 5: 685–692

Uhlmann H, Kraft R, Bernhardt R (1994) C-terminal region of adrenodoxin affects its structural integrity and determines differences in its electron transfer function to cytochrome P450. J Biol Chem 269: 22557–22564

Ullah AJ, Murray RI, Bhattacharyya PK, Wagner GC, Gunsalus IC (1990) Protein components of a cytochrome P450 linalool 8-methyl hydroxylase. J Biol Chem 265: 1345–1351

Usanov SA, Turko IV, Chashchin VL, Akhrem AA (1985) Cross-linking studies of steroidogenic electron transfer: covalent complex of adrenodoxin reductase with adrenodoxin. Biochim Biophys Acta 832: 288–296

Usanov SA, Chashchin VL, Akhrem AA (1990) Cytochrome P450-dependent pathways of the biosynthesis of steroid hormones. In: Ruckpaul K, Rein H (eds) Frontiers in biotransformation, vol 3. Akademie, Berlin, pp 1–57

Uvarov VY, Bachmanova GI, Archakov AI, Sukhomudrenko AG (1980) Conformation and thermostability of soluble cytochrome P450 incorporated into liposomal membrane and cytochrome P450 reductase (in Russian). Biochemistry 45: 1463–1469

Van de Straat R (1987) Role of hepatic microsomal and purified cytochrome P450 in one-electron reduction of two quinone imines and concomitant reduction of molecular oxygen. Biochem Pharmacol 36: 613–619

Van de Straat R, Vromans RM, Bosman P, de Vries J, Vermeulen NPE (1988) Cytochrome P450-mediated oxidation of substrates by electron-transfer; role of oxygen radicals and of 1- and 2-electron oxidation of paracetamol. Chem Biol Interact 64: 267–280

Van der Hoeven TA, Coon MJ (1974) Preparation and properties of partially purified cytochrome P450 and reduced nicotinamide adenine dinucleotide phosphate-cytochrome P450 reductase from rabbit liver microsomes. J Biol Chem 249: 6302–6310

Van Dyke RA, Baker MT, Jansson I, Schenkman J (1988) Reductive metabolism of halothane by purified cytochrome P450. Biochem Pharmacol 37: 2357–2361

Vergeres G, Winterhalter K, Richter C (1991) Localization of the N-terminal methionine of rat liver cytochrome P450 in the lumen of the endoplasmic reticulum. Biochim Biophys Acta 1063: 235–241

Vermilion JL, Coon MJ (1978) Purified liver microsomal NADPH-cytochrome P450 reductase (spectral characterization of oxidation-reduction states). J Biol Chem 253: 2694–2704

Vermilion JL, Ballou DP, Massey V, Coon MJ (1981) Separate roles for FMN and FAD in catalysis by liver microsomal NADPH-cytochrome P450 reductase. J Biol Chem 256: 266–277

Vickery LE (1993) Cholesterol side chain cleavage cytochrome P450 (P450scc). In: Schenkman JB, Greim H (eds) Cytochrome P450. Springer, Berlin Heidelberg New York (Handbook of experimental pharmacology, vol 105), pp 651–665

Vilgrain I, Defaye G, Chambaz EM (1984) Adrenocortical cytochrome p450 responsible for cholesterol side chain cleavage (P450scc) is phosphorylated by the calcium-activated, phospholipid-sensitive protein kinase (protein kinase C). Biochem Biophys Res Commun 125: 554–561

Voznesensky AI, Schenkman JB (1992) The cytochrome P4502B4-NADPH cytochrome P450 reductase electron transfer complex is not formed by charge-pairing. J Biol Chem 267: 14669–14676

Voznesensky AI, Schenkman JB (1994) Quantitative analyses of electrostatic interactions between NADPH-cytochrome P450 reductase and cytochrome P450 enzymes. J Biol Chem 269: 15724–15731

Voznesensky AI, Schenkman JB, Pernecky SJ, Coon MJ (1994) The NH_2-terminal region of rabbit CYP2E1 is not essential for interaction with NADPH-cytochrome P450 reductase. Biochem Biophys Res Commun 203: 156–161

Wada A, Waterman MR (1992) Identification by site-directed mutagenesis of two lysine residues in cholesterol side chain cleavage cytochrome P450 that are essential for adrenodoxin binding. J Biol Chem 267: 22877–22882

Wade AE, Norred WP, Evans JS (1978) Lipids in drug detoxication. In: Hathcock JN, Coon J (eds) Nutrition and drug interrelations. Academic, New York, pp 475–503

Wade RC (1990) Solvation of the active site of cytochrome P450cam. J Comput Aided Mol Des 4: 199–204

Wang J, Rousseau DL, Abu-Soud HM, Stuehr DJ (1994) Heme coordination of NO in NO synthase. Proc Natl Acad Sci USA 91: 10512–10516

Waterman MR, Estabrook RW (1983) The induction of microsomal electron transport enzymes. Mol Cell Biochem 53/54: 267–278

Waterman MR, Simpson ER (1990) Mechanisms of regulation of steroid hydroxylase gene expression. In: Ruckpaul K, Rein H (eds) Molecular mechanisms of adrenal steroidogenesis and aspects of regulation and application, Akademie, Berlin (Frontiers in biotransformation, vol 3), pp 101–126

Wendoloski JJ, Matthew JB, Salemme FR (1987) Molecular dynamics of a cytochrome c-cytochrome b5 electron transfer complex. Science 238: 794–797

White KA, Marletta MA (1992) Nitric oxide synthase is a cytochrome P450 type hemoprotein. Biochemistry 31: 6627–6631

White KA, Marletta MA (1993) Nitric oxide synthase (NOS). In: Schenkman JB, Greim H (eds) Cytochrome P450. Springer, Berlin Heidelberg New York (Handbook of experimental pharmacology, vol 105), pp 719–728

White PC, Curnow KM, Pascoe L (1993) Steroid 11β-hydroxylase isozymes (CYP11B1 and CYP11B2). In: Schenkman JB, Greim H (eds) Cytochrome P450. Springer, Berlin Heidelberg New York (Handbook of experimental pharmacology, vol 105), pp 641–650

White PC, Dupont J, New MI, Leiberman E, Hochberg Z, Rosler A (1991) A mutation in CYP11B1 (Arg-448—His) associated with steroid 11 beta-hydroxylase deficiency in Jews of Moroccan origin. J Clin Invest 87: 1664–1667

White RE (1991) The involvement of free radicals in the mechanisms of monooxygenases. Pharmacol Ther 49: 21–42

White RE (1994) The importance of one-electron transfers in the mechanism of cytochrome P450. In: Lechner MC (ed) Cytochrome P450. 8th international conference. Libbey Eurotext, Paris, pp 333–340

White RE, Coon MJ (1980) Oxygen activation by cytochrome P450. Annu Rev Biochem 49: 315–356

White RE, McCarthy M (1986) Active site mechanics of liver-microsomal cytochrome P450. Arch Biochem Biophys 246: 19–32

Whitlock JP Jr (1990) Genetic and molecular aspects of 2,3,7,8-tetrachlorodibenzo-p-dioxin action. Annu Rev Pharmacol Toxicol 30: 251–277

Whysner JA, Ramseyer J, Harding BW (1970) Substrate-induced changes in visible absorption and electron spin resonance properties of adrenal cortex mitochondria P450. J Biol Chem 245: 5441–5449

Williams CH, Kamin H (1962) Microsomal triphosphopyridine nucleotide-cytochrome c reductase of liver. J Biol Chem 237: 587–595

Willie A, McLean M, Liu RQ, Hilgen-Willis S, Saunders AJ, Pielak GJ, Sligar SG, Durham B, Millett F (1993) Intracomplex electron transfer between ruthenium-65-cytochrome b_5 and position 82 variants of yeast iso-1-cytochrome c. Biochemistry 32, 7519–7525

Wu DA, Chung BC (1991) Mutations of P450c21 (steroid 21-hydroxylase) at Cys428, Val281, and Ser268 result in complete, partial, or no loss of enzymatic activity, respectively. J Clin Invest 88: 519–523

Yasukochi Y, Masters BSS (1976) Some properties of a detergent-solubilized NADPH-cytochrome c (cytochrome P450) reductase purified by biospecific affinity chromatography. J Biol Chem 2515337–5344

Yasukochi Y, Peterson JA, Masters BSS (1979) NADPH-cytochrome c (P450) reductase (spectrophotometric and stopped flow kintic studies on the formation of reduced flavoprotein intermediates). J Biol Chem 254: 7097–7104

Yasukochi T, Okada O, Hara T, Sagara Y, Sekimizu K, Horiuchi T (1994) Putative functions of phenylalanine-350 of *Pseudomonas putida* cytochrome P-450_{cam}. Biochim Biophys Acta 1204: 84–90

Zhukov AA, Archakov AI (1982) Complete stoichiometry of free NADPH oxidation in liver microsomes. Biochem Biophys Res Commun 109: 813–818

Zhukov AA, Blanck J, Ristau O, Ruckpaul K, Archakov A (1989) Stoichiometry of cytochrome P450 catalysed oxygenase and oxidase reactions. Correlations with the spin state of the ferric heme. In: Schuster I (ed) Cytochrome P450: biochemistry and biophysics. Proceedings of the 6th international conference on biochemistry and biophysics of Cytochrome P450, Vienna, 3–8 July 1988, pp 85–88

Editor-in-Charge: Professor H. Grunicke and Professor M. Schweiger

Rev. Physiol. Biochem. Pharmacol., Vol. 127

DNA Damage Profiles Induced by Oxidizing Agents

B. EPE

Contents

1 Introduction

The formation of reactive oxygen species (ROS) inside cells constitutes a serious threat to the integrity of the cellular genome. This is best indicated by the fact that apparently all types of cells contain a number of repair endonucleases which specifically recognize DNA modifications induced by ROS (for reviews, see Friedberg 1985; Wallace 1988; Lindahl 1990; Boiteux 1993; Demple and Harrison 1994). These repair endonucleases act independently from and in addition to the unspecific nucleotide excision repair system, which is represented in *Escherichia coli* by the UvrABC endonuclease.

Institute of Pharmacy, University of Mainz, Staudinger Weg 5, 55099 Mainz, Germany

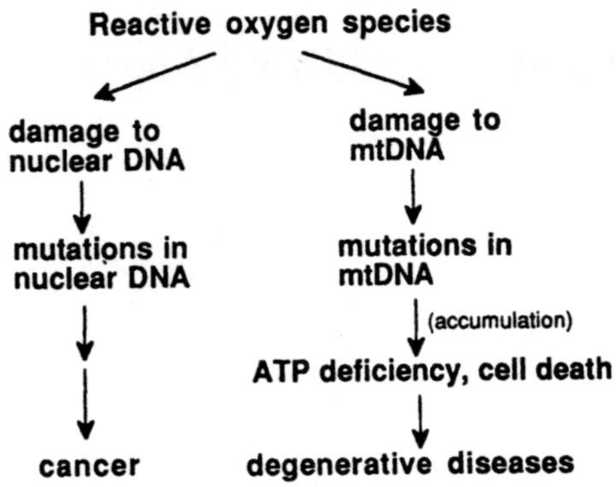

Fig. 1. Hypothetic role of nuclear and mitochondrial DNA damage induced by ROS in the development of cancer and degenerative diseases

The term ROS is usually used to describe intermediates of the reduction of molecular oxygen to water, such as superoxide (O_2^-) and hydroxyl radicals (OH), several other oxygen-centered radicals, such as alkoxy radicals, peroxy radicals and triplet excited states of carbonyl compounds, and singlet oxygen ($^1O_2$) (Pryor 1986). ROS are generated inside cells not only under the influence of exogenous agents such as visible light, UV radiation, ionizing radiation and several xenobiotics (e.g., quinones, peroxides), but also endogenously as by-products of oxygen metabolism under natural conditions (Halliwell and Gutteridge 1986; Sies 1986, 1991; Joenje 1989; Clayson et al. 1994). The relevance of this endogenous ROS production as a high risk factor for genetic stability is indicated by the considerable (up to 1000-fold) increase in spontaneous mutation rates that is associated with certain defects in the specific repair of oxidative DNA modifications (Michaels et al. 1992a).

In recent years, oxidative DNA modifications and the mutations resulting from them have been implicated in several diseases and physiologic processes of clinical relevance. The list ranges from age-correlated disorders such as arteriosclerosis, Parkinson's disease, diabetes and cancer to the process of aging itself (Ames, 1983; Halliwell and Gutteridge 1989; Breimer 1990; Wallace, 1992; Frenkel 1992; Gutteridge 1993; Lindahl 1993; Shigenaga et al. 1994; Halliwell 1994). In the case of cancer, oxidative damage to chromosomal DNA is thought to induce mutations that can activate oncogenes or inactivate tumor supressor genes (Breimer 1990). In other diseases, oxidative damage to mitochondrial DNA (mtDNA) may be of particular importance (Wallace 1992; Richter 1992; Shigenaga et al. 1994). ROS generated during oxidative (respiratory chain) phosphorylation in the mitochondria may result in mtDNA damage and mtDNA mutations. On the assumption that there is no selection against mutated and therefore malfunctioning mitochondria during cell propagation, the percentage of defective mitochondria would be expected to increase during the lifetime of an organism and result in energy (adenosine triphosphate, ATP) shortage in the cells and ultimately in tissue defects (Fig. 1). Vicious circles are well possible, since certain mutations in the chromosomal DNA (nDNA) or mtDNA may increase ROS formation, and therefore the rate of further oxidative DNA damage (Bandy and Davison 1990).

To obtain further insight into the role of oxidative DNA damage as a causal or ancillary risk factor for carcinogenesis and other diseases, it is necessary to characterize and quantify the oxidative DNA damage under various conditions and correlate it with its direct genotoxic consequences, i.e., with the numbers and types of mutation induced. Two different approaches, which complement each other, are feasible. In the first approach, oxidative DNA modifications with the highest mutagenic potential are identified and then serve as marker lesions that can be correlated with the effects of interest, e.g., the incidence of mutations and disease (see Breimer 1990; Essigmann and Wood 1993; Feig et al. 1994). Major problems associated with this approach are the large number of different types of DNA modifications that are induced by ROS (see below) and the fact that the mutagenic potential of a defined modification is very dependent on the sequence context. In the second approach, oxidative DNA damage spectra ("damage profiles") rather than single, defined DNA modifications are correlated with the mutagenic consequences. In this case, for the conditions of oxidative stress which are of interest (a) the type of DNA damage profile induced and (b) the extent of any suitable marker lesion has to be determined. This second approach is particularly valuable as it may be assumed that the number of relevant mechanisms that are directly responsible for oxidative DNA damage in the cells – and therefore the number of oxidative damage profiles that has to be investigated – is lower than the number of relevant DNA modifications.

Here recent data for the genotoxic consequences of (a) defined oxidative DNA modifications and (b) oxidative damage profiles induced by various ROS will be summarized. Then, oxidative damage profiles that are generated under various conditions of oxidative stress in cells will be described.

2 Defined Oxidative DNA Modifications and Their Consequences

The number of different oxidative DNA modifications that are generated by highly reactive species such as hydroxyl radicals is very large; approximately 100 different modifications have been identified (von Sonntag 1987; Dizdaroglu 1992). They include DNA single-strand and double-strand breaks, base modifications, sites of base loss (apurinic/apyrimidinic sites, AP sites) and cross-links. Sites of base loss comprise both regular AP sites, which are generated by a mere hydrolysis of the *N*-glycosylic bond, and oxidized AP sites, which frequently result from radical attack at the positions 1', 2', or 4' of the sugar moiety. The term "strand break" also covers several defined modifications which differ in the phosphate and sugar residues at the 5'and 3' terminus of the strand break. Among the modifications of pyrimidine bases are several 5,6-dihydropyrimidine derivatives (e.g., thymine glycols and cytosine hydrates) and fragmentation products, e.g., 5-hydroxy-5-methylhydantoin, a ring-contracted thymine (Breimer and Lindahl 1985). Modifications of purines include 8-hydroxypurines and ring fragmentation products such as an oxazolone (Cadet et al. 1994) and formamidopyrimidines (Fapy lesions). The formation of the last type of modification involves opening of the imidazole ring, but no net oxidation takes place. DNA protein cross-links can be formed between various DNA bases and amino acids; however, thymine to tyrosine cross-links seem to prevail in the presence of oxygen (Nackerdien et al. 1991). Some chemical structures of oxidative base modifications and AP sites are shown in Fig. 2.

Only a small fraction of the known oxidative DNA modifications has been analyzed for its genotoxic consequences so far. Since different target genes have been used and since both the repair and the miscoding properties of DNA modifications are sequence dependent (Tornaletti and Pfeifer 1994; Gao et al. 1994; Retèl et al. 1993), data obtained for the various lesions

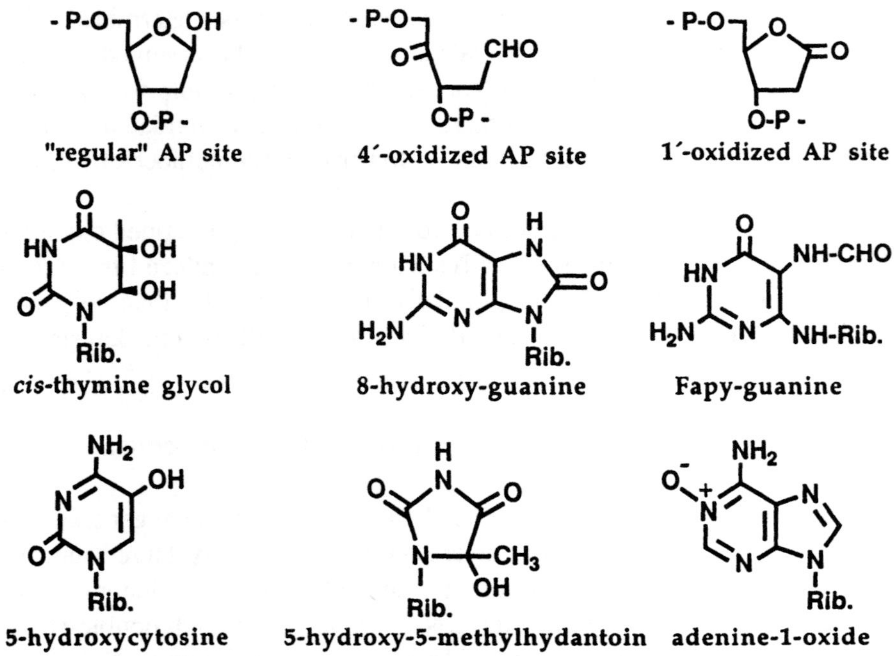

Fig. 2. Chemical structures of some types of AP sites (sites of base loss) and oxidative DNA base modifications mentioned in the text

cannot be compared directly and give only a rough estimate of the mutagenic risk resulting from the lesions.

8-Hydroxyguanine (7,8-dihydro-8-oxoguanine, 8-oxoG) (see Fig. 2) is most probably the most frequent base modification after reaction of DNA with hydroxyl radicals, singlet oxygen, excited photosensitizers and other species (see below) and has received greatest interest in recent years. Research has been facilitated by the sensitive and relatively easy determination of 8-oxoG by HPLC with an electrochemical detector (Floyd et al. 1986). 8-OxoG mispairs with adenine during DNA replication (Shibutani et al. 1991) and therefore gives rise to G:C→T :A transversions in bacteria (Wood et al. 1990; Moriya et al. 1991; Cheng et al. 1992) and mammalian cells (Klein et al. 1992, Moriya 1993). The mutation frequencies observed at single 8-oxoG residues in a single-stranded vector in mammalian cells were 2.5%–4.8% (Moriya 1993). The relatively low mutagenicity is the result of several specific protective mechanisms. In *E. coli*, Fpg protein (formamidopyrimidine-DNA glycosylase) removes 8-oxoG from 8-oxoG:C pairs, but not from oxoG:A mismatches that result from a misincorporation of adenine opposite 8-oxoG. MutY protein, on the other hand, removes the unmodified adenine from 8-oxoG:A mismatches (Michaels et al. 1992a,b). A third protein, MutT, is a phosphatase that removes 8-oxoGTP from the nucleotide pool and thereby prevents a misincorporation of 8-oxoG opposite adenine residues in the template (Maki and Sekiguchi 1992). A homologue of MutT is active in human cells (Mo et al. 1992). Furthermore, the unspecific nucleotide excision repair systems in *E. coli* (Czeczot et al. 1991) and in mammalian cells (Klein et al. 1992) can most probably also remove 8-oxoG residues. 8-OxoG residues apparently do not block DNA replication (Shibutani et al. 1991; Moriya 1993) and transcription (Chen and Bogenhagen 1993) nor induce significant cell-cycle arrest (Ballmaier and Epe 1995). Nevertheless, in Xp-A cells (deficient in nucleotide excision repair) the replication of 8-oxoG-containing plasmids was impaired (Klein et al. 1992).

8-Hydroxyadenine (8-oxoA) is generated in DNA by hydroxyl radicals in somewhat lower yields than 8-oxoG (Fuciarelli et al. 1990; Dizdaroglu 1992). In a direct comparison, a single 8-oxoA in a vector was at least an order of magnitude less mutagenic than an 8-oxoG in the same position (Wood et al. 1992). Interestingly, 8-oxoA is not a good substrate for Fpg protein (Boiteux et al. 1992; Tchou et al. 1994).

Formamidopyrimidines (Fapy lesions) (Fig. 2) are formed in significant amounts in the reaction of hydroxyl radicals with DNA (Fuciarelli et al. 1990; Dizdaroglu 1992). They apparently block DNA synthesis rather than induce mutations (Tudek et al. 1992).

2-Hydroxy-adenine (2-oxoA) is a minor purine modification found after reaction with hydroxyl radicals (Dizdaroglu et al. 1993). It has recently been shown to give rise to misincorporation of dAMP during *in vitro* replication, which could result in A→T transversions (Kamiya et al. 1995).

Among the oxidative modifications of pyrimidines, *cis*-thymine glycol is the best studied. It acts as a blocking lesion. (Ide et al. 1985; Clark and Beardsley 1987; Evans et al. 1993). When placed at a unique site in a single-stranded vector, *cis*-thymine glycol caused T→C transitions in *E. coli* with low frequency (0.3%); no mutagenicity was observed with double-stranded DNA (Basu et al. 1989; Hayes et al. 1988). The specific repair of thymine glycol and other 5,6-dihydropyrimidine derivatives by endonuclease III in *E. coli* and similar enzymes in many other organisms is one clear reason for the low mutagenic potential (Boiteux 1993; Demple and Harrison 1994). In addition, thymine glycols in bacteria can also be repaired by nucleotide excision repair (Lin and Sancar 1989; Kow et al. 1990).

5,6-Dihydrocytosines (e.g., cytosine hydrates and glycols) are of interest since their spontaneous rates of deamination are probably increased relative to that of unmodified cytosine (O'Donnell et al. 1994). At sites of cytosine methylation, deamination yields thymine derivatives which pair with adenine upon replication and thus induce G:C→A:T transitions. This transition is the most frequent type of base subsitution found in the p53 gene of tumors (Hollstein et al. 1991). Furthermore, defects in the mismatch repair of G:T pairs, which increase the risk for this transition, are responsible for a hereditary form of colon cancer (Papadopoulos et al. 1994; Bronner et al. 1994).

5-Hydroxymethyuracil is an example of an oxidative base modification that probably causes few or no genetic problems, as indicated by the fact that in the genome of *Bacillus subtilis* phage SP8 thymine is completely substituted by 5-hydroxymethyluracil (Kallen et al. 1962).

In a new approach to identifying the cytosine modification(s) with the highest mutagenic potential, Loeb and coworkers treated 2'-deoxycytidine-5'-triphosphate (dCTP) with H_2O_2 in the presence of $FeSO_4$ and ascorbic acid, separated the reaction products by high-performance liquid chromatography (HPLC) and incorporated them into a gene that allows analysis of the formation of mutations in *E. coli* (Feig et al. 1994). 5-Hydroxycytosine (5-OH-Cyt) was identified as one of the premutagenic modifications that give rise to G:C→A:T transitions. Under cell-free conditions, in addition to guanine both adenine and cytosine were found to be incorporated opposite 5-OH-Cyt, depending on the sequence context (Purmal et al. 1994). 5-OH-Cyt can be formed by dehydration of cytosine glycols, but is probably also a direct product in the reaction of DNA with oxidants. In DNA of rat liver and kidney, it was detected in amounts similar to those of 8-oxoG (Wagner et al. 1992). Dizdaroglu et al. (1993) reported that 5-OH-Cyt was not excised by endonuclease III from γ-irradiated DNA. In contrast, Hatahet et al. (1994)

observed surprisingly that 5-OH-Cyt in a defined sequence context was recognized by both endonuclease III and Fpg protein.

DNA double-strand breaks are both highly toxic and mutagenic and generally are assumed to be responsible for (some of) the chromosomal aberrations and large deletions observed with ionizing radiation and other types of oxidative damage. DNA single-strand breaks, in contrast, are readily repaired. The repair after exposure to oxidants, however, is biphasic (Churchill et al. 1991) and the possibility cannot be excluded that a minor subtype of single-strand breaks contributes significantly to the mutagenicity of some oxidants. In V79 cells treated with various photosensitizers plus light, however, the mutation frequency (HPRT locus) was not correlated with the number of single-strand breaks induced (Noodt et al. 1993).

Sites of base loss (AP sites) are noninstructive lesions and block DNA replication, but they have been shown to be premutagenic too. Guanine was most frequently inserted opposite regular (i.e., unoxidized) AP sites in a shuttle vector replicated in mammalian cells (Neto et al. 1992; Klinedinst and Drinkwater 1992). In addition to base substitutions, deletions were frequently observed, possibly because AP sites are easily converted into strand breaks, both enzymatically and spontaneously. In contrast, AP sites in bacteria with induced SOS response most frequently code for A (Loeb and Preston 1986; Lawrence et al. 1990). A yeast strain deficient in its AP endonuclease Apn-1 had increased spontaneous mutation rates (Kunz et al. 1994). The mutagenic potential of oxidized AP sites versus regular AP sites is not known. Two ring fragmentation products of thymine, urea and β-ureidoisobutyric acid, which can be regarded as noninstructive modifications similar to AP sites, have recently been shown to code for cytosine and adenine, respectively, in a single-stranded vector replicated in SOS-induced *E. coli* cells (Maccabee et al. 1994).

3 Oxidative DNA Damage Profiles In Vitro and Their Consequences

The damage profile gives the (absolute or relative) extents to which various types of modification are formed in DNA. On the assumption that different ROS (e.g., hydroxyl radicals and singlet oxygen) generate different types of modification or at least different ratios of common modifications, a damage profile can serve as a fingerprint of the ultimate (i.e., directly) DNA damaging species. Provided that the mutagenic potentials of the most relevant defined oxidative DNA modifications are known, the damage profile can in principle also be used to calculate an overall mutagenic risk. Last, but not least, it may often be more appropriate to correlate defined DNA damage profiles with their mutagenic consequences than to do that for defined modifications.

Table 1. Recognition of oxidative DNA modifications by repair endonucleases[a]

Repair endonuclease	Gene locus	Recognition spectrum			
		Sites of base loss (AP sites)			Base modifications
		regular[b]	1'-oxid[c]	4'-oxid[d]	
Fpg protein	*fpg*	+	–	+	8-oxoG[e], Fapy[f]
Endonuclease III	*nth*	+	–	+	5,6-dihydropyrimidines; hyd[g]
T4 endonuclease V	*denV*	+	–	+	Py<>Ph[h]
Endonuclease IV	*nfo*	+	+	+	—
Exonuclease III	*xth*	+	–	(+)[i]	—

[a] See Friedberg (1985); Wallace (1988); Lindahl (1990); Boiteux (1993); Tchou et al. (1994); Häring et al (1994); Demple and Harrison (1995). At very high enzyme concentrations, additional modifications can be incised (Hatahet et al. 1994).
[b] Unmodified desoxyribose moiety.
[c] Desoxyribose oxidized in the 1' position.
[d] Desoxyribose oxidized in the 4' position.
[e] 7,8-Dihydro-8-oxoguanine (8-hydroxyguanine).
[f] Formamidopyrimidines (imidazole ring-opened purines).
[g] 5-Hydroxy-5-methylhydantoin.
[h] Cyclobutane pyrimidine photodimers.
[i] Recognition requires high enzyme concentrations (200 U/ml).

Few DNA damage profiles have been described so far since many techniques allow the quantification of only one type of modification and it is difficult to combine the results from different assay systems. Two techniques are particularly suitable for quantifying several types of oxidative DNA modifications in parallel: gas chromatography coupled with mass spectrometry in the selected ion-monitoring mode (GC/MS/SIM) can determine a large number of base modifications and also DNA-protein cross-links (Dizdaroglu 1992; 1994). In the second technique, the recognition of oxidative DNA modifications by purified repair endonucleases is exploited to quantify various types of modification (Epe et al. 1993a; Epe and Hegler 1994). The repair endonucleases incise the DNA at the substrate modifications, generating single-strand breaks. These breaks can be detected and quantified with high sensitivity by a variety of techniques, e.g., the alkaline elution assay (for nuclear DNA of mammalian cells) or the relaxation assay (for supercoiled mitochondrial DNA or plasmids). The number of single-strand breaks generated directly by the damaging agent is obtained in all cases from tests without repair endonucleases. Thus, several types of base modification, regular and oxidized AP sites and single-strand breaks can be determined in parallel. A major limitation of the assay is the relatively broad (and not completely established) substrate specificity of some repair endonucleases (Table 1). The methods described above can complement each other when DNA base modifications excised by repair endonucleases are analyzed by GC/MS/SIM (Dizdaroglu et al. 1993). It is very likely that other newly developed techniques, e.g., HPLC coupled with thermospray mass spectrometry and ^{32}P postlabeling methods will also allow determination DNA damage profiles in the near future (Cadet and Weinfeld 1993).

In the following paragraphs, DNA damage profiles induced by ROS that react directly with DNA will be discussed. For each damage profile, our knowledge of its mutagenic consequences will be summarized. Then, data on the occurrence in vivo of the various damage profiles will be discussed.

3.1 The DNA Damage Profile Induced by Hydroxyl Radicals

Hydroxyl radicals are considered to be the most reactive ROS. A DNA damage profile generated by the direct action of hydroxyl radicals is obtained by exposure of DNA (in diluted solution) to ionizing radiation in the presence of oxygen or N_2O (von Sonntag 1987). Under these conditions the GC/MS/SIM technique revealed the formation of many types of base modifications (Fuciarelli et al. 1990; Dizdaroglu et al. 1993). Purine derivatives were approximately twice as frequent as pyrimidine derivatives. 8-OxoG was the most frequent modification; 4,6-diamino-5-formamidopyrimidine (Fapy-Ade), 2,6-diamino-4-hydroxy-5-formamidopyrimidine (Fapy-Gua) and

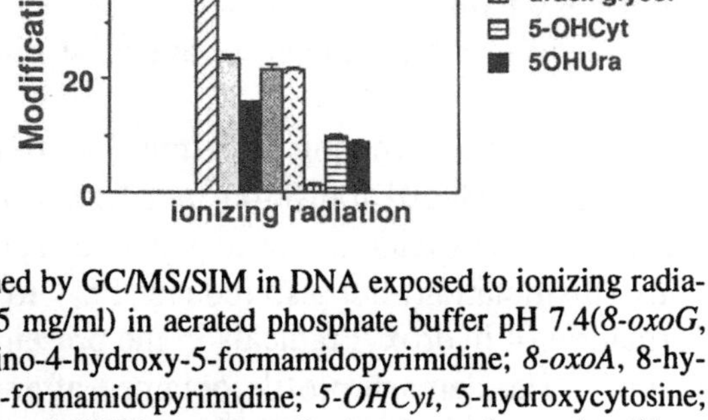

Fig. 3. DNA damage profile determined by GC/MS/SIM in DNA exposed to ionizing radiation (55 Gy; DNA concentration 0.25 mg/ml) in aerated phosphate buffer pH 7.4(*8-oxoG*, 8-hydroxyguanine; *FapyG*, 2,6-diamino-4-hydroxy-5-formamidopyrimidine; *8-oxoA*, 8-hydroxyadenine; *FapyA*, 4,6-diamino-5-formamidopyrimidine; *5-OHCyt*, 5-hydroxycytosine; *5-OHUra*, 5-hydroxyuracil). 5-OHCyt, 5-OHUra and uracil glycol are (in part) artifacts generated from cytosine glycol during the experimental procedure. Data are taken from Dizdaroglu et al. 1993

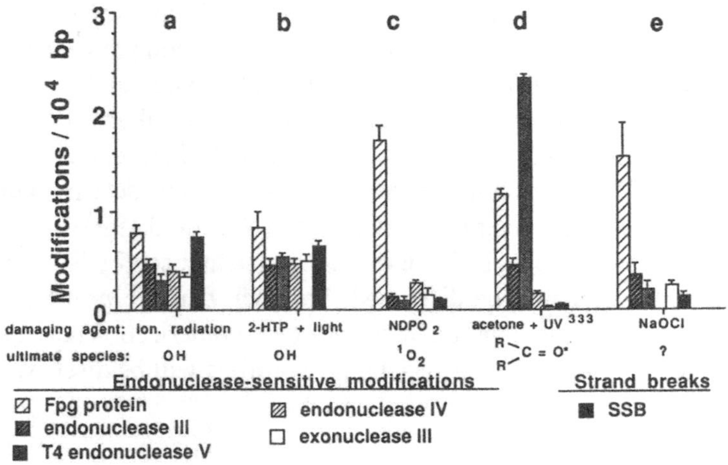

Fig. 4. DNA damage profiles determined by means of repair endonucleases in DNA exposed in phosphate buffer to the following oxidizing agents: *a*, ionizing radiation (20 Gy), *b* N-hydroxy-2-thiopyridone (2-HTP) (1 m*M*) plus light (225 kJ/m^2), *c* $NDPO_2$ (3.5 m*M* in D_2O buffer), *d* acetone (2.6 *M*) plus UV^{333} (8.6 J/m^2) and *e* NaOCl (28 m*M*) Data are taken from Epe et al. (1993a,b,c; and unpublished results in the case of 2-HTP and NaOCl)

thymine glycol were each found at approximately one third of the level of 8-oxoG (Fig. 3). The analysis by repair endonucleases (Fig. 4) revealed a 2:1 ratio of single-strand breaks and AP sites (recognized by endonuclease IV or exonuclease III from *E. coli*). Oxidative purine modifications such as 8-oxoG, Fapy-Gua and Fapy-Ade that are recognized by Fpg protein were formed at a level similar to that of AP sites, while oxidative pyrimidine modifications recognized by endonuclease III (thymine glycol and other 5,6-dihydropyrimidine derivatives) were not more than half as frequent, in agreement with the results obtained by GC/MS/SIM. By means of the different potentials of various repair enzymes to recognize regular and oxidized AP sites it was shown that both regular and 1'-oxidized AP sites represent only a minor fraction of the AP sites induced by hydroxyl radicals; 4'-oxidized AP sites therefore are most probably dominant (Häring et al. 1994).

As well as by ionizing radiation, hydroxyl radicals are generated in the reaction of reduced transition metal ions (Fe^{2+}, Cu^+) with H_2O_2 (Fenton reaction). Superoxide not only generates H_2O_2 spontaneously or enzymatically (dismutation) but also reduces Fe^{3+} to Fe^{2+} so that it gives rise to the formation of hydroxyl radicals in the presence of catalytic amounts of Fe^{2+} or Fe^{3+}. The damage profile obtained after treatment of DNA with superoxide in the presence of Fe^{3+}-ethylenediaminetetraacetic acid (EDTA) or other agents that generate hydroxyl radicals via the Fenton reaction did not differ significantly from that obtained after exposure to ionizing radiation, according to damage analysis both by GC/MS/SIM (Aruoma et al. 1989a,b) and by repair endonucleases (Epe et al. 1993a). This result was not necessarily expected since the generation of hydroxyl radicals from H_2O_2 involves the formation of reactive intermediates such as a peroxo complex $Fe^{II}(OOH)$ and an iron(IV)oxo complex (FeO^{2+}) which could induce a spectrum of modifications different from that induced by hydroxyl radicals (Wink et al. 1994). Furthermore, the Fenton reaction may involve metal ions bound to DNA and therefore the hydroxyl radicals attack DNA from a steric orientation different from that observed with "free" hydroxyl radicals. Indeed, the Fenton-analogous reaction of a copper-phenanthroline complex in the presence of H_2O_2 apparently gave rise to a damage profile that significantly differed from that observed with free hydroxyl radicals. It was characterized by a high number of AP sites oxidized at C-1' (Häring et al. 1994).

Another interesting source of hydroxyl radicals is the photoinduced decomposition of certain hydroperoxides and *N*-oxides ("photo-Fenton reagents"; Saito 1992; Hess and Dix 1992). As expected, the DNA damage profiles induced upon illumination of certain furocoumarin hydroperoxides (Epe et al. 1993b) or of *N*-hydroxy-2-thiopyridone in the presence of DNA were very similar to the damage profile induced by ionizing radiation (Fig. 4).

The mutagenic consequences associated with the DNA damage spectrum induced by hydroxyl radicals can be studied when DNA after exposure to hydroxyl radicals under cell-free conditions is allowed to replicate in host cells. In a double-stranded vector (pZ189) modified by Fe-EDTA plus H_2O_2 and propagated in simian LV-1 cells, several types of mutations were observed, including a relatively high number of small deletions and base substitutions at A:T sites (Moraes et al. 1989). In another system, however, G:C→A:T transitions and G:C→T:A transversions were observed with highest incidence, in addition to deletions (Akman et al. 1991). In a single-stranded vector replicated in bacteria, G→C transversion and C→T transitions were most frequent (McBride et al. 1991). As described above, DNA strand breaks could be responsible for the formation of the deletions, while deamination of 5,6-dihydrocytosines or the specific mispairing of 5-hydroxycytosine with adenine (Feig et al. 1994) could explain the G:C→A:T transitions. The formation of G:C→T:A transversions could indicate that 8-oxoG contributes to the mutagenicity of hydroxyl radicals as well. When hydroxyl radicals were generated by ionizing radiation rather than by a Fenton reaction, G:C→T:A transitions were observed again with the highest incidence (Waters et al. 1991). Also in bacteria, deletions and substitutions at G:C sites were more frequent than substitutions at T:A sites. The type of substitution at G:C, however, was quite variable and sequence dependent and included many G:C to C:G transversions (Retèl et al. 1993).

3.2 The DNA Damage Profile Induced by Singlet Oxygen

Singlet oxygen, the lowest excited state of molecular oxygen, can be generated both in photoreactions and in dark reactions. Typically in the former case, light absorption by a so-called photosensitizer molecule is followed by an energy transfer reaction of the excited photosensitizer with molecular oxygen (type II reaction). The transfer reaction with oxygen is very efficent and often competes favorably with reactions of the photosensitizer with other molecules, including DNA, which are called type I reactions. Among the dark reactions, the formations of singlet oxygen from peroxy radicals (Russell mechanism) and from hypochlorite and peroxynitrite plus H_2O_2 are well-established and of potential biological significance under conditions of oxidative stress, in particular during inflammation (Sies 1986, 1991; Naqui et al. 1986; Kanofsky 1989; Di Mascio et al. 1994).

In phosphate buffer containing 10 mg/ml DNA, singlet oxygen is approximately 1000-fold less reactive with DNA than are hydroxyl radicals (Epe et al. 1988). Therefore, the reaction is much more selective. Accordingly, the DNA damage profile induced by singlet oxygen is quite different from that caused by hydroxyl radicals (Müller et al. 1990) (Fig. 4). Guanine modifications (sensitive to Fpg protein) are generated in high excess of single-strand breaks, AP sites and pyrimidine modifications sensitive to endonuclease III. Analysis by GC/MS/SIM supports the assumption that most of the Fpg-sensitive guanine modifications are actually 8-oxoG (Boiteux et al. 1992), although the formation of so far unknown guanine modifications that are also sensitive to Fpg protein is difficult to exclude. The spectrum of mutations that is observed when singlet oxygen-modified DNA is replicated in bacteria or mammalian host cells is in agreement with the assumption that singlet oxygen generates predominantly 8-oxoG: as expected from the known miscoding properties of 8-oxoG (see above), G:C→T:A transversions were the most frequent type of mutation, no matter whether $NDPO_2$, which is a chemical source of singlet oxygen (DiMascio and Sies 1989), or type II photosensitizers were used to damage DNA (Decuyper-Debergh et al. 1987; McBride et al. 1992; Costa de Oliveira et al. 1992; Tudek et al. 1993; Retèl et al. 1993). The mechanism of 8-oxoG formation in DNA by singlet oxygen is not completely understood. Interestingly, the reaction of deoxyguanosine as an isolated nucleoside with singlet oxygen gives rise to quite different modifications, in particular the formation of 4,8-dihydro-4-hydroxy-8-oxo-deoxyguanosine (Cadet et al. 1986).

3.3 DNA Damage Profiles Induced by Other Excited-State Species

Several photosensitizers that modify DNA not via singlet oxygen, but in a direct (type I) reaction, gave rise to a DNA damage profile very similar to that induced by singlet oxygen, i.e., 8-oxoG is formed in high excess of single-strand breaks, AP sites and other base modifications (Epe et al. 1993c,d). The reaction most probably proceeds via a one-electron-transfer or hydrogen abstraction from DNA, and the selective modification of guanine is in agreement with the fact that this is the most easily oxidized base (Steenken 1989).

The formation of 8-oxoG by electron transfer or hydrogen abstraction from DNA also takes place with excited carbonyl species such as triplet acetone (Epe et al. 1993d), which are potential secondary products during lipid oxidation (Sies 1986). The excitation energy of these species is high enough to allow energy transfer to thymine residues as well, and therefore the damage profile consists of pyrimidine photodimers at levels greater than those of 8-oxoG (Fig. 4). It is not yet known whether the mutation spectrum associated with this kind of damage is of the type characteristic for 8-oxoG or for pyrimidine dimers. The question is of some interest since the formation of both pyrimidine dimers and 8-oxoG is also expected in the case of cellular DNA damage by solar radiation in the UVB range of the spectrum (see below).

3.4 DNA Damage Profiles Induced by H_2O_2 and Superoxide

H_2O_2 and superoxide are primary products during oxidative stress, i.e., they are generated directly in many enzymatic and nonenzymatic reactions of molecular oxygen in the cells (Halliwell and Gutteridge 1989). However, it is often assumed that they do not react directly with DNA. Indeed, DNA damage observed after exposure to superoxide was inhibited by iron chelators and therefore most probably was mediated by traces of iron present in the incubation medium (Fenton reaction) (Lesco et al. 1980; Aruoma et al. 1989a). When isolated DNA was treated with H_2O_2 alone, none of the oxidative base modifications described above were detected (Aruoma et al. 1989b) and no mutations were found when DNA exposed to H_2O_2 under cell-free conditions was replicated in mammalian cells (Moraes et al. 1989). On the other hand, the formation of adenine-1-oxide residues in both isolated and cellular DNA exposed to H_2O_2 has been described recently (Mouret et al. 1990; Cadet et al. 1992). It seems that this modification has little effect on DNA replication.

3.5 DNA Damage Profiles Induced by Hypochlorite and Related Species

Hyperchlorite (ClO^-) is generally not considered to be a ROS, but is generated together with ROS by the oxidative burst of activated leukocytes (Halliwell and Gutteridge 1989). DNA damage by hypochlorite therefore could play a role during inflammation and other reactions involving the immune system.

The DNA damage profile determined with repair endonucleases after exposure of isolated DNA to hypochlorite is shown in Fig. 4. The profile is of the same type as that observed after exposure to singlet oxygen or type I photosensitizers, i.e., base modifications sensitive to Fpg protein (most probably 8-oxoG), are formed in high excess of single-strand breaks, AP sites, and other modifications that are characteristic for the DNA damage by hydroxyl radicals. The same type of damage profile is also observed when DNA is exposed to bromate (BrO_3^-) in the presence of the reduced form of glutathione (GSH) as reducing agent. As concluded from the effects of radical scavengers, D_2O and other agents, bromine radicals ($Br^{\bullet}$) or oxidized bromine radicals ($BrO^{\bullet}$, $BrO_2^{\bullet}$) are likely to be the ultimate species responsible for the DNA damage. The results suggest that DNA damage profiles dominated by 8-oxoG residues are frequently observed when DNA is attacked by mildly oxidizing species and that modifications at other sites in the DNA (strand breaks, AP sites, pyrimidine modifications) are formed in higher yields only with very reactive radicals such as hydroxyl radicals. Further DNA damage profiles have to be determined to confirm this assumption.

3.6 DNA Damage Profiles Induced by Other ROS

DNA damage profiles that could be assigned to the ultimate action of alkoxy radicals ($RO^{\bullet}$) or peroxy radicals ($ROO^{\bullet}$) have not yet been described. Peroxy radicals are formed during lipid peroxidation. They have relatively long half-lives (possibly of the order of seconds), but is is not known whether they react directly with DNA under physiologically relevant conditions (Marnett 1987). Alkoxy radicals are more reactive (Gray and Williams 1959) and should be able to abstract hydrogen atoms from DNA. They can be formed in the reduction of (lipid) peroxides by transition metals in a Fenton-analogous reaction. Indeed, the production of *tert*-butoxy radicals from *tert*-butylhydroperoxide in rat liver mitochondria has been reported (Kennedy et al. 1992).

Nitric oxide (NO), although not a classical ROS, has received much interest as a damaging species since its endogenous production from arginine by NO synthetases was discovered in 1987 (reviewed by Bredt and Snyder 1994). The reaction of NO with DNA is expected to result in a deamination of the DNA bases adenine, guanine and cytosine and – as a consequence – in transition mutations. Indeed, C→T transitions were observed in *Salmonella* strains treated with NO-releasing compounds and transitions at both G:C and A:T sites in a shuttle vector treated with nitric oxide under cell-free conditions and replicated in mammalian cells or *E. coli*. (Wink et al. 1991; Routledge et al. 1993).

Peroxynitrite ($ONOO^-$) is efficiently generated from NO and superoxide; it decomposes spontaneously into hydroxyl radicals and nitric dioxide (NO_2). The DNA damage profiles for the various nitric oxides still have to be determined.

4 Cellular Oxidative DNA Damage Profiles

Various types of oxidative DNA modifications have been detected in cellular DNA, and there is no doubt that oxidative DNA damage takes place inside cells. However, the species directly responsible for the oxidative DNA damage in the cells and therefore the damage spectrum is unknown for most conditions of oxidative stress. The major problem comes from the fact that in a cellular environment the various ROS can be converted into one another by a number of reactions and the formation of one type of species will trigger the generation of many others. An excited photosensitizer molecule, for example, can not only damage DNA directly or via singlet oxygen as under cell-free conditions, but can also give rise to the generation of hydroxyl radicals in a Fenton reaction, since superoxide and H_2O_2 are by-products in photosensitization reactions and transition metals are present in all cells. Yet another possibility is the ultimate DNA damage by cellular nucleases which may be activated by a release of Ca^{2+} from intracellular stores as a result of membrane damage by ROS (Cantoni et al. 1989; Halliwell and Aruoma 1991; Weis et al. 1994; Zhivotovsky et al. 1994). As described above, cellular DNA damage profiles not only give quantitative information about certain types of oxidative DNA modifications, but can also help to identify the DNA damaging mechanisms and species that are actually relevant under given conditions of oxidative stress.

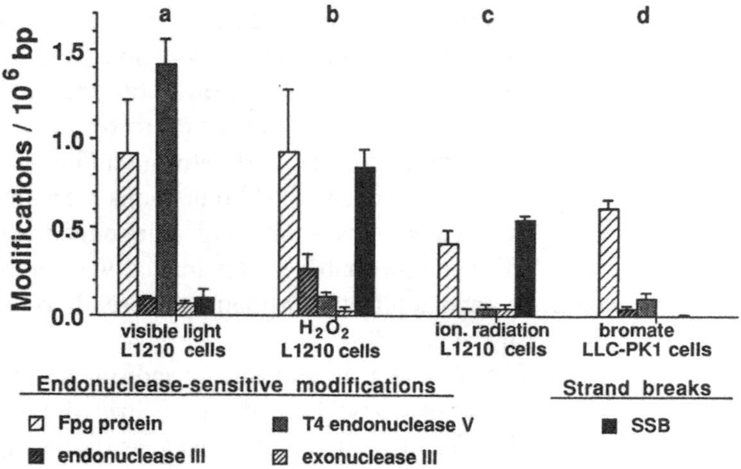

Fig. 5. DNA damage profiles determined by means of repair endonucleases in cultured mammalian cells exposed to the following oxidizing agents: *a* visible light (450 kJ/m^2, measured between 400 and 800 nm) at 0°C, *b* H_2O_2 (500 m*M*) at 0°C, *c* ionizing radiation (3 Gy) at 0°C and *d* potassium bromate (7.5 m*M*) 15 min at 37°C. Data are taken from Pflaum et al. 1994, Epe and Hegler 1994, Ballmaier and Epe 1995 and unpublished results (ionizing radiation)

4.1 Cellular DNA Damage Profiles Induced by Endogenous and Xenobiotic Photosensitizers

Cellular constituents such as flavins and porphyrins absorb light and UV radiation and are potential photosensitizers, i.e., the life times of the excited states of the molecules are long enough for them to react with oxygen (type II reaction, generating singlet oxygen and superoxide) or directly with DNA (type I reaction). The DNA damage profile that was revealed by repair endonucleases when L1210 mouse leukemia cells were illuminated with visible light (400–500 nm) at 0°C is shown in Fig. 5. It is of the same type as that observed under cell-free conditions after reaction of DNA with singlet oxygen or type I photosensitizers, i.e., Fpg-sensitive base modifications (probably 8-oxoG) are generated in high excess of strand breaks, AP sites and endonuclease III-sensitive pyrimidine modifications. This indicates that the cellular DNA damage is not induced by hydroxyl radicals or cellular nucleases (which would generate a much higher relative number of DNA strand breaks), but most probably results from a direct reaction of DNA with an excited photosensitizer or with singlet oxygen. The result predicts that mutations characteristic for 8-oxoG are induced by visible light. The number of mutations to be expected from the level of damage shown in Fig. 5, however, is quite low: a mutation frequency of 1% at Fpg-sensitive modifications (8-oxoG) (Cheng et al. 1992) would generate only 30 genotype mutations per genome (3 x10^9 bp).

DNA damage profiles similar to that described above were observed when mammalian cells were illuminated in the presence of photosensitizers such as acridine orange (Epe et al. 1993c). The conclusion that here too hydroxyl radical are not responsible for the DNA damage is supported by an analysis of the types of chromosomal aberrations induced under these conditions in Syrian hamster embryo cells (Uggla 1990). The treatment of cells with very high doses of light in the presence of lipophilic photosensitizers which localize in the membrane gave rise to a nonrandom distribution of DNA strand breaks (Kvam et al. 1992). The results were considered to demonstrate that the diffusion path of singlet oxygen generated in the nuclear membrane is limited. However, the results could also indicate that Ca^{2+}-dependent nucleases are activated at such high doses and induce DNA fragmentation similar to that observed in apoptosis (Agarwal et al. 1993).

4.2 Cellular DNA Damage Profiles Induced by Peroxides

The spectra of DNA base modifications detected by GC/MS/SIM in mammalian cells exposed to H_2O_2 (Dizdaroglu 1992) or *tert*-butanol (Altman et al. 1994) both resembled that induced by H_2O_2 in the presence of Fenton catalysts under cell-free conditions (Aruoma et al. 1989b). These results support the suggestion that the cellular DNA damage is caused by hydroxyl radicals in both cases. The damage profile obtained by means of repair endonucleases when L1210 cells were exposed to H_2O_2 at 0°C (to exclude DNA repair) led to the same conclusion (Epe and Hegler 1994; Fig. 5): similar amounts of DNA single-strand breaks and Fpg sensitive modifications were detected, as was the case in the cell-free damage profiles attributed to the reaction of DNA with hydroxyl radicals. The relatively low number of AP sites in the cellular damage profile may be a consequence of the alkaline assay conditions, which convert oxidized AP sites into strand breaks. The damage profile supports the proposal that the cellular DNA damage caused by H_2O_2 is mediated by a Fenton reaction, in agreement with earlier suggestions (Meneghini 1988). An activation of intracellular endonucleases is not an important mechanism at 0°C. It probably plays a role only at 37°C, as has been suggested earlier (Cantoni et al. 1989).

The involvement (not necessarily the ultimate responsibility) of hydroxyl radicals in the cellular DNA damage caused by H_2O_2 was previously concluded from findings that H_2O_2-induced strand breaks in CHO cells (Cantoni et al. 1989) and mutations in V79 cells (Nassi-Calò et al. 1989) can be prevented by 1,10-phenanthroline, which inhibits an intracellular Fenton reaction.

4.3 Cellular DNA Damage Profiles Induced by Ionizing Radiation

The spectrum of DNA base modification characteristic for hydroxyl radicals was also found in γ-irradiated mammalian cells (von Sonntag 1987; Halliwell and Aruoma 1991; Nackerdien et al. 1992), in aggreement with the earlier suggestions based on results obtained with radical scavengers that approximately 70% of the cellular DNA damage induced by γ-irradiation is mediated by hydroxyl radicals (Roots and Okada 1975). Analysis of the cellular DNA damage by means of repair endonucleases confirmed that result (Fig. 5): the ratio of single-strand breaks and Fpg-sensitive modifications in cells treated with γ-rays (3 Gy) was similar to that in the cells treated with H_2O_2 at 0°C. This would predict that the mutation spectrum induced by ionizing radiation in cells would be similar to that induced by H_2O_2. However, exposure to H_2O_2 generates exclusively chromatid-type (S-phase dependent) aberrations, while ionizing radiation gives rise to chromosome-type aberrations as well (Duell et al. 1995). For gene mutations the results are less clear. The treatment of mammalian cells transfected with the shuttle vector pZ189 with H_2O_2 gave rise to deletions and base substitutions at G:C pairs, which were predominantly G:C → T:A transitions (Moraes et al. 1990). This is the same type of mutation spectrum that is observed in various shuttle vectors exposed under cell-free conditions to hydroxyl radicals (see above). Somewhat conflicting results, however, were observed for mammalian cells exposed to ionizing radiation, while a high incidence of transversions at A:T sites was reported in some studies (Kimura et al. 1993; Yuan et al. 1995), substitutions at G:C sites were more frequent in others (Nelson et al. 1994; Giver et al. 1995).

4.4 Cellular DNA Damage Profiles Induced by Potassium Bromate

Potassium bromate is a renal carcinogen in rodents (Kurokawa et al. 1990) Increased levels of 8-oxoG were observed in kidney DNA from the treated animals (Kasai et al. 1987). The cellular DNA damage profile was measured by means of repair endonucleases after exposure of LLC-PK1 porcine kidney cells (Ballmaier and Epe 1995) (Fig. 5). Interestingly, it was not of the type characteristic for hydroxyl radicals, but resembled that observed in DNA exposed under cell-free conditions to bromate in the presence of GSH. This indicates that the cellular DNA damage caused by bromate is produced by the same species that is responsible for the cell-free DNA damage seen in the presence of GSH. Indeed, the cellular DNA damage caused by bromate is inhibited in GSH-depleted cells. Besides excited photosensitizers, bromate is therefore a second type of agent that apparently induces 8-oxoG formation in cells quite selectively.

4.5 Spontaneous Oxidative DNA Damage in Cells

Both the type and the extent of oxidative DNA damage induced in cells by normal oxygen metabolism are of particular interest. The steady-state levels of the various oxidative DNA modifications in repair-proficient cells, however, cannot be used as a fingerprint of the ultimate damaging species, since under steady-state conditions the genuine damage profile is modified by the different rates of DNA repair. Therefore, direct information about the species or mechanism(s) relevant for spontaneous oxidative DNA damage cannot be obtained from the damage profiles.

Steady-state levels of 8-oxoG in DNA from mammalian cells have been frequently determined in various types of cells by means of GC/MS/SIM and in particular HPLC/electrochemical detection (ECD). It has been demonstrated that values determined by GC/MS/SIM are often too high due to artifactual generation of 8-oxoG from guanine during the derivatization reaction (Ravanat et al. 1995). Values determined by HPLC/ECD are usually between 2×10^{-6} and 10×10^{-6} 8-oxoG residues per bp (Table 2). Values as low as 5×10^{-7} 8-oxoG residues per bp were obtained by means of Fpg protein and the alkaline elution technique (Pflaum et al. 1994). The apparent discrepancies remain to be resolved. So far, neither an underestimation of Fpg-sensitive modifications by the alkaline elution technique nor an artifactual formation of 9-oxoG during cell lysis and work-up procedures for HPLC analysis can be excluded. Assuming an average mutation frequency of 1% at 8-oxoG residues (Cheng et al. 1992), a steady-state level of 10^{-5} 8-oxoG residues per bp would result in 10^{-7} mutations per bp per generation. This high value is difficult to reconcile with total "spontaneous" mutation frequencies of only approximately 10^{-9}/bp, which have been estimated.

Table 2. Steady state levels of 8-oxoG determined in rat liver DNA by means of HPLC and electrochemical detection

8-oxoG / 10^6bp	Reference
10.5 ± 4	Fraga et al. (1994)
5.3 ± 4.5	Mizumoto et al. (1994)
3	Umegaki et al. (1993)
5.8 ± 0.4	Wagner et al. (1992)
6 ± 2	Conway et al. (1991)
20 ± 6	Denda et al. (1991)
37 ± 5	Hinrichsen et al. (1990)
40	Hegi et al. (1990)

HPLC, High-performance liquid chromatography.

5 Conclusions

It is obvious that our present knowledge of oxidative DNA damage and its consequences does not allow estimation of the (additional) risk of cancer or other diseases that results from a given level of oxidative DNA modifications in the cells. Risk assessment not only requires establishment of the mutagenic risk associated with oxidative DNA damage, but also knowledge of which types of mutation in nuclear or mitochondrial DNA (base substitutions, deletions or chromosomal mutations) are most relevant for a particular health risk. The data described above indicate, however, that our knowledge of both the types and the extent of oxidative DNA damage has increased considerably in recent years. Two major types of oxidative DNA damage apparently prevail both in cells and under cell-free conditions: while hydroxyl radicals give rise to a complex damage profile, mild (less reactive) agents such as singlet oxygen or hypochlorite seem to induce mostly 8-oxoG and only few pyrimidine modifications, AP sites, and strand breaks. The observation that a number of different species ultimately induces the same type of damage suggests that the determination of damage profiles is a valuable step in the investigation of oxidative DNA damage and its consequences.

Both hydroxyl radicals and singlet oxygen seem to induce similar damage profiles in cells and under cell-free conditions. Apparently, the differences in the reaction conditions [e.g., in ion concentrations, the presence of various proteins, the oxygen concentration (Joenje 1989) and the degree of DNA supercoiling] have only a moderate influence on the damage profiles. Although this may not hold true for all types of modification, it allows the use of DNA damage profiles as fingerprints in the identification of the ultimate DNA damaging species.

It is now well established that ROS are not only genotoxic, but also modulate gene expression via activation of transcription factors such as NF-κB and AP-1 and influence second messengers such as Ca^{2+} (Abate et al. 1990; Schreck et al. 1991; Remacle et al. 1995). These "epigenetic" effects could be relevant for the development of cancer, degenerative diseases, and aging as well as the genotoxic effects discussed above. This is best indicated by the long-established tumor-promoting activity of ROS (Zimmerman and Cerutti 1984; Troll and Wiesner 1985; Cerutti 1985). The relative relevance of the two modes of action of ROS for the various adverse effects is not known. Again, the determination of DNA damage profiles may help to distiguish between the genotoxic and epigenetic effects and provide some of the information required to understand the consequences of a normal or enhanced generation of ROS in cells and tissues.

References

Abate C, Patel L, Rauscher F, Curran T (1990) Redox regulation of fos and jun DNA binding activity *in vitro*. Science 249: 1157–1161

Agarwal ML, Larkin HE, Zaidi SIA, Mukhtar H, Oleinick NL (1993) Phospholipase activation triggers apoptosis in photosensitized mouse lymphoma cells. Cancer Res 53: 5897–5902

Akman SA, Forrest GP, Doroshow JH, Dizdaroglu M (1991) Mutation of potassium permanganate- and hydrogen peroxide-treated plasmid pZ189 replicating in CV-1 monkey kidney cells. Mutat Res 261: 123–130

Altman SA, Zastawny TH, Randers L, Lin Z, Lumpkin JA, Remacle J, Dizdaroglu M, Rao G (1994) *tert*-Butyl hydroperoxide-mediated DNA base damage in cultured mammalian cells. Mutat Res 306: 35–44

Ames BN (1983) Dietary carcinogens and anticarcinogens. Oxygen radicals and degenerative diseases. Science 221: 1256–1264

Aruoma OI, Halliwell B, Dizdaroglu M (1989a) Iron ion dependent modifications of bases in DNA by the superoxide radical generating system hypoxanthine/xanthine oxidase. J Biol Chem 264: 13024–13028

Aruoma OI, Halliwell B, Gajewski E Dizdaroglu M (1989b) Damage to the bases in DNA induced by hydrogen peroxide and ferric ion chelates. J Biol Chem 264: 20509–20512

Ballmaier D, Epe B (1995) Oxidative DNA damage induced by potassium bromate under cell-free conditions and in mammalian cells. Carcinogenesis 16: 335–342

Bandy B, Davison AJ (1990) Mitochondrial mutations may increase oxidative stress implications for carcinogenesis and aging? Free Radic Biol Med 8: 523–539

Basu AK, Loechler EL, Leadon SA, Essigman JM (1989) Genetic effects of thymine glycol site specific mutagenesis and molecular modeling studies Proc Natl Acad Sci USA 86: 7677–7681

Boiteux S (1993) Properties and biological functions of the NTH and FPG proteins of *Escherichia coli*: two DNA glycosylases that repair oxidative damage in DNA. Photochem Photobiol B 19: 87–96

Boiteux S, Gajewski E, Laval J, Dizdaroglu M (1992) Substrate specificity of the *Escherichia coli* Fpg protein (formamidopyrimidine-DNA glycosylase) excision of purine lesions in DNA produced by ionizing radiation or photosensitization. Biochemistry 31: 106–110

Bredt DS and Snyder SH (1994) Nitric oxide a physiologic messenger molecule. Annu Rev Biochem 63: 175–195

Breimer (1990) Molecular mechanisms of oxygen radical carcinogenesis and mutagenesis the role of base damage. Mol Carcinog 3: 188–197

Breimer LH, Lindahl T (1985) Thymine lesions produced by ionizing radiation in double-stranded DNA. Biochemistry 24: 4018–4022

Bronner CE, Baker SM, Morrison PT, Warren G, Smith LG, Lescoe MK, Kane M, Earabino C, Lipford J, Lindblom A, Tannergård P, Bollag RJ, Godwin AR, Ward DC, Nordenskjøld M, Fishel R, Kolodner R, Liskay M (1994) Mutation in the DNA mismatch repair gene homologue hMLH1 is associated with hereditary nonpolyposis colon cancer. Nature 368: 258–261

Cadet J, Weinfeld M (1993) Detecting DNA damage. Anal Chem 65: 675A–682A

Cadet J, Berger M, Decarroz C, Wagner JR, van Lier JE, Ginot YM, Vigny P (1986) Photosensitized reactions of nucleic acids. Biochimie 68: 813–834

Cadet J, Odin F, Mouret JF, Polverelli M, Audic A, Giacomoni P, Favier A Richard MJ (1992) Chemical and biochemical postlabeling methods for singling out specific oxidative DNA lesions. Mutat Res 275: 343–354

Cadet J, Berger M, Buchko GW, Joshi PC, Raoul S, Ravanat J-L (1994) 2,2-Diamino-4-[(3, 5di-O-acetyl-2-deoxy-β-D-erythro-pentafuranosyl)amino]-5-(2H)-oxazolone a novel and predominant radical oxidation product of 3',5'-di-O-acetyl-2'-deoxyguanosine. J Am Chem Soc: 116: 7403–7404

Cantoni O, Sestili P, Cattabeni F, Bellomo G, Pou S, Cohen M, Cerutti P (1989) Calcium chelator quin 2 prevents hydrogen-peroxide-induced DNA breakage and cytotoxicity. Eur J Biochem 182: 209–212

Cerutti PA (1985) Prooxidant states and tumor promotion. Science 227: 375–381

Chen YH, Bogenhagen DF (1993) Effects of DNA lesions on transcription elongation by T7 RNA polymerase. J Biol Chem 268: 5849–5855

Cheng KC, Cahill DS, Kasai H, Nishimura S, Loeb LA (1992) 8-Hydroxyguanine, an abundant form of oxidative DNA damage, causes G→T and A→C substitutions. J Biol Chem 267: 166–172

Churchill ME, Peak JG Peak MJ (1991) Repair of near-visible and blue-light-induced DNA single-strand breaks by the CHO cell lines AA8 and EM9. Photochem Photobiol 54: 639–644

Clark JM, Beardsley GP (1987) Functional effects of cis-thymine glycol lesion on DNA synthesis in vitro. Biochemistry 26: 5398–5403

Clayson DB, Mehta R, Iverson F (1994) Oxidative DNA damage – the effects of certain genotoxic and operationally non-genotoxic carcinogens. Mutat Res 317: 25–42

Conway CC, Nie G, Huesain NS, Fiala ES (1991) Comparision of oxidative damage to rat liver DNA and RNA by primary nitroalkanes, secondary nitroalkanes, cyclopentanone oxime and related compounds. Cancer Res 51: 3143–3147

Costa de Oliveira R, Ribeiro DT, Nigro RG, Di Mascio P, Menck CFM (1992) Singlet oxygen induced mutation spectrum in mammalian cells. Nucleic Acids Res 20: 4319–4323

Czeczot H, Tudek B, Lambert B, Laval J, Boiteux S (1991) *Escherichia coli* FPG protein and UvrABC endonuclease repair DNA damages induced by methylene blue plus visible light in vitro and in vivo. J Bacteriol 173: 3419–3424

Decuyper-Debergh D, Piette J, Van de Vorst A (1987) Singlet oxygen-induced mutations in M13 lacZ phage DNA. EMBO J 6: 3155–3161

Demple B, Harrison L (1994) Repair of oxidative damage to DNA enzymology and biology. Annu Rev Biochem 63: 915–948

Denda A, Sai K, Tang Q, Tsujuchi T, Tsutsumi M, Amanuwa T, Murata Y, Nakoe D, Maruyama H, Kurokawa Y, Konishi Y (1991) Induction of 8-hydroxydeoxyguanosine but not initiation of carcinogenesis by redox enzyme modulations with or without menadione in rat liver. Carcinogenesis 12: 719–726

Di Mascio P, Sies H (1989) Quantification of singlet oxygen generated by thermolysis of 3,3'-(1,4-naphthylidene)dipropionate. Monomol and dimol photoemission and the effects of 1,4-diazabicyclo[2.2.2]octane. J Am Chem Soc 111: 2909–2914

Di Mascio P, Bechera EJH, Medeiros MHG, Briviba K, Sies, H (1994) Singlet molecular oxygen production in the reaction of peroxynitrite with hydrogen peroxide. FEBS Lett 355: 287–289

Dizdaroglu M (1992) Oxidative damage to DNA in mammalian chromatin. Mutat Res 275: 331–342

Dizdaroglu M (1994) Chemical determination of oxidative DNA damage by gas chromatography-mass spectrometry. Methods Enzymol 234: 3–16

Dizdaroglu M, Laval J, Boiteux S (1993) Substrate specificity of the *Escherichia coli* endonuclease III excision of thymine- and cytosine-derived lesions in DNA produced by radiation-generated free radicals. Biochemistry 32: 12105–12111

Duell T, Lengfelder E, Fink R, Giesen R, Bauchinger M (1995) Effect of activated oxygen species in human lymphocytes. Mutat Res 336: 29–38

Epe B, Hegler J (1994) Oxidative DNA damage: endonuclease fingerprinting. Methods Enzymol 234: 122–131

Epe B, Mützel P, Adam W (1988) DNA damage by oxygen radicals and excited state species: a comparative study using enzymatic probes in vitro. Chem Biol Interact 67: 149–165

Epe B, Pflaum M, Häring M, Hegler J, Rüdiger H (1993a) Use of repair endonucleases to characterize DNA damage induced by reactive oxygen species in cellular and cell-free systems. Toxicol Lett 67: 57–72

Epe, B, Häring M, Ramaiah D, Stopper H, Adam W, Abou-Elzahab MM, Saha-Möller CR (1993b) DNA damage induced by furocoumarin hydroperoxides plus UV (360 nm). Carcinogenesis 14: 2271–2276

Epe B, Pflaum M, Boiteux S (1993c) DNA damage induced by photosensitizers in cellular and cell-free systems. Mutat Res 299: 135–145

Epe B, Henzl H, Adam W, Saha-Möller CR (1993d) Endonuclease-sensitive DNA modifications induced by acetone and acetophenone as photosensitizers. Nucleic Acids Res 21: 863–869

Essigmann JM, Wood ML (1993) The relationship between the chemical structures and mutagenic specificities of the DNA lesions formed by chemical and physical mutagens. Toxicol Lett 67: 29–39 (1993)

Evans J, Maccabee M, Hatahet Z, Courcelle J, Bockrath R, Ide H, Wallace S (1993) Thymine ring saturation and fragmentation products: lesion bypass, misinsertion and implications for mutagenesis. Mutat Res 299: 147–156

Feig DI, Sowers LC, Loeb LA (1994) Reverse chemical mutagenesis Identification of the mutagenic lesions resulting from reactive oxygen species-mediated damage to DNA. Proc Natl Acad Sci USA 91: 6609–6613

Floyd RA, Watson JJ, Wong PK, Altmiller DH, Rickard RC (1986) Hydroxyl free radical adduct of deoxyguanosine sensitive detection and mechanisms of formation. Free Radic Res Commun 1: 163–172

Fraga CG, Onuki J, Lucesoli F, Bechara EJH, Di Mascio P (1994) 5-Aminolevulinic acid mediates the in vivo and in vitro formation of 8-hydroxy-2'-deoxyguanosine in DNA. Carcinogenesis 15: 2241–2244

Frenkel K (1992) Carcinogen-mediated oxidant formation and oxidative DNA damage. Pharmacol Ther 53: 127–166

Friedberg EC (1985) DNA Repair. Freeman, New York

Fuciarelli AF, Wegher BJ, Blakely WF, Dizdaroglu M (1990) Yields of radiation-induced base products in DNA effects of DNA conformation and gassing conditions. Int J Radiat Biol 58: 397–415

Gao S, Drouin R, Holmquist GP (1994) DNA repair rates mapped along the human pgk1 gene at nucleotide resolution. Science 263: 1438–1440

Giver CR, Nelson Jr SL, Cha MY, Pongsaensook P, Grosovsky AJ (1995) Mutational spectrum of X-ray induced TK- human cell mutants. Carcinogenesis 16: 267–275

Gray P, Williams A (1959) The thermochemistry and reactivity of alkoxyl radicals. Chem Rev 59: 239–328

Gutteridge JMC (1993) Free radicals in disease processes: a compilation of cause and consequences. Free Radic Res Commun 19: 141–158

Halliwell B (1994) Free radicals, antioxidants, and human disease: curiosity, cause, or consequence? Lancet 344: 721–724

Halliwell B, Aruoma OI (1991) DNA damage by oxygen-derived species. Its mechanism and measurement in mammalian cells. FEBS Lett 281: 9–19

Halliwell B, Gutteridge JM (1986) Oxygen free radicals and iron in relation to biology and medicine, some problems and concepts. Arch Biochem Biophys 246: 501–514

Halliwell B, Gutteridge JM (1989) Free radicals in biology and medicine, 2nd edn. Oxford University Press, Oxford

Häring M, Rüdiger H, Demple B, Boiteux S, Epe B (1994) Recognition of oxidized abasic sites by repair endonucleases. Nucleic Acids Res 22: 2010–2015

Hatahet Z, Kow YW, Purmal AA, Cunningham RP, Wallace SS (1994) New substrates for old enzymes. 5-Hydroxy-2'-deoxycytidine and 5-hydroxy-2'-deoxyuridine are substrates for Escherichia coli endonuclease III and formamidopyrimidine DNA glycosylase, while 5-hydroxy-2'-uridine is a substrate for uracil DNA N-glycosylase. J Biol Chem 269: 18814–18820

Hayes RC, Petrullo LA, Huang H, Wallace SS, LeClerc JE (1988) Oxidative DNA damage in DNA. Lack of mutagenicity by thymine glycol lesions. J Mol Biol 201: 239–246
Hegi ME, Ulrich D, Sagelsdorff P, Richter C, Lutz WK (1990) No measurable increase in thymidine glycol or 8-hydroxydeoxyguanosine in liver DNA of rats treated with nafenopin or choline-devoid low-methionine diet. Mutat Res 238: 325–329
Hess KM, Dix TA (1992) Evaluation of N-hydroxy-2-thiopyridone as a nonmetal dependent source of the hydroxyl radical (HO) in aqueous systems. Anal Biochem 206: 309–314
Hinrichsen LI, Floyd RA, Sudilovsky O (1990) Is 8-hydroxydeoxyguanosine a mediator of carcinogenesis by a coline-devoid diet in the rat liver? Carcinogenesis 11: 1879–1881
Hollstein M, Sidransky D, Vogelstein B, Harris CC (1991) p53 Mutations in human cancers. Science 253: 49–53
Ide H, Kow YW, Wallace SS (1985) Thymine glycol and urea residues in M13 DNA constitute replicative blocks in vitro. Nucleic Acids Res 13: 8032–8052
Joenje H (1989) Genetic toxicology of oxygen. Mutat Res 219: 193–208
Kallen RG, Simon M, Marmur J (1962) The occurrence of a new pyrimidine base replacing thymine in a bacteriophage DNA: 5-hydroxymethyl uracil. J Mol Biol 5: 248–250
Kamiya H, Ueda T, Ohgi T, Matsukage A, Kasai H (1995) Misincorporation of dAMP opposite 2-hydroxyadenine, an oxidative form of adenine. Nucleic Acids Res 23: 761–766
Kanofsky, JR (1989) Singlet oxygen production by biological systems. Chem-Biol Interact 70: 1–28
Kasai H, Nishimura S, Kurokawa, Y Hayashi Y (1987) Oral administration of the renal carcinogen, potassium bromate, specifically produces 8-hydroxydeoxyguanosine in rat target organ. Carcinogenesis 8: 1959–1961
Kennedy CH, Church DF, Winston GW, Pryor WA (1992) tert-Butyl hydroperoxide induced radical production in rat liver mitochondria. Free Radic Biol Med 12: 381–387
Kimura H, Higuchi H, Iychara-Ogawa H, Kato T (1993) Sequence analysis of X-ray induced mutations occuring in a cDNA of the human hprt gene integrated into mammalian chromosomal DNA. Radiat Res 134: 202–208
Klein JC, Bleeker MJ, Saris CP, Roelern HCPF, Brugghe HF, Van den Elst H, Van der Marel GA, Van Boom JH, Westra JG, Kriek E, Berns AJM (1992) Repair and replication of plasmids with site-specific 8-oxodG and AAF-dG residues in normal and repair-deficient human cells. Nucleic Acids Res 20: 4437–4443
Klinedinst DK, Drinkwater NR (1992) Mutagenesis by apurinic sites in normal and ataxia teleangiectasia human lymphoblastoid cells. Mol Carcinogenesis 6: 32–42
Kow YW, Wallace SS, Van Houten B (1990) UvrABC nuclease complex repairs thymine glycol, an oxidative DNA damage. Mutat Res 235: 147–156
Kunz BA, Henson ES, Roche H, Ramotar D, Nunoshiba T, Demple B (1994) Specifity of the mutator caused by deletion of the yeast structural gene (APN1) for the major apurinic endonuclease. Proc Natl Acad Sci USA 91: 8165–8169
Kurokawa Y, Maekawa A, Takahashi M, Hayashi Y (1990) Toxicity and carcinogenicity of potassium bromate – a new renal carcinogen. Environ Health Perspect 87: 309–335
Kvam E, Stocke T, Moan J, Steen HB (1992) Plateau distributions of DNA fragment lengths produced by extended light exposure of extranuclear photosensitizers in human cells. Nucleic Acids Res 20: 6687–6693
Lawrence CW, Borde A, Banerjee SK, LeClerk JR (1990) Mutation frequency and spectrum resulting from a single abasic site in a single-stranded vector. Nucleic Acids Res 18: 2153–2157
Lesco SA, Lorentzen RJ, Ts'o PO (1980) Role of superoxide in desoxyribonucleic acid strand scission. Biochemistry 19: 3023–3028
Lin J-J, Sancar A (1989) A new mechanism for repairing oxidative damage to DNA: (A)BC excinuclease removes AP sites and thymine glycol from DNA. Biochemistry 28: 7979–7984
Lindahl T (1990) Repair of intrinsic DNA lesions. Mutat Res 238: 305–311
Lindahl T (1993) Instability and decay of the primary structure of DNA. Nature 362: 709–715

Loeb LA, Preston BD (1986) Mutagenesis by apurinic/apyrimidinic sites. Annu Rev Genet 20: 201–230

Maccabee M, Evans JS, Glackin MP, Hatatet Z, Wallace SS (1994) Pyrimidine ring fragmentation products Effects of lesion structure and sequence context on mutagenesis. J Mol Biol 236: 514–530

Maki H, Sekiguchi M (1992) Mut T protein specifically hydrolyses a potent mutagenic substrate for DNA synthesis. Nature 355: 273–275 (1992)

Marnett LJ (1987) Peroxy free radicals potential mediators of tumor initiation and promotion. Carcinogenesis 8: 1365–1373

McBride TJ, Preston BD, Loeb LA (1991) Mutagenic spectrum resulting from DNA damage by oxygen radicals. Biochemistry 30: 207–213

McBride TJ, Schneider JE, Floyd RA, Loeb LA (1992) Mutations induced by methylene blue plus light in single-stranded M13mp2. Proc Natl Acad Sci USA 89: 6866–6870

Meneghini R (1988) Genotoxicity of active oxygen species in mammalian cells. Mutat Res 195: 215–230

Michaels ML, Cruz C, Grollman AP, Miller JH (1992a) Evidence that MutY and MutM combine to prevent mutations by an oxidative damaged form of guanine. Proc Natl Acad Sci USA 89: 7022–7025

Michaels ML, Tchou J, Grollman AP, Miller JH (1992b) A repair system for 8-oxo-7, 8-dihydrodeoxyguanine. Biochemistry 31: 10964–10968

Mizumoto, Y, Nakae D, Yoshiji H, Andoh N, Horiguchi K, Endoh T, Kobayashi E, Tsujiuchi T, Shimoji N, Denda A, Tsujii T, Nagao M, Wakabayashi K, Konishi Y (1994) Inhibitory effects of 2-O-octadecylascorbic acid and other vitamin C and E derivatives on the induction of enzyme-altered putative preneoplastic lesions in the livers of rats fed a choline-deficient, L-amino acid-defined diet. Carcinogenesis 15: 241–246

Mo JY, Maki H, Sekiguchi M (1992) Hydrolytic elimination of a mutagenic nucleotide, 8oxodGTP, by human 18-kilodalton protein: sanitization of nucleotide pool. Proc Natl Acad Sci USA 89: 11021–11025

Moraes EC, Keyse SM, Pidoux M, Tyrrell RM (1989) The spectrum of mutations generated by passage of a hydrogen peroxide damaged shuttle vector plasmid through a mammalian host. Nucleic Acids Res 17: 8301–8312

Moraes EC, Keyse SM, Tyrrell RM (1990) Mutagenesis by hydrogen peroxide treatment of mammalian cells a molecular analysis. Carcinogenesis 11: 283–293

Moriya M (1993) Single-stranded shuttle phagemid for mutagenesis studues in mammalian cells 8-oxoguanine in DNA induces targeted GC→TA transversions in simian kidney cells. Proc Natl Acad Sci USA 90: 1122–1126

Moriya M, Ou C, Bodepudi V, Johnson F, Takeshita M, Grollman AP (1991) Site-specific mutagenesis using a gapped duplex vector: A study of translesion synthesis past 8-oxodeoxyguanosine in *E coli*. Mutation Res 254: 281–288

Mouret JF, Odin F, Polverelli M, Cadet J (1990) ^{32}P-Postlabeling measurement of adenine N-1 oxide in cellular DNA exposed to hydrogen peroxide. Chem Res Toxicol 3: 102–110

Müller E, Boiteux S, Cunningham RP, Epe B (1990) Enzymatic recognition of DNA modifications induced by singlet oxygen and photosensitizers. Nucleic Acids Res 18: 5969–5973

Nackerdien Z, Rao G, Cacciuttolo MA, Gajewski E, Dizdaroglu M (1991) Chemical nature of DNA-protein-cross-links produced in mammalian chromatin by hydrogen peroxide in the presence of iron or copper ions. Biochemistry 30: 4873–4879

Nackerdien Z, Olinski R, Dizdaroglu M (1992) DNA base damage in chromatin of γ-irradiated cultured human cells. Free Radic Res Commun 16: 259–273

Naqui A, Chance B, Cadenas E (1986) Reactive oxygen intermediates in biochemistry. Annu Rev Biochem 55: 137–166

Nassi-Calò L, Mello-Filho AC, Meneghini R (1989) o-Phenanthroline protects mammalian cells from hydrogen peroxide-induced gene mutation and morphological transformation. Carcinogenesis 10: 1055–1057

Nelson SL, Giver CR, Grosovsky AJ (1994) Spectrum of X-ray-induced mutations in the human hprt gene. Carcinogenesis 15: 495–502

Neto JB, Gentil A, Cabral RE, Sarasin A (1992) Mutation spectrum of heat-induced abasic sites on a single-stranded shuttle vector replicated in mammalian cells. J Biol Chem 267: 19718–19723

Noodt BB, Kvam E, Steen HB, Moan J (1993) Primary DNA damage, HPRT mutation and cell inactivation photoinduced with various sensitizers in V79 cells. Photochem Photobiol 58: 541–547

O'Donnell RE, Boorstein RJ, Cunningham RP, Teebor GW (1994) Effect of pH and temperature on the stability of UV-induced repairable pyrimidine hydrates in DNA. Biochemistry 33: 9875–9880

Papadopoulos N, Nicolaides NC, Wei Y-F, Ruben SM, Carter KC, Rosen CA, Haseltine WA, Fleischmann RD, Fraser CM, Adams MD, Venter JC, Hamilton SR, Petersen GM, Watson P, Lynch HT, Peltomäki P, Mecklin J-P, de la Chapelle A, Kinzler KW, Vogelstein B (1994) Mutation of a mutL homolog in heriditary colon cancer. Science 263: 1625–1629

Pflaum M., Boiteux S, Epe B (1994) Visible light generates oxidative DNA base modifications in high excess of strand breaks in mammalian cells. Carcinogenesis 15: 297–300

Pryor WA (1986) Oxy-radicals and related species: their formation, lifetimes, and reactions. Annu Rev Physiol 48: 657–667

Purmal AA, Kow YW, Wallace SS (1994) Major oxidative products of cytosine, 5-hydroxycytosine and 5-hydroxyuracil, exhibit sequence context-dependent mispairing in vitro. Nucleic Acids Res 22: 72–78

Ravanat JL, Turesky RJ, Gremaud E, Trudesl JL, Stadler R (1995) Determination of 8-oxoguanine in DNA by gas chromatography – mass spectrometry and HPLC – electrochemical detection: overestimation of the background level of the oxidized base by the gas chromatography – mass spectrometry assay. Chem Res Toxicol (in press)

Remacle J, Raes M, Toussaint O, Renard P, Rao G (1995) Low levels of reactive oxygen species as modulators of cell function. Mutat Res 316: 103–122

Retèl J, Hoebee B, Braun JEF, Lutgerink JT, Van der Akker E, Wanamarta AH, Joenje H, Lafleur MVM (1993) Mutational specificity of oxidative DNA damage. Mutat Res 299: 165–182

Richter C (1992) Reactive oxygen and DNA damage in mitochondria. Mutat Res 275: 249–255

Roots R, Okada S (1975) Estimation of life times and diffusion distances of radicals involved in X-ray-induced DNA strand breaks or killing of mammalian cells. Radiat Res 64: 306–320

Routledge MN, Wink DA, Keefer LK, Dipple A (1993) Mutations induced by saturated aqueous nitric oxide in the pSP189 *supF* gene in human Ad293 and *E coli* MBM7070 cells. Carcinogenesis 14: 1251–1254

Saito I (1992) Photochemistry of highly organized biomolecules: sequence-selective photoreaction of DNA. Pure Appl Chem 64: 1305–1310

Schreck R, Rieber P, Baeuerle P (1991) Reactive oxygen species are apparently widely used messengers in the activation of NFκB transcription factor and HIV-1. EMBO J 10: 2247–2258

Shibutani S, Takeshita M, Grollman AP (1991) Insertion of specific bases during DNA synthesis past the oxidation-damaged base 8-oxodG. Nature 349: 431–434

Shigenaga MK, Hagen TM, Ames B (1994) Oxidative damage and mitochondrial decay in aging. Proc Natl Acad Sci USA 91: 10771–10778

Sies H (1986) Biochemistry of oxidative stress. Angew Chem Int Ed Engl 25: 1058–1071

Sies H (1991) Oxidative stress oxidants and antioxidants. Academic, London

Steenken S (1989) Purine bases, nucleosides, and nucleotides aquous solution redox chemistry and transformation reactions of their radical cations and e^- and OH adducts. Chem Rev 89: 503–520

Tchou J, Bodepudi V, Shibutani S, Antoshechkin I, Miller J, Grollman AP, Johnson F (1994) Substrate specificity of Fpg protein. J Biol Chem 269: 15318–15324
Tornaletti S, Pfeifer GP (1994) Slow repair of pyrimidine dimers at p53 mutation hotspots in skin cancer. Science 263: 1436–1438
Troll W, Wiesner R (1985) The role of oxygen radicals as a possible mechanism of tumor promotion. Annu Rev Pharmacol Toxicol 25: 509–528
Tudek B, Boiteux S, Laval J (1992) Biological properties of imidazole ring-opended N7-methylguanine in M13mp18 phage DNA. Nucleic Acids Res 20: 3079–3084
Tudek B, Laval J, Boiteux S (1993) SOS-independent mutagenesis in lacZ induced by methylene blue plus visible light. Mol Gen Genet 236: 433–439
Uggla AH (1990) The induction of chromosomal aberrations and SCEs by visible light in combination with dyes II. Cell cycle dependence, and the effect of hydroxyl radical scavengers during light exposure in cultures of Chinese hamster ovary cells sensitized with acridine orange. Mutat Res 231: 233–242
Umegaki K, Ikegami S, Ichikawa T (1993) Influence of dietary vitamin E on the 8-hydroxy-deoxyguanosine levels in rat liver DNA. J Nutr Sci Vitaminol Tokyo 39: 303–310
von Sonntag C (1987) The chemical basis of radiation biology. Taylor and Francis, London
Wagner JR, Hu C-C, Ames BN (1992) Endogenous oxidative damage of deoxycytidine in DNA. Proc Natl Acad Sci USA 89: 3380–3384
Wallace SS (1988) AP endonucleases and DNA glycosylases that recognize oxidative DNA damage. Environ Mol Mutagen 12: 431–477
Wallace DG (1992) Mitochondrial genetics a paradigm for aging and degenerative diseases? Science 256: 628–632
Waters LC, Sikpi MO, Preston RJ, Mitra S, Jaberaboansari A (1991) Mutations induced by ionizing radiation in a plasmid replicated in human cells. Radiat Res 127: 190–201
Weis M, Kass GE, Orrenius S (1994) Further characterization of the events involved in mitochondrial Ca^{2+} release and pore formation by prooxidants. Biochem Pharmacol 47: 2147–2156
Wink DA, Kasprzak KS, Maragos CM, Elespuru RK, Misra M, Dunams TM, Cebula TA, Koch WA, Andrews AW, Allen JS, Keefer LK (1991) DNA deaminating ability and genotoxicity of nitric oxide and its progenitors. Science 254: 1001–1003
Wink DA, Nims RW, Saavedra JE, Utermahlen Jr WE, Ford PC (1994) The Fenton oxidation mechanism reactivities of biologically relevant substrates with two oxidizing intermediates differ from those predicted for the hydroxyl radical. Proc Natl Acad Sci USA 91: 6604–6608
Wood, ML, Dizdaroglu M, Gajewski E, Essigmann JM (1990) Mechanistic studies of ionizing radiation and oxidative mutagenesis genetic effects of a single 8-hydroxyguanine (7-hydro-8-oxoguanine) residue inserted at a unique site in a viral genome. Biochemistry 29: 7024–7032
Wood ML, Esteve A, Morningstar ML, Kuziemko GM, Essigmann JM (1992) Genetic effects of oxidative DNA damage comparative mutagenesis of 7,8-dihydro-8-oxoguanine and 7,8-dihydro-8-oxoadenine in *Escherichia coli*. Nucleic Acids Res 20: 6023–6032
Yuan J, Yeasky TM, Rhee MC, Glazer PM (1995) Frequent TA→GC transversions in X-irradiated mouse cells. Carcinogenesis 16: 83–88
Zhivotovsky B, Wade D, Gahm A, Orrenius S, Nicotera P (1994) Formation of 50 kbp chromatin fragments in isolated liver nuclei is mediated by protease and endonuclease activation. FEBS Lett 351: 150–154
Zimmerman R, Cerutti P (1984) Active oxygen acts as a promoter of transformation in mouse embryo C3H/10$T_{1/2}$/C18 fibroblasts. Proc Natl Acad Sci USA 81: 2085–2087

Editor-in-Charge: Professor H. Grunicke and Professor M. Schweiger

Rev. Physiol. Biochem. Pharmacol., Vol. 127

Depolarization – Transcription Coupling in Excitable Cells

J. SCHMIDT

Contents

1 Introduction

Changes in the electrical potential of the plasma membrane accompany synaptic transmission and axonal conduction and constitute the elements of rapid signaling in the nervous system. Such signals are transmitted via transmembrane ionic fluxes and, depending on whether they involve the activity of ion channels alone or include the operation of second messenger

Department of Biochemistry and Cell Biology, State University of Stony Brook, Stony Brook, NY 11794, USA systems, require milliseconds to seconds and minutes. Ionic fluxes also have long-term consequences, affecting the biochemistry of nerve and effector cells over periods of hours and days. Compared to the mechanisms that mediate fast information processing, still very little is known about the processes underlying activity-dependent long-term or adaptive changes. It has long been suspected that plastic changes of excitable cells that are caused by changes in electrical membrane activity must involve alterations in gene expression (Goelet et al. 1986; Black et al. 1987; Curran and Morgan 1987); more specifically, the activity-controlled epigenetic flexibility of the neuronal phenotype has been attributed to the selection of one of several cellular programs characterized by different rates of transcription of specific subsets of genes (Laufer and Changeux 1989). Advances in the methodology of signal transduction research and of gene activity measurements have vindicated this view. While previously changes in protein composition, enzyme activities, and physiological properties could be measured and monitored in response to electrical activity, it has now become possible to infer transcriptional control from manipulations of promoters and quantification of message levels or even to measure gene activities directly; in addition, the use of transgenic animals and the generation of transgenic tissues have become more widespread, and research in intracellular messengers and signaling cascades involving kinases has progressed rapidly.

There is, accordingly, a large and growing literature on how depolarization affects the structure, function, and development of excitable cells. Evidence is accumulating for the electrical activity-driven regulation of genes coding for proteins involved in synaptic transmission and functional maturation of the nervous system. These include, but are not limited to, enzymes of neurotransmitter biosynthesis and metabolism, such as tyrosine hydroxylase (Kilbourne et al. 1992) and glutamate decarboxylase (Feldblum et al. 1990); receptors for neurotransmitters, including glutamate receptors (Gall et al. 1990; Bessho et al. 1994) and nicotinic receptors in skeletal muscle (this review); voltage-gated ion channels such as the Kv1.2 and Kv4.2 potassium channels (Tsaur et al. 1992); trophic factors such as nerve growth factor (NGF; Lu et al. 1991) and brain-derived neurotrophic factor (BDNF; Isackson et al. 1991; Ghosh et al. 1994a) and their receptors (p140trk; Birren et al. 1992); and neuropeptides such as proenkephalin (Morris et al. 1988; MacArthur et al. 1993), prodynorphin (Morris et al. 1988), vasoactive intestinal peptide (Adler and Fink 1993), gastrin (Brand and Wang 1988), and neuropeptide Y (Higuchi et al. 1990). In addition, electrical activity influences more complex intercellular interactions, such as neural sprouting, long-term potentiation and depression, morphological changes, maturation of pathways, and map refinement and rearrangement (Constantine-Paton et al. 1990). Although in these instances molecular mechanisms are largely unknown, it is plausible that proteins such as tissue plasminogen activator, induced as an immediate-early gene (IEG) in various activation paradigms (Qian et al. 1993), and neural cell adhesion molecule (N-CAM), upregulated by denervation in skeletal muscle (Covault et al. 1986), participate in remodeling processes.

No attempt will be made in this review to cover this entire literature. Instead, the focus will be on two intensively studied paradigms of the coupling of membrane depolarization and gene activity: (1) the regulation of the proto-oncogene c-*fos* in neurons and (2) the control of genes coding for subunits of the acetylcholine receptor (AChR) in skeletal muscle. At the same time the emphasis will be shifted to the intracellular signaling cascades involved. These two examples were selected not only for their physiological relevance, but also for the relative abundance of currently available information. The expectation is that at least some of the observations made in these systems will shed light on other cases of depolarization-transcription coupling.

2 Depolarization-Dependent c-*fos* Stimulation in Neurons: Actuation of a Transcriptional "On" Switch

2.1 Neuronal Activity and Immediate-Early Genes

Stimulation of neurons with neurotransmitters and neuromodulators causes changes in gene expression that are critical for neuronal survival, differentiation, and the plasticity, particularly of chemical and electrical excitability, of mature neuronal cells (Black et al. 1987; Morgan et al. 1987). Much of the early work on the signaling pathway connecting the plasma membrane and gene activation was devoted to a class of genes designated "immediate-early genes" (IEG). Characteristically, these genes are expressed at low levels in quiescent cells and upon stimulation are induced rapidly, transiently, and independently of protein synthesis. Many IEG are thought to encode nuclear proteins that govern genomic responses to extracellular stimuli by coordinating expression of downstream genes. These late-onset or late-response genes are induced and repressed more slowly, and they differ mechanistically from IEG in that their response depends on new protein synthesis (Curran and Morgan 1987). IEG were first identified as genes that are rapidly and transiently activated when quiescent fibroblasts are stimulated with serum or growth factors to reenter the cell cycle (Greenberg and Ziff 1984). In the nervous system several IEG have been identified and analyzed; they include the following: c-*fos* (the *fos* proto-oncogene) and its dimerization partner c-*jun*; *jun*-B and *jun*-D, which also dimerize with c-*fos* and *fos*-related factors; and zif/268, a transcription factor containing zinc finger motifs, and proteins related to the steroid receptor family members nur77 and nurr1 (for review of IEG in the nervous system, see Sheng and Greenberg 1990; Morgan and Curran 1991).

2.2 Neuronal Activation Induces c-*fos* Transcription

Among the IEG present in the nervous system, c-*fos* has been most thoroughly studied. Its protein product Fos, a nuclear 62-kDa phosphoprotein, and similar factors (fra, *fos*-related antigens) form heterodimers with members of the Jun protein family (c-Jun, Jun-B, Jun-D). These complexes are held together by a leucine zipper (Landschulz et al. 1988) and, on binding to an activating protein (AP)-1 site (consensus, TGACTCA), activate target genes (for review see Morgan and Curran 1991). In the nervous system, potential examples of target genes downstream of c-*fos* include proenkephalin (Sonnenberg et al. 1989c) and genes responsible for the apoptotic program (Smeyne et al. 1993). Much of the work on c-*fos* activation in neurons was done with rat pheochromocytoma PC12 cells, which upon treatment with NGF acquire a sympathetic neuron-like phenotype. Treatment with NGF not only induces neuronal differentiation, but also stimulates, within minutes, c-*fos* transcription by over 100-fold (Greenberg et al. 1985; Kruijer et al. 1985). Initially the expression of the gene was analyzed by measuring levels of the Fos protein, which is present at low levels in unstimulated cells. Due to the short half-life of the protein (about 2 h; Müller et al. 1984), the rapid turnover of the mRNA coding for it (half-life, 10–15 min; Sheng and Greenberg 1990), and the transient nature of gene activation (onset of activity within 5 min and duration of about 15 min; Greenberg et al. 1985), the amount of protein reasonably well reflects the activity of the gene. For the same reasons, the steady state level of c-*fos* mRNA is an even better indicator of gene activity. To directly demonstrate increased rates of transcription nuclear run-on or transcript elongation experiments are necessary; using such assays, it was shown that changes in mRNA levels in response to depolarization, calcium ionophores, and barium ions are a result of transcriptional activation (Sheng et al. 1988). In addition, the profound and well-documented effects of promoter manipulation in transfection protocols strongly suggest that c-*fos* regulation occurs at least in part at the transcriptional level. The transient nature of the activation may be related to the ability of Fos to inhibit its own expression (Morgan and Curran 1991).

The analogy with effects of mitogenic polypeptide growth factors on fibroblasts prompted studies with nerve cells which led to the conclusion that in these cells likewise upregulation of c-*fos* and other IEG is triggered by a large number of stimuli. These include the following: agonists of neurotransmitter receptors such as nicotine (Greenberg et al. 1986), carbachol (Smeyne et al. 1992), glutamate (Smeyne et al. 1992; Ghosh et al. 1994b), kainic acid (Sonnenberg et al. 1989b; Smeyne et al. 1992; Lerea and McNamara 1993), and *N*-methyl-d-aspartate (NMDA; Lerea and McNamara 1993); neurotrophic and growth factors (other than NGF) such as platelet-derived growth factor (PDGF; Smeyne et al. 1992), epidermal growth factor (EGF; Greenberg et al. 1985), and fibroblast growth factor (FGF; Curran and Morgan 1985; Greenberg et al. 1985); insulin (Greenberg et al. 1985); drugs that stimulate various intracellular signaling pathways such as phorbol esters (Greenberg et al. 1985) and cyclic adenosine monophosphate (cAMP) analogues (Greenberg et al. 1985); the calcium ionophore A23187 (Sheng et al. 1988, 1990); the L channel agonist Bay-K8644 (Murphy et al. 1991; Morgan and Curran 1986); and elevated potassium (Greenberg et al. 1985; Morgan and Curran 1986; Smeyne et al. 1992).

2.3 c-*fos* is Stimulated by Depolarization

The most characteristic type of neuronal excitation is membrane depolarization, which is an integral part of synaptic activation and axonal conduction. It was therefore important to examine whether depolarization of neuronal cells might be one of the stimuli to trigger activation of IEG and particularly c-*fos*. Several depolarizing agents were shown to cause c-*fos* activation, among them KCl (Greenberg et al. 1985; Morgan and Curran 1986; Bartel et al. 1989; Smeyne et al. 1992), the sodium channel activator veratridine (Morgan and Curran 1986), and the calcium mimic barium (Curran and Morgan 1986; Sheng et al. 1988), while the sodium channel blocker tetrodotoxin antagonized these effects (Morgan and Curran 1986). IEG other than c-*fos* also have been investigated; depolarization, in contrast to the action of growth factors, affects c-*jun* and zif/268 little, while very strongly activating nur77 (Bartel et al. 1989; Morgan and Curran 1991) as well as nurr1 (Law et al. 1992). In addition to PC12 cells, primary cortical and hippocampal neurons have been utilized in these investigations (Bading and Greenberg 1991; Bading et al. 1993). Effects of membrane activity on c-*fos* expression have been studied in vivo as well. An elegant demonstration that c-*fos* expression is a physiological consequence of neuronal activity is the appearance of c-*fos* immunoreactivity in dorsal horn neurons after peripheral sensory stimulation (Hunt et al. 1987). Direct electrical stimulation was shown to induce c-*fos* as well as several other IEG (Dragunow and Robertson 1987; Sonnenberg et al. 1989a; Wisden et al. 1990). Long-term potentiation was shown not to induce c-*fos*, but instead rather specifically to upregulate zif/268 (Wisden et al. 1990; Qian et al. 1993).

Curran and Morgan and colleagues, among others, have investigated chemically induced seizures which are likely to bring about electrical activity of a large number of neurons (Morgan et al. 1987; Sonnenberg et al. 1989a; Saffen et al. 1988, Watson and Millbrandt 1989). c-*fos* mRNA as well as Fos protein transiently appear after seizures with kinetics similar to those seen in cultured cells, suggesting that the upregulation is likewise based on gene activation; levels of c-*fos* mRNA peak between 30 min and 1 h after administration of the convulsant, while Fos protein reaches a maximum at 1–2 h and then gradually declines. AP-1 binding activity follows a more protracted time course, a fact that finds its explanation in the delayed expression of fra (*fos*-related antigens; Morgan et al. 1987; Sonnenberg et al. 1989a). Massive upregulation of c-*fos* is observed in cortex, basal ganglia, and the olfactory bulb, with more limited responses seen in other parts of the brain and very little in non-nervous tissue (lung; Morgan et al. 1987). It is difficult in these experiments to separate the effects of depolarization from secondary synaptic effects due to release of neurotransmitters and neuromodulators. For instance, the fact that metrazol seizure effects are largely blocked by amino-phosphonovalerate (APV) and MK-801 (Sonnenberg et al. 1989b) suggests that depolarization as such, though undoubtedly occurring, is less important than the activation of NMDA receptors. Similarly, the effects seen with various kinds of other stimulatory compounds are not completely separable from the consequences of depolarization; such compounds may themselves cause depolarization, as in the case of veratridine (Morgan and Curran 1986) and the cholinergic agonist nicotine (Greenberg et al. 1986; Bartel et al. 1989), or require extracellular calcium, as in the case of activators of NMDA receptors (Bading et al. 1993; Lerea and McNamara 1993) or target transcription factors that are also utilized by depolarization signals, as would be the case with cAMP analogues and forskolin (Sheng et al. 1990; Liu et al. 1993).

2.4 *Cis* Elements in the c-*fos* Regulatory Region

The signals leading to the activation of the c-*fos* gene ultimately must impinge on specific *cis* elements in control regions of the gene. As a result, the promoter region of this gene has been thoroughly characterized, and several regulatory elements have been identified. Among these a serum response element (SRE) at – 310 upstream of the transcription start site (Treisman 1985) plays a central role in the growth factor-dependent activation of the gene. From the point of view of depolarization-transcription coupling, the important motif is the sequence TGACGTTT at – 60, reminiscent of the cAMP reponse element (CRE; Montminy et al. 1986), whose consensus sequence is TGACGTCA. The CRE mediates cAMP effects; since it responds to depolarization and calcium influx as well, it has also been designated a CaRE (calcium response element; Sheng et al. 1988). The CRE/CaRE thus has the capacity of integrating signals from both second messengers (Sheng et al. 1990). CRE-like motifs have been identified in the regulatory regions of several activity-dependent genes. These include other IEG; nur77, for example, contains four CRE-like elements (Watson and Millbrandt 1989). In addition, genes coding for tyrosine hydroxylase (Kim et al. 1993) and the neuropeptides somatostatin (Montminy et al. 1986), vasoactive intestinal polypeptide (Tsukuda et al. 1987), and proenkephalin (Comb et al. 1986) contain CRE (for review of neuropeptide genes, see Goodman 1990). While early findings (Sheng et al. 1988) suggested that the SRE is unimportant for calcium activation, later analysis has led to a more differentiated view. Thus it is now appreciated that CRE-binding protein (CREB; see below) can bind to at least two sites flanking the SRE, the sis-inducible elements (SIE) at –340 and the c-*fos* AP-1/CRE site (FAP) at –290; these elements contain the CGTCA CREB consensus recognition sequence (Ginty et al. 1994). In transgenic animals, conservation of the SRE and the neighboring SIE and FAP elements are crucial for all regulatory influences, while the CRE/CaRE is essential only for the transmission of the cAMP signal (Robertson et al. 1994). At any rate there is at present little evidence that factors which directly interact with the SRE are crucially involved in the depolarization response.

2.5 Transacting Factors

The CRE is recognized and activated by the transcription factor CREB (Montminy and Bilezikjian 1987; Sheng et al. 1990), a 43-kDa nuclear protein and member of a family of transcription factors that, like Fos and Jun, dimerize via a leucine zipper; this family includes CREB B (Sheng et al. 1991) and activating transcription factor (ATF)-1 (Liu et al. 1993), both of which respond to cAMP as well as to depolarization signals. CREB is thought to be active as a homodimer. Since AP-1 binding activity does not change during c-*fos* induction – based on gel mobility shift assays (Sheng et al. 1990) and interaction with affinity sorbents (Sheng et al. 1991) – it has been proposed that CREB homodimers are constitutively bound to the CRE motif and activated in situ. Activation is effected by phosphorylation at a specific residue, Ser-133, in response to increases in cAMP or to the influx of calcium into stimulated cells. The integration of these two signaling pathways targeting the CRE/CaRE element seemingly occurs at the level of the CREB protein; this was confirmed in experiments in PC12 cells in which

GAL-4-CREB fusion constructs drive GAL-4-containing reporter genes in response to calcium influx (Greenberg et al. 1992).

2.6 Role of Calcium

Intracellular calcium functions as a second messenger in all cells; in nerve cells it triggers a variety of physiological events such as neurotransmitter release and modulation of synaptic transmission. There seems to be little doubt that influx of extracellular calcium also effects long-term, adaptive changes in neurons, including synaptic remodeling, and morphological changes such as neurite extension, and finally alterations in gene expression. It might be postulated on theoretical grounds that external stimuli should affect neuronal plasticity via the influx of calcium; as Bertil Hille observed in 1984, calcium channels "serve as the *only* link to transduce depolarization into all the nonelectrical activities controlled by excitation. Without Ca channels our nervous system would have no outputs." The notion that electrical stimulation induces c-*fos* expression via an influx of extracellular calcium has been corroborated by many studies involving a variety of calcium-specific reagents. Removal of extracellular calcium with chelating agents eliminates the effect of KCl-induced depolarization (Morgan and Curran 1986; Bading et al. 1993; Ghosh et al. 1994b). The signaling cascade can be triggered with reagents that activate ion channels and thereby directly permit the influx of calcium ions, such as BayK-6844, an activator of the L-type calcium channel (Murphy et al. 1991). Alternatively, treatments that cause depolarization, such as KCl and the sodium channel agonist veratridine, can be employed; they act through activation of voltage-gated channels. Blockers of the L-type voltage-gated calcium channel, such as dihydropyridines, block c-*fos* induction by KCl and veratridine (Morgan and Curran 1986; Greenberg et al. 1986); they also suppress c-*fos* expression in cortical neurons in culture (Murphy et al. 1991) and depolarization-induced expression of c-*fos* in hippocampal neurons (Bading et al. 1993) and of BDNF in cortical neurons (Ghosh et al. 1994a). L channels differ in biophysical and pharmacological properties from other types of voltage-gated calcium channels. Although a systematic investigation of the involvement of these other channel subtypes seems not to have been undertaken, it is noteworthy that L channels carry the calcium current in PC12 cells, a cell line widely used in these investigations; furthermore, L channels are suitably positioned over the soma and dendrites of neurons to play a role in transsynaptic signaling (Westenbroek et al. 1990); finally, the N-channel blocker omega conotoxin does not interfere with depolarization-transcription coupling in primary cortical neurons (Murphy et al. 1991). Since the calmodulin antagonists chlorpromazine and trifluoperazine abolish the response (Morgan and Curran 1986), it has been assumed that calmodulin acts as the calcium sensor in the cytosol; these somewhat nonspecific drugs were subsequently replaced with the more selective drugs calmidazolium and the calcium-calmodulin kinase antagonist KN-62, which in hippocampal neurons partly inhibit depolarization-triggered c-*fos* activation (Bading et al. 1993). In PC12 cells, A23187 mimics the effects of KCl-induced depolarization (Sheng et al. 1988, 1990, 1991). Thapsigargin, which blocks reuptake of calcium into intracellular stores, was shown to moderately activate *ras*, which may link depolarization to c-*fos* activation via mitogen-activated protein (MAP) kinase (Rosen et al. 1994). Barium, which readily passes through calcium channels, has been shown to mimic the effect of depolarization (Curran and Morgan 1986; Sheng et al. 1988). In hippocampal neurons, even NMDA receptor-mediated c-*fos* stimulation requires calcium influx (Bading and Greenberg 1991), and it is interesting that in these cells, depending on the port of entry (L-type channel or NMDA receptor channel), calcium differentially stimulates gene transcription (Bading et al. 1993); in the same cells, calcium entering through NMDA and non-NMDA glutamate receptors has been shown to utilize distinct signaling pathways (Lerea and McNamara 1993). Similarly, in cortical neurons, expression of BDNF is stimulated by voltage-gated calcium channels, but not NMDA receptor channels, although both conduct calcium (Ghosh et al. 1994a).

2.7 Transcytosolic Signaling

The strong evidence from hippocampal and cortical neurons suggesting that there are at least two distinct calcium-triggered pathways from the plasma membrane to the nucleus makes it highly unlikely that calcium itself could serve as the transcytosolic messenger. Instead the assumption must be made that calcium acts in a well-circumscribed microdomain at its site of entry to initiate distinct signals, most likely via a phosphorylation cascade. In PC12 cells, it has long been known that several distinct kinase signaling systems can induce c-*fos*: the diacylglycerol (DAG)/protein kinase C (PKC) pathway (Greenberg et al. 1986; Sheng et al. 1988), the cAMP pathway (Montminy et al. 1986; Sheng et al. 1988), and the calcium/calmodulin pathway (Morgan and Curran 1986). Of these kinases, protein kinase A (PKA) and a calcium/calmodulin-dependent kinase (CamK) can phosphorylate CREB on Ser-133 and thereby activate it; recently, the novel CREB kinase has been added to this group (Ginty et al. 1994). Tracing the depolarization transcription coupling pathway to completion requires linking one of these candidate kinases to calcium influx.

There is little evidence that PKA mediates depolarization effects, since cAMP does not rise in response to depolarization (Sheng et al. 1990). A better case can be made for calmodulin: calmodulin inhibitors (phenothiazines as well as the more specific calmidazolium) were shown to disrupt the signaling cascade, at least partly. Calmodulin-dependent kinase is not only the most plausible enzyme to provide a simple link between calcium influx and gene activation; its candidacy is also supported by considerable experimental evidence. CamKII is activated by depolarization in PC12 cells (Griffith and Schulman 1988); CREB fused to the DNA-binding domain of the yeast transactivating factor GAL-4 is responsive to calcium influx (Greenberg et al. 1992); both CamKI and CamKII act on CREB in vitro (Greenberg et al. 1992), resulting in the phosphorylation of Ser-133, the same residue that is targeted in vivo by the depolarization signal (Sheng et al. 1991). CREB is a very efficient substrate for CamKII (Greenberg et al. 1992). Significantly, CREB phosphorylated at Ser-133 is a more potent activator in an in vitro transcription system (Dash et al. 1991), and mutation of Ser-133 to Ala significantly inhibits the depolarization response (Sheng et al. 1991). The evidence, however, at present is still circumstantial and rests to a large extent on the specificity of calmodulin and CamK inhibitors; it remains to be established that CamK is localized to the nucleus and that the enzyme is required for CREB phosphorylation in vivo (Ghosh et al. 1994b).

The calcium signal may, to some extent, target elements other than the CRE/CaRE in the c-*fos* promoter. This may seem surprising considering promoter investigations carried out in vitro; however, more recent work with *fos-lac*Z transgenic animals suggests that the CRE/CaRE is crucial for regulation only by cAMP, and not by membrane depolarization. Mutation of any of the c-*fos* regulatory elements (SIE, SRE, FAP, CRE/CaRE) inhibited KCl-induced c-*fos* upregulation (Robertson et al. 1994). This would indicate that, in vivo, integrity of the promoter is required because either CREB interacts with elements other than, and in addition to, the CRE/CaRE or because other transcription factors, such as the SRF, are regulated by calcium influx to a greater extent than has been appreciated. At any rate, components of such an alternative calcium/c-*fos* pathway have been identified. NMDA receptor activation has been shown to activate MAP kinase (Bading and Greenberg 1991). Activation of MAP kinase by electroconvulsive shock (Baraban et al. 1993) and generalized seizures (Gass et al. 1993) is likely to be a consequence of NMDA receptor activation. In PC12 cells, membrane depolarization and the resulting flow of calcium ions through activated L channels likewise leads to activation of MAP kinase; the latter pathway has been elucidated in some detail more recently and has been shown to involve the monomeric G-protein p21ras and the dual-function kinase MEK1 (Rosen et al. 1994). The link between the influx of calcium and the activation of *ras* remains to be found. Greenberg and colleagues (Rosen et al. 1994) speculate that PKC might act as the calcium target, since activation of cholinergic receptors had previously been shown to cause PKC translocation in these cells (Messing et al. 1989); PKC could then lead to the inhibition of a GAP (GTPase-activating protein), thereby leading to *ras* activation, as phorbol esters activate *ras* in PC12 cells (Thomas et al. 1992). This model, which does not address the origin of the lipid activator probably also involved in PKC activation, implies that the level of cytosolic calcium rather than the amount of a lipid activator is limiting in the resting cell. Another possibility that these authors consider involves the tyrosine phosphorylation of calcium channels in response to calcium influx, eventually leading to *ras* activation via adaptor proteins harboring SH2 domains; this would still leave the primary calcium target undefined. CREB kinase has recently been identified as a kinase in the *ras* pathway linking NGF stimulation to CREB phosphorylation. It may be activated by a number of extracellular stimuli, including elevated KCl (Ginty et al. 1994), but too little is known at present to speculate on how a depolarization signal might reach it.

3 Downregulation of the γ-Isoform of the Acetylcholine Receptor by Electrical Activity: Actuation of a Transcriptional "Off" Switch

3.1 Effect of Electrical Activity on Gene Expression in Skeletal Muscle

Skeletal muscle displays phenotypic plasticity; in response to external stimuli and changing functional demands it is capable of switching fiber type by expressing functionally different variants of membrane (sarcolemma; sarcoplasmic reticulum, SR) and sarcomeric proteins and of enzymes of energy metabolism. Much of our knowledge of these activity-driven changes has been obtained in experiments using chronic, low-frequency stimulation of mammalian muscle. The majority of these studies involved measurement of protein levels, enzyme activities, or physiological parameters and did not provide much insight into regulatory mechanisms. Thus changes in protein levels could be the result of transcriptional or post-transcriptional events; for example, increased protein turnover accounts for the drop in the amount of myofibrillar protein in inactive muscle (Strohman et al. 1981); similarly, tetrodotoxin treatment has been shown to cause impaired assembly and enhanced secretion of acetylcholinesterase in cultured muscle cells (Fernandez-Valle and Rotundo 1989). The functional state of muscle probably also affects gene activity, since electrostimulation leads to the appearance of proteins that were previously present at low or undetectable levels. In many cases a correlation between protein and mRNA levels has been detected, suggesting that protein synthesis is affected by membrane activity and, in addition, that the gene coding for the protein in question might be the site of control. Nevertheless, none of the numerous studies on activity-driven phenotypic plasticity of skeletal muscle have yet been backed up by direct gene activity analysis (for a review, see Purves 1976; Pette and Vrbová 1992).

3.2 Elimination of the Embryonic Form of the Acetylcholine Receptor

Among the many genes whose expression is influenced by electrical activity, those coding for the subunits of the AChR are probably the most thoroughly studied. The junctional AChR in the mature muscle fiber is a pentameric membrane protein composed of α-, β-, δ-, and ε-subunits in the stoichiometry 2:1:1:1 (for a recent review of AChR structure, see Devillers-Thiery et al. 1993). Analysis of the formation of the neuromuscular junction has shown that the appearance of this junctional AChR is preceded by the ubiquitous expression of an embryonic form in which a γ-subunit substitutes for the ε-chain of the adult AChR. These embryonic or extrajunctional AChR appear when myoblasts fuse to form myofibers. In differentiated and uninnervated myofibers such as arise in cultures of primary myogenic cells, the γ-isoform of the receptor is expressed at high rates. In the embryo, innervation first leads to clustering of receptor at the point where the nerve contacts the fiber; subsequently, extrasynaptic receptors are lost, and finally ε-type receptors replace the γ-isoform.

Denervation leads to the reappearance of the γ-subtype in extrajunctional areas, a change that is reversed by reinnervation or direct electrical stimulation. Denervated muscle thus serves as a convenient preparation in which to study AChR suppression (for a brief review, see Hall and Sanes 1993). The neurally mediated suppression of the γ-isoform could be mediated by a chemical signal or might be a simple consequence of neuromuscular transmission. Numerous experiments have shown that electrical stimulation of the denervated muscle fiber, even in the absence of reinnervation, leads to downregulation (Jones and Vrbová 1974; Lømo and Rosenthal 1972; Drachman and Witzke 1972; Lømo and Westgaard 1975; for an excellent review of the older literature, see Fambrough 1979); contractile activity does not seem to be required for this effect (Gruener et al. 1974). In 1976, Hall and collaborators (Hogan et al. 1976), using the electrostimulated rat diaphragm, showed that the activity-driven elimination of the γ-type AChR proceeds without enhanced turnover, suggesting inhibition of receptor expression. Subsequently, the same group (Hall and Reinness 1977) demonstrated that the decline of AChR levels is the result of a block in new receptor synthesis.

Activity-dependent gene regulation can also be studied in cultured muscle cells displaying spontaneous activity. Inhibition of the voltage-gated sodium channel with tetrodotoxin (Cohen and Fischbach 1973; Betz and Changeux 1979) and of voltage-gated calcium channels with the phenylalkylamines verapamil (Klarsfeld et al. 1989) and D600 (Shainberg et al. 1976; Pezzementi and Schmidt 1981) leads to increased receptor expression, while the effect of depolarization can be mimicked by exposure to the sodium channel agonist veratridine (Betz and Changeux 1979; Shieh et al. 1988; for additional information, see Fambrough 1979; Salpeter and Loring 1985; Laufer and Changeux 1989).

3.3 Electrical Activity Blocks Acetylcholine Receptor Subunit Gene Transcription

That muscle membrane depolarization blocks AChR expression at the level of individual subunit genes was first suggested by mRNA measurements. Heinemann and collaborators observed that electrical stimulation, begun at the time of denervation, blocks the increase in the amount of mRNA coding for α-, β-, γ-, and δ- subunits in rat muscle (Goldman et al. 1988); this was confirmed for the α-subunit by Eftimie et al. (1991). Witzemann et al. (1991) found that mRNA encoding the α-, γ-, and δ-subunits are strongly reduced by exogenous stimulation of chronically denervated rat soleus muscle. Later, Neville et al. (1992) showed that in denervated chick muscle, upon onset of electrostimulation, the α-subunit mRNA disappears with a half-life of 4 h, indistinguishable from the turnover rate of the mRNA in unstimulated muscle. This observation strongly suggested a rapid shutdown of the α-subunit gene. Direct confirmation of this inference came when Huang et al. (1992) found that electrostimulation of denervated chick muscle in situ blocks the transcription of the α-, γ-, and δ-subunit genes within minutes. In addition, the delineation, however preliminary, of a depolarization response element in receptor promoters (see below) confirms the gene as the site of activity-dependent regulation.

3.4 Effect of Cessation of Electrical Activity on Receptor Expression

Just as treatments causing depolarization of the sarcolemma inhibit receptor expression, manipulations leading to the elimination of neurally evoked or spontaneous electrical activity of muscle fibers cause an upregulation of AChR. This phenomenon was originally discovered as an increase in acetylcholine sensitivity of skeletal muscle following nerve section (Axelsson and Thesleff 1959). Denervation supersensitivity was first correctly attributed to an increase in the number of AChR by Miledi and Potter (1971), who, using radiolabeled α-bungarotoxin, established that denervation is followed by an increase in receptor density that easily accounts for the observed increase in transmitter sensitivity. It was subsequently shown that other treatments resulting in the block of action potentials within the muscle fiber such as tetrodotoxin cuffs around the motor nerve, treatment of muscle with botulinum toxin (which inhibits acetylcholine release), treatment with α-bungarotoxin (which blocks the postsynaptic response to the transmitter), and treatment of the muscle with tetrodotoxin or local anesthetics that block action potentials directly all cause an upregulation of the AChR (for review, see Fambrough 1979). Brockes and Hall (1975) established that the denervation effect was a consequence of increased receptor synthesis (rather than of a stabilization of the receptor in the plasma membrane); this in turn was later attributed to an increase in the availability of subunit transcripts (Merlie et al. 1984; Evans et al. 1987; Moss et al. 1987; Shieh et al. 1988). Similar observations were made with cultured cells; Klarsfeld and Changeux (1985) found that tetrodotoxin treatment of chick primary muscle cells raises the level of α-subunit mRNA by about an order of magnitude, similar to the effect of denervation (17-fold increase); comparable results were obtained by Shieh et al. (1988). That the rise in mRNA levels in turn might result from increased synthesis rather than decelerated turnover was suggested by the appearance of a putative splicing intermediate of the α-subunit transcript in response to denervation (Shieh et al. 1987) and by the detection of intronic transcripts (Klarsfeld et al. 1989). Eventually gene activity was measured directly in nuclei isolated from muscle at various times after denervation, confirming a significant and coordinate increase in the activity of the α-, γ-, and δ-subunit genes (Tsay and Schmidt 1989). In vitro treatment with tetrodotoxin and local anesthetics had previously been known to cause AChR upregulation (see above). These treatments also result in an increase in receptor messages including the putative α-subunit mRNA precursor (Klarsfeld and Changeux 1985; Shieh et al. 1988; Harris et al. 1988).

As in many other situations in cellular metabolism, up- and downregulation of the AChR are not simply mirror images of each other. While depolarization-driven receptor gene inactivation is a fast, post-translational event that happens on a time scale of minutes, cessation of electrical activity leads to gene activation over a period of 10–20 h and requires both RNA and protein synthesis. As early as 1970 Fambrough showed that denervation effects can be prevented by treating organ-cultured rat diaphragm with actinomycin D or cycloheximide. Later, Changeux' group demonstrated that protein synthesis is required for the tetrodotoxin-stimulated increase in AChR expression (Duclert et al. 1990), an observation confirmed for the denervation-triggered activation of AChR genes in chick skeletal muscle by Tsay et al. (1990).

3.5 Search for the Depolarization Response Element

The regulatory regions of many AChR subunit genes have been analyzed; all available evidence points to a subset of E boxes in the AChR subunit gene 5' flanking regions as the element responsive to electrical activity of the plasma membrane. Thus, every promoter fragment known to be responsive to depolarization contains at least one E box: chick α850 (Merlie and Kornhauser 1989; Sanes et al. 1991; Salmon and Changeux 1992); chick α842 (Bessereau et al. 1994); chick α829 (Gundersen et al. 1993); chick α829 and α111 (Merlie et al. 1994); mouse δ1850 (Simon et al. 1992); rat δ102 (Chahine et al. 1992); mouse δ181 (Dutton et al. 1993); mouse myogenin MYG3700 (Buonanno et al. 1993); mouse myogenin MYG1565 and MYG335 (Merlie et al. 1994); mouse ε3500 (Sanes et al. 1991); and mouse ε3500, ε830, and ε280 (Gundersen et al. 1993). As a rule these E boxes are parts of enhancer-like elements that impart tissue and developmental specificity to the receptor genes. The most direct evidence for the involvement of E boxes in the activity response was obtained by Bessereau et al. (1994), who showed that in transgenic animals carrying *lac*Z under the control of an α-subunit region of 850 bp a denervation response of the reporter enzyme is observed, which is significantly diminished upon mutation of one of the two E boxes present in the α-enhancer; however, a shortened α-enhancer construct which still contains both intact E boxes lost its ability to respond to denervation. Thus specific E boxes seem to be required, but are not necessarily sufficient for the depolarization response.

3.6 Search for Transacting Factors – Possible Involvement of Myogenic Factors

The tentative identification of E boxes as mediators of the depolarization response is plausible as well as perplexing: plausible, because E boxes serve as targets for the MyoD family of transcription regulators, and perplexing, because clearly genes that are little affected by activity as well as those exceedingly sensitive to it contain E boxes as functional elements in their promoters. Examples are the myosin light chain (MLC) and muscle creatine kinase (MCK) genes, which are little affected by denervation, but contain E boxes as important functional elements in their promoters. Thus, Merlie and Kornhauser used MLC as a negative control in a denervation experiment (Merlie and Kornhauser 1989), and MCK mRNA levels, in contrast to messages for the AChR subunits, gradually decline in response to denervation (Duclert et al. 1991). The solution to this E box dilemma must lie in the multiplicity of E boxes and of transactivator complexes associating with them. The CANNTG motif occurs every 256 nucleotide (nt) in a random sequence; as a result there are many more such motifs than are likely to have physiological significance within the genome or even within promoter sequences. Therefore, only a subset of these sequences are probably regulatory elements. A subset of these again reside in muscle-specific promoters and are targets of myogenic determination factors or "M proteins." Since there are four such proteins (MyoD, myogenin, myf5, and herculin/MRF4; for a review of myogenic factors, see Weintraub et al. 1991; Edmondson and Olson 1993; Weintraub 1993), it is conceivable that one of them, possibly in conjunction with a specific E protein as dimerization partner, could serve as the transactivating factor responsive to, i.e., inhibited by, membrane depolarization. This presupposes that individual E boxes are preferentially recognized by a specific transactivator. That individual M proteins have subtle predilections for flanking and central nucleotides not specified in the basic CANNTG motif has been shown using cycles of amplification and selection of appropriate target sequences ("CASTing"; Blackwell and Weintraub 1990; Wright et al. 1991; Funk and Wright 1992).

Of the four known myogenic determination factors or M proteins, myogenin is most likely to mediate the depolarization response. It is noteworthy that an antisense oligonucleotide directed against the 5' end of the myogenin message was shown to block the expression of the AChR δ-subunit in BC3H-1 cells (Brunetti and Goldfine 1989); however, since myogenin in this system is required for differentiation of myogenic cells, a distinction between general developmental effects and specific receptor regulation is difficult. Analysis of M protein expression during development and after denervation in mouse muscle has shown that both myf-5 and herculin/MRF4 are unlikely candidates, the former because it is not strongly upregulated by denervation and the latter because, unlike AChR genes, it is not downregulated in the postnatal phase (Duclert et al. 1991). MyoD and myogenin, on the other hand, in general parallel the expression of receptor subunits; in particular, denervation-induced increases in MyoD and myogenin protein have been documented (Weis 1994). In rat muscle, Witzemann and Sakmann (1991) found that myogenin also is more strongly affected by denervation and reinnervation than is MyoD and proposed that it could be involved in the regulation of α-, γ-, and δ-subunits. An mRNA analysis in denervated chick muscle revealed that only myogenin resembled receptor subunit expression in the extent and kinetics of activation (Neville et al. 1992). This also holds at the level of gene activity: myogenin transcription rate goes up by approximately an order of magnitude and rapidly drops in response to electrostimulation (Huang et al. 1993). The slow upregulation of receptor in response to denervation and the quick inactivation of receptor genes upon stimulation is best explained by invoking an autocatalytic transcription factor (Neville et al. 1991); such a factor would be assumed to activate its own promoter after cessation of membrane signals, thereby giving rise to increased amount of factor protein, in agreement with the finding that denervation-induced activation of receptor genes requires ongoing protein synthesis (Duclert et al. 1990; Tsay et al. 1990). Indeed myogenin promoters have been studied in human, mouse, and chick muscle, and a conserved functional E box has been identified near the TATA box. It has also been shown that this E box serves as a target for the myogenin protein (S. Malik and J. Schmidt, unpublished). Another argument for myogenin comes from the analysis of the inactivation kinetics: the myogenin gene in chick muscle is rapidly inactivated by electrostimulation, as rapidly in fact as the genes coding for receptor subunits, while genes for the M proteins respond much more slowly (Huang et al. 1993). Finally, the state of phosphorylation of myogenin depends on the electrical activity of the muscle cell (Mendelzon et al. 1994). The most convincing experiment would be to block denervation supersensitivity by elimination of myogenin. This is not easily accomplished, as muscle differentiation completely fails in myogenin null mutants during early development (Hasty et al. 1993). Perhaps this aim should be pursued using antisense technology in the postnatal period.

3.7 The Signaling Pathway Proper – Role of Protein Kinase C

When Laufer and Changeux reviewed activity-dependent receptor expression in 1989, they emphasized that the slow progress in the field was due to insufficient specificity of the drugs employed, a preponderance of measurements of events downstream of gene activity (mRNA levels, protein levels), and the complexity of signaling pathways and their possible crosstalk. These difficulties have been at least partly solved in the meantime: gene activities can be measured directly in cultured cells and in nuclei isolated from skeletal muscle; since such measurements are possible within seconds and minutes after the onset of electrostimulation, gene regulation can therefore be isolated not only mechanistically, but also kinetically, from regulatory events affecting post-transcriptional processes; and more specific reagents are becoming available, such as antisense oligonucleotides and expression vectors for specific inhibitors of potential cascade elements.

The rapid inactivation of the transcription factor driving AChR gene expression is likely to come about through a post-translational modification. A commonly encountered covalent modification is phosphorylation; attempts at identifying specific kinase pathways have involved the use of specific activating drugs. The first kinase to be investigated was PKA.

Moderate increases in receptor expression were observed in primary muscle cells upon stimulation of PKA (Betz and Changeux 1979). Subsequent investigations revealed that cAMP levels do not change upon treatment with tetrodotoxin and local anesthetics; in addition tetrodotoxin and cholera toxin have additive effects, suggesting independent mechanisms (Betz 1980; McManaman et al. 1982). More recently, Goldman and collaborators reported that forskolin counteracts tetrodotoxin effects on receptor subunit messages in rat primary muscle cells in culture; these investigators have proposed that depolarization blocks receptor gene activity via inactivation of PKA (Chahine et al. 1993). It is difficult to reconcile their findings with the experiments performed by other investigators in the chick and mouse, especially since possible contributions of post-transcriptional effects have not been ruled out.

The case for PKC is altogether more convincing. Phorbol esters were observed to reduce the rate of AChR synthesis in cultured chick myotubes as early as 1978 (Miskin et al. 1978); these findings were confirmed by Bursztajn et al. (1988) and by Changeux' group, who, in addition, demonstrated that phorbol esters enhance receptor expression upon chronic exposure, thought to lead to a depletion of PKC from the treated cells (Klarsfeld et al. 1989). Huang et al. (1992) were then able to connect depolarization of skeletal muscle in vivo to the stimulation of nuclear PKC and the resulting shutoff of receptor genes. This mechanism is compatible with the notion that one of the M proteins is the immediate target of kinase activity. Even a minimal chick AChR α-subunit promoter, containing little beside the enhancer and its two E boxes, is responsive to phorbol esters (Laufer et al. 1991). Olson and collaborators (Li et al. 1992b) have shown that all M proteins can be inactivated by PKC through phosphorylation of a conserved threonine residue in the basic DNA-binding domain. However, Huang et al. (1994b) demonstrated that of all four factors only myogenin was inactivated rapidly enough to account for the kinetics of receptor gene shutdown in vivo. This discrepancy can be resolved by considering that the threonine studied by the Olson group is likely to be inaccessible in the DNA-bound factor and that consequently a different target, present in myogenin but absent from other M proteins, mediates depolarization-transcription coupling. A promising candidate is Ser-80 in chick myogenin, which resides within a PKC target motif common to all myogenins sequenced so far but absent from the other M proteins. This residue, although near the basic region and only two α-helical turns away from the threonine in the DNA-binding region, is probably not occluded even in the DNA-bound factor, since the equivalent position remains accessible in the MyoD-DNA complex whose structure was recently elucidated by Pabo and colleagues (Ma et al. 1994). It should be pointed out that Goldman and collaborators recently reported that stimulation of PKC in primary rat muscle cells does not downregulate receptor subunit messages (Walke et al. 1994); this would suggest that the depolarization-transcription coupling in the rat differs from that in the chick and mouse studied by other groups.

3.8 Calcium and Calcium Channels

Depolarization activates voltage-dependent calcium channels, which in turn triggers the release of calcium from the SR. Therefore, cytosolic calcium has long been thought to play a role in blocking receptor expression (Lømo and Westgaard 1975). The first experimental support for this hypothesis came from investigations of changes in receptor levels in cultured muscle cells. Exposure to the calcium channel blockers D600 (Shainberg et al. 1976; Pezzementi and Schmidt 1981) and verapamil (Klarsfeld et al. 1989) resulted in enhanced expression of AChR. Similarly, dantrolene, a drug that inhibits the release of calcium from the SR, increased receptor levels (Birnbaum et al. 1980; Pezzementi and Schmidt 1981; Klarsfeld et al. 1989). Ryanodine, at the low concentrations at which it activates the SR calcium release channel, reduced receptor expression (Pezzementi and Schmidt 1981; Rotzler et al. 1991), but enhanced it at high concentrations that cause calcium depletion from the muscle fiber (Pezzementi and Schmidt 1981). Quantification of mRNA levels has also been used to study the effects of calcium drugs. The calcium channel blockers D600 (Shieh et al. 1988) and verapamil (Klarsfeld et al. 1989) proved to be potent stimulators of receptor subunit mRNA expression, while dantrolene was found to be considerably weaker (Klarsfeld et al. 1989). Since message measurements provide a reliable picture of gene activity only when combined with technically difficult mRNA turnover studies, it is preferable to analyze gene activity directly. Treatments thought to stimulate calcium influx caused inactivation of AChR genes in denervated chick mucle in vivo (Huang and Schmidt 1994). Using the mouse muscle cell line C2C12, Huang et al. (1994a) measured receptor gene activity under various conditions which alter cytoplasmic free calcium concentrations and observed that calcium entering the cytosol from the extracellular medium acts as a signal inhibiting receptor gene activity; unexpectedly, calcium released from the SR does not trigger inactivation of AChR subunit genes.

3.9 From the Plasma Membrane to the Nucleus

For the depolarization-transcription cascade to be triggered, influx of extracellular calcium ions through L-type calcium channels is required. How the signal from the activated L channel reaches the nucleus is not known. Depolarization activates nuclear PKC within 2–3 min (in intact chick muscle; Huang et al. 1992). There is no measurable delay between PKC activation and AChR gene inactivation, and a plausible pathway has been delineated from the activation of PKC to the phosphorylation of a myogenic factor to gene inactivation. However, the pathway from the plasma membrane to the nucleus leads through largely uncharted territory. Based on the effects of calcium drugs on receptor levels in cultured muscle cells, the original expectation had been that depolarization-transcription coupling shares elements with excitation-contraction coupling, up to and including the excitation-triggered release of calcium from the SR. This idea, however, did not survive the demonstration that depolarization-transcription coupling requires external calcium, while excitation-contraction coupling ostensibly does not. The independence of the two pathways is further supported by the demonstration that release of calcium from intracellular stores by caffeine and thapsigargine does not affect membrane-to-genome signaling. What the two pathways do share is: (a) the need for the activation of L-type channels, which in depolarization-transcription coupling direct the influx of calcium ions and in excitation-contraction coupling appear to function as electromechanic transducers (Catterall 1991), and (b) the flow of calcium through a membrane bounding the cytosol, where it interacts with an unknown sensor in the vicinity of the cytoplasmic mouth of the channel in the case of depolarization-transcription coupling and with troponin C in the case of excitation-contraction coupling. More work is necessary to elucidate the failure of SR-derived calcium to initiate depolarization-transcription coupling from the inside of the sarcolemma; one possible explanation is that an active L channel is required for triggering the pathway and that SR-derived calcium, via calmodulin/calcineurin, dephosphorylates and inactivates the L channel before the ion reaches the sensor. It has been shown in cultured rat muscle cells that the L channel in the sarcolemma is activated via stimulation of PKA (Sculptoreanu et al. 1993); phosphorylation of the channel protein in response to PKA has also been demonstrated (Mundiña-Weilenmann et al. 1991).

At present virtually nothing is known about the link between the hypothetical calcium sensor and PKC in the nucleus. Most likely, PKC is stimulated by diacylglycerol or a related compound, but the site at which this lipid activator is generated is unknown. A phospholipase C (PLC) probably participates in the cascade; however, what type of phospholipase it is, how it is linked to the activation of the calcium channels in the sarcolemma, and even whether it resides in the plasma membrane or in the nucleus or in some other compartment remains to be elucidated.

4 Concluding Remarks

Many genes are likely to be influenced by electrical membrane activity, but only a few such cases have been investigated in any detail, including the ones considered in this review. The stimulation of c-*fos* gene activity in neurons and the silencing of AChR genes in skeletal muscle, both in response to depolarization of the plasma membrane, differ in important aspects. Depolarization activates c-*fos* transcription but inactivates AChR receptor genes. CREB, which is an efficient substrate for PKA, CamKII, CREB kinase, and possibly other kinases, integrates signals from several converging pathways; the transcription factor associated with the depolarization response element in muscle, on the other hand, is likely to be targeted only by PKC, at least during residence at the depolarization response element, although whether directly or via another kinase has not been firmly established yet. Like all M proteins it can be inactivated by PKA (Li et al. 1992a) as well as by PKC (Li et al. 1992b) through phosphorylation of a conserved threonine reside in the DNA-binding region, but this is unlikely to play a role in depolarization-transcription coupling, as the rapid inactivation is difficult to reconcile with a kinase target buried in the DNA-binding site. Nevertheless, modification in the DNA-binding domain could contribute to the inactivation of receptor genes in the long term. Another difference between the two fields of research is related to the scope of the pathways under investigation. Although in both cases physiological observations preceded and eventually prompted molecular investigations, the ultimate target of the c-*fos* pathway has not come into focus, possibly because several downstream or late-response genes exist and/or because the several cell preparations being utilized differ in this respect, whereas in the pathway linking depolarization and the reduction of acetylcholine sensitivity it was soon realized (changes in acetylcholinesterase levels notwithstanding) that the AChR represents the major object of regulation or at least one well-defined end point of the signaling cascade.

It is remarkable, nevertheless, that the two examples covered here, though seemingly disparate, have common elements. In both pathways depolarization leads to the activation of L channels and calcium influx and eventually to the phosphorylation of a transcription factor that is already associated with the proper promoter element. In both paradigms strong evidence argues for the coexistence of at least two distinct signaling pathways that commence with the influx of calcium ions into the cytosol, but nevertheless reach different targets. In muscle the calcium released from the SR in the course of excitation–contraction coupling remains without effect on AChR gene transcription, while influx through L channels in the plasma membrane triggers receptor gene block even in the absence of excitation-contraction coupling. The mechanistic detail of this pathway separation remains to be worked out in either case. It appears most likely that the existence of calcium microdomains account for it. In neurons the plausible argument has been put forward that the calcium channels in question, the L channel and NMDA receptors, are topologically separated enough. In principle this explanation can also be invoked for skeletal muscle; there, however, the dimension of the microdomain in question cannot exceed the dimensions of the channel protein involved since the dihydropyridine receptor, which serves as the port of entry of extracellular calcium, and the calcium release channel in the SR are believed to be physically linked. It is possible to estimate the size of calcium ion diffusion ranges by means of intracellular chelators with different calcium association rate constants (Robitaille et al. 1993). Clearly it would be useful to employ such compounds, perhaps in combination with calcium imaging, approaches that do not seem to have been utilized much in the exploration of either pathway.

Sheng et al. (1990) proposed that it is likely that mechanisms defined for gene activation by membrane activity in PC12 cells would apply also to neurons in the intact central nervous system (CNS). Similarly, it may be hoped that the analysis of the two rather disparate systems summarized here will eventually pave the way to the elucidation of other depolarization-driven regulatory schemes. The discovery of amazing parallels in the two signaling pathways is certainly grounds for optimism.

References

Adler EM, Fink JS (1993) Calcium regulation of vasoactive intestinal polypeptide mRNA abundance in SH-SY5Y human neuroblastoma cells. J Neurochem 61: 727–737

Axelsson J, Thesleff S (1959) A study of supersensitivity in denervated mammalian skeletal muscle. J Physiol (Lond) 147: 178–193

Bading H, Greenberg M (1991) Stimulation of protein tyrosine phosphorylation by NMDA receptor activation. Science 253: 912–914

Bading H, Ginty DD, Greenberg ME (1993) Regulation of gene expression in hippocampal neurons by distinct calcium signaling pathways. Science 260: 181–186

Baraban JM, Fiore RS, Sanghera JS, Paddon HB, Pelech SL (1993) Identification of p42 mitogen-activated protein kinase as a tyrosine kinase substrate activated by maximal electroconvulsive shock in hippocampus. J Neurochem 60: 330–336

Bartel DP, Sheng M, Lau LF, Greenberg ME (1989) Growth factors and membrane depolarization activate distinct programs of early response gene expression: dissociation of fos and jun induction. Genes Dev 3: 304–313

Bessereau JL, Stratford-Perricaudet LD, Piette J, LePoupon C, Changeux J-P (1994) In vivo and in vitro analysis of electrical activity-dependent expression of muscle acetylcholine receptor genes using adenovirus. Proc Natl Acad Sci USA 91: 1304–1308

Bessho Y, Nawa H, Nakanishi S (1994) Selective up-regulation of an NMDA receptor subunit mRNA in cultured cerebellar granule cells by K^+-induced depolarization and NMDA treatment. Neuron 12: 87–95

Betz H (1980) Effects of drug-induced paralysis and depolarization on acetylchopline receptor and cyclic nucleotide levels of chick muscle cultures. FEBS Lett 118: 289–292
Betz H, Changeux J-P (1979) Regulation of muscle acetylcholine receptor synthesis in vitro by derivatives of cyclic nucleotides. Nature 278: 749–752
Birnbaum M, Reis MA, Shainberg A (1980) Role of calcium in the regulation of acetylcholine receptor synthesis in cultured muscle cells. Eur J Physiol 385: 37–43
Birren SJ, Verdi JM, Anderson DJ (1992) Membrane depolarization induces p140trk and NGF responsiveness, but not p75LNGFR, in MAH cells. Science 257: 395–397
Black IB, Adler JE, Dreyfus CF, Friedman WF, LaGamma EF, Roach A (1987) Biochemistry of information storage in the nervous system. Science 236: 1263–1268
Blackwell TK, Weintraub (1990) Differences and similarities in DNA-binding preferences of MyoD and E2A protein complexes revealed by binding site selection. Science 250: 1104–1110
Brand SJ, Wang TC (1988) Gastrin gene expression and regulation in rat islet cell lines. J Biol Chem 263: 16597–16603
Brockes JP, Hall ZW (1975) Synthesis of acetylcholine receptor by denervated rat diaphragm. Proc Natl Acad Sci USA 72: 1368–1372
Brunetti A, Goldfine ID (1990) Role of myogenin in myoblast differentiation and its regulation by fibroblast growth factor. J Biol Chem 265: 5960–5963
Buonanno A, Edmondson DG, Hayes WP (1993) Upstream sequences of the myogenin gene convey responsiveness to skeletal muscle denervation in transgenic mice. Nucleic Acids Res 21: 5684–5693
Bursztajn S, Schneider LW, Jong YJ, Berman SA (1988) Phorbol esters inhibit synthesis of acetylcholine receptors in cultured muscle cells. Biol Cell 63: 57–65
Catterall WA (1991) Excitation-contraction coupling in vertebrate skeletal muscle: a tale of two calcium channels. Cell 64: 871–874
Chahine KG, Walke W, Goldman D (1992) A 102-basepair sequence of the nicotinic acetylcholine receptor δ-subunit gene confers regulation by muscle electrical activity. Development 115: 213–219
Chahine KG, Baracchini E, Goldman D (1993) Coupling muscle electrical activity to gene expression via a cAMP-dependent second messenger system. J Biol Chem 268: 2893–2898
Cohen SA, Fischbach GD (1973) Regulation of muscle acetylcholine sensitivity by muscle activity in cell culture. Science 181: 76–78
Comb MC, Birnberg NC, Seasholtz A, Herbert E, Goodman HM (1986) A cyclic AMP- and phorbol ester-inducible DNA element. Nature 323: 353–356
Constantine-Paton M, Cline HT, Debski E (1990) Patterned activity, synaptic convergence, and the NMDA receptor in developing visual pathways. Annu Rev Neurosci 13: 129–154
Covault J, Merlie JP, Goridis C, Sanes JR (1986) Molecular forms of N-CAM and its RNA in developing and denervated skeletal muscle. J Cell Biol 102: 731–738
Curran T, Morgan JI (1987) Memories of fos. Bioessays 7: 255–258
Curran T, Morgan JI (1985) Superinduction of c-fos by NGF in the presence of peripherally active benzodiazepines. Science 229: 1265–1268
Curran T, Morgan JI (1986) Barium modulates c-fos expression and post-translational modification. Proc Natl Acad Sci USA 83: 8521–8524
Dash PK, Karl KA, Colicos MA, Prywes R, Kandel EB (1991) cAMP response element binding protein is activated by Ca^{2+}/calmodulin as well as cAMP-dependent protein kinase. Proc Natl Acad Sci USA 88: 5061–5065
Devillers-Thiery A, Galzi JL, Eisele JL, Bertrand S, Bertrand D, Changeux J-P (1993) Functional architecture of the nicotinic acetylcholine receptor: a prototype of ligand-gated ion channels. J Membrane Biol 136: 97–112
Drachman DB, Witzke F (1972) Trophic regulation of acetylcholine sensitivity of muscle: effect of electrical stimulation. Science 176: 514–516
Dragunow M, Robertson HA (1987) Kindling stimulation induces c-fos protein in granule cells of the rat dentate gyrus. Nature 329: 441–442

Duclert A, Piette J, Changeux J-P (1990) Induction of acetylcholine receptor α-subunit gene expression in chicken myotubes by blocking electrical activity requires ongoing protein synthesis. Proc Natl Acad Sci USA 87: 1391–1395

Duclert A, Piette J, Changeux J-P (1991) Influence of innervation on myogenic factors and acetylcholine receptor α-subunit mRNAs. NeuroReport 2: 25–28

Dutton EK, Simon AM, Burden SJ (1993) Electrical activity-dependent regulation of the acetylcholine receptor δ subunit gene, MyoD, and myogenin in primary myotubes. Proc Natl Acad Sci USA 90: 2040–2044

Edmondson DG, Olson E (1993) Helix-loop-helix proteins as regulators of muscle-specific transcription. J Biol Chem 268: 755–758

Eftimie R, Brenner HR, Buonanno A (1991) Myogenin and MyoD join a family of skeletal muscle genes regulated by electrical activity. Proc Natl Acad Sci USA 88: 1349–1353

Evans S, Goldman D, Heinemann S, Patrick J (1987) Muscle acetylcholine receptor biosynthesis. Regulation by transcript availability. J Biol Chem 262: 4911–4916

Fambrough DM (1970) Acetylcholine sensitivity of muscle fiber membranes: mechanism of regulation by motoneurons. Science 168: 372–373

Fambrough DM (1979) Control of acetylcholine receptors in skeletal muscle. Physiol Rev 59: 165–266

Feldblum S, Ackerman RF, Tobin AJ (1990) Long-term increase of glutamate decarboxylase mRNA in a rat model of temporal lobe epilepsy. Neuron 5: 361–371

Fernandez-Valle C, Rotundo RL (1989) Regulation of acetylcholinesterase synthesis and assembly by muscle activity: effects of tetrodotoxin. J Biol Chem 264: 14043–14049

Funk WD, Wright WE (1992) Cyclic amplification and selection of targets for multicomponent complexes: myogenin interacts with factors recognizing binding sites for basic helix-loop-helix, nuclear factor 1, myocyte-specific enhancer-binding factor 2, and COMP1 factor. Proc Natl Acad Sci USA 89: 9484–9488

Gall C, Sumikawa K, Lynch G (1990) Levels of mRNA for a putative kainate receptor are affected by seizures. Proc Natl Acad Sci USA 87: 7643–7647

Gass P, Kiessling M, Bading H (1993) Regionally selective stimulation of mitogen activated protein (MAP) kinase tyrosine phosphorylation after generalized seizures in the rat brain. Neurosci Lett 162: 39–42

Ghosh A, Carnahan J, Greenberg ME (1994a) Requirement for BDNF in activity-dependent survival of cortical neurons. Science 263: 1618–1623

Ghosh A, Ginty DD, Bading H, Greenberg ME (1994b) Calcium regulation of gene expression in neuronal cells. J Neurobiol 25: 294–303

Ginty DD, Bonni A, Greenberg ME (1994) Nerve growth factor activates a ras-dependent protein kinase that stimulates c-fos transcription via phosphorylation of CREB. Cell 77: 1–20

Goelet P, Castellucci VF, Schacher S, Kandel ER (1986) The long and short of short-term memory – a molecular framework. Nature 322: 419–422

Goldman D, Brenner HR, Heinemann S (1988) Acetylcholine receptor α, β-, γ-, and δ-subunit mRNA levels are regulated by muscle activity. Neuron 1: 329–333

Goodman RH (1990) Regulation of neuropeptide gene expression. Annu Rev Neurosci 13: 111–127

Greenberg ME, Ziff EB (1984) Stimulation of 3T3 cells induces transcription of the c-fos proto-oncogene. Nature 311: 433–438

Greenberg ME, Greene LA, Ziff EB (1985) Nerve growth factor and epidermal growth factor induce rapid transient changes in proto-oncogene transcription in PC12 cells. J Biol Chem 260: 14101–14110

Greenberg ME, Ziff EB, Greene LA (1986) Stimulation of neuronal acetylcholine receptors induces rapid gene transcription. Science 234: 80–83

Greenberg ME, Thompson MA, Sheng M (1992) Calcium regulation of immediate early gene transcription. J Physiol (Paris) 86: 99–108

Griffith LC, Schulman H (1988) The multifunctional calcium/calmodulin-dependent protein kinase mediates calcium-dependent phosphorylation of tyrosine hydroxylase. J Biol Chem 263: 9542–9549

Gruener R, Baumbach N, Coffee D (1974) Reduction of denervation sensitivity of muscle by submechanical threshold stimulation. Nature 248: 68–69

Gundersen K, Sanes JR, Merlie JP (1993) Neural regulation of muscle acetylcholine receptor ε- and α-subunit gene promoters in transgenic mice. J Cell Biol 123: 1535–1544

Hall ZW, Reinness CG (1977) Electrical stimulation of denervated muscles reduces incorporation of methionine into acetylcholine receptor. Nature 268: 655–657

Hall ZW, Sanes JR (1993) Synaptic structure and development: the neuromuscular junction. Neuron 10: 99S-121S

Harris DA, Falls DL, Dill-Devor RM, Fischbach GD (1988) Acetylcholine receptor-inducing factor from chicken brain increases the level of mRNA encoding the receptor α subunit. Proc Natl Acad Sci USA 85: 1983–1987

Hasty P, Bradley A, Morris JH, Edmondson DG, Venuti JM, Olson EN, Klein WH (1993) Muscle deficiency and neonatal death in mice with a targeted mutation in the myogenin gene. Nature 364: 501–506

Higuchi H, Iwasa A, Yoshida H, Miki N (1990) Long-lasting increase in neuropeptide Y gene expression in rat adrenal gland with reserpine treatment: positive regulation of transsynaptic regulation and membrane depolarization. Mol Pharmacol 38: 614–623

Hille B (1984) Ionic channels of excitable membranes. Sinauer, Sunderland, pp 1–426

Hogan PG, Marshall JM, Hall ZW (1976) Muscle activity decreases rate of degradation of α-bungarotoxin bound to extrajunctional receptors. Nature 261: 328–330

Huang C-F, Schmidt J (1994) Calcium influx blocks the skeletal muscle acetylcholine receptor α subunit genes in vivo. FEBS Lett 338: 277–280

Huang C-F, Tong J, Schmidt J (1992) Protein kinase C couples excitation to acetylcholine receptor gene inactivation in chick skeletal msucle. Neuron 9: 671–678

Huang C-F, Neville CM, Schmidt J (1993) Control of myogenic factor genes by the membrane depolarization/protein kinase C cascade in chick skeletal muscle. FEBS Lett 319: 21–25

Huang C-F, Flucher BE, Schmidt MM, Stroud SK, Schmidt J (1994a) Depolarization-transcription signals in skeletal muscle use calcium flux through L channels, but bypass the sarcoplasmic reticulum. Neuron 13: 167–177

Huang C-F, Lee Y-S, Schmidt M, Schmidt J (1994b) Rapid inhibition of myogenin-driven acetylcholine receptor subunit gene transcription. EMBO J 13: 634–640

Hunt SP, Pini A, Evan G (1987) Induction of c-fos-like protein in spinal cord neurons following sensory stimulation. Nature 328: 632–634

Isackson PJ, Huntsman MM, Murray KD, Gall CM (1991) BDNF mRNA expression is increased in adult rat forebrain after limbic seizures: temporal patterns of induction distinct from NGF. Neuron 6: 937–948

Jones R, Vrbová G (1974) Two factors responsible for the development of denervation supersensitivity. J Physiol (Lond) 236: 517–538

Kilbourne EJ, Nanlova BB, Lewis EJ, McMahon A, Osaka H, Sabban DB, Sabban EL (1992) Regulated expression of the tyrosine hydroxylase gene by membrane depolarization. Identification of the responsive element and possible second messengers. J Biol Chem 267: 7563–7569

Kim K-S, Lee MK, Carroll J, Joh TH (1993) Both the basal and inducible transcription of the tyrosine hydroxylase gene are dependent upon a cAMP response element. J Biol Chem 268: 15689–15695

Klarsfeld A, Changeux J-P (1985) Activity regulates the levels of acetylcholine receptor α-subunit mRNA in cultured chicken myotubes. Proc Natl Acad Sci USA 82: 4558–4562

Klarsfeld A, Laufer R, Fontaine B, Devillers-Thiery A, Dubreuil C, Changeux J-P (1989) Regulation of muscle AChR α-subunit gene expression by electrical activity: involvement of protein kinase C and Ca^{2+}. Neuron 2: 1229–1236

Kruijer W, Schubert D, Verma IM (1985) Induction of the proto-oncogene fos by nerve growth factor. Proc Natl Acad Sci USA 82: 7330–7334

Landschulz WH, Johnson PF, McKnight SL (1988) The leucine zipper: a hypothetical structure common to a new class of DNA binding proteins. Science 240: 1759–1764

Laufer R, Changeux J-P (1989) Activity-dependent regulation of gene expression in muscle and neuronal cells. Mol Neurobiol 3: 1–53

Laufer R, Klarsfeld A, Changeux J-P (1991) Phorbol esters inhibit the activity of the chicken acetylcholine receptor α-subunit gene promoter. Eur J Biochem 202: 813–818

Law SW, Conneely OM, DeMayo FJ, O'Malley BW (1992) Identification of a new brain-specific transcription factor, NURR1. Mol Endocrinol 6: 2129–2135

Lerea LS, McNamara JO (1993) Ionotropic glutamate receptor subtypes activate c-fos transcription by distinct calcium-requiring intracellular signaling pathways. Neuron 10: 31–41

Li L, Heller-Harrison R, Czech M, Olson EN (1992a) Cyclic AMP-dependent protein kinase inhibits the activity of myogenic helix-loop-helix proteins. Mol Cell Biol 12: 4478–4485

Li L, Zhou J, James G, Heller-Harrison R, Czech MP, Olson EN (1992b) FGF inactivates myogenic helix-loop-helix proteins through phosphorylation of a conserved protein kinase C site in their DNA binding domain. Cell 71: 1181–1194

Liu F, Thompson MA, Wagner S, Greenberg ME, Green MR (1993) Activating transcription factor-1 can mediate Ca^{2+}- and cAMP-inducible transcriptional activation. J Biol Chem 268: 6714–6720

Lømo T, Rosenthal J (1972) Control of ACh sensitivity by muscle activity in the rat. J Physiol (Lond) 221: 493–513

Lømo T, Westgaard RH (1975) Further studies on the control of acetylcholine sensitivity by muscle activity in the rat. J Physiol (Lond) 252: 603–626

Lu B, Yokoyama M, Dreyfus CF, Black IB (1991) Depolarizing stimuli regulate nerve growth factor gene expression in cultured hippocampal neurons. Proc Natl Acad Sci USA 88: 6289–6292

Ma PCM, Rould MA, Weintraub H, Pabo CO (1994) Crystal structure of MyoD bHLH domain-DNA complex: perspectives on DNA recognition and implications for transcription activation. Cell 77: 451–459

McArthur L, Koller KJ, Eiden LE (1993) Enkephalin gene transcription in bovine chromaffine cells is regulated by calcium and protein kinase A signal transduction pathways: identification of DNAse I hypersensitive sites. Mol Pharmacol 44: 545–551

McManaman JL, Blosser JC, Appel SH (1982) Inhibitors of membrane depolarization regulate acetylcholine receptor synthesis by a calcium-dependent, cyclic nucleotide-independent mechanism. Biochim Biophys Acta 720: 28–35

Mendelzon D, Changeux J-P, Nghiem H-O (1994) Phosphorylation of myogenin in chick myotubes: regulation by electrical activity and by protein kinase C. Implications for acetylcholine receptor gene expression. Biochemistry 33: 2568–2575

Merlie JP, Kornhauser JM (1989) Neural regulation of gene expression by an acetylcholine receptor promoter in muscle of transgenic mice. Neuron 2: 1295–1300

Merlie JP, Isenberg KE, Russell SD, Sanes JR (1984) Denervation supersensitivity in skeletal muscle: analysis with a cloned cDNA probe. J Cell Biol 99: 332–335

Merlie JP, Mudd J, Cheng T-C, Olson EN (1994) Myogenin and acetylcholine receptor α gene promoters mediate transcriptional regulation in response to motor innervation. J Biol Chem 269: 2461–2467

Messing RO, Stevens AM, Kiyasu E, Sneade AB (1989) Nicotinic and muscarinic agonists stimulate rapid protein kinase C translocation in PC12 cells. J Neurosci 9: 507–512

Miledi R, Potter LT (1971) Acetylcholine receptors in muscle fibers. Nature 233: 599–600

Miskin R, Easton TG, Maelicke A, Reich E (1978) Metabolism of acetylcholine receptor in chick embryo muscle cells: effects of RSV and PMA. Cell 15: 1287–1300

Montminy MR, Bilezikjian LM (1987) Binding of a nuclear protein to the cyclic AMP response element of the somatostatin gene. Nature 328: 175–178

Montminy MR, Sevarino KA, Wagner JA, Mandel G, Goodman RH (1986) Identification of cAMP response element within the rat somatostatin gene. Proc Natl Acad Sci USA 83: 6682–6686

Morgan JI, Curran T (1986) Role of ion flux in the control of c-fos expression. Nature 322: 552–555

Morgan JI, Curran T (1991) Stimulus-transcription coupling in the nervous system: involvement of the inducible proto-oncogenes fos and jun. Annu Rev Neurosci 14: 421–451

Morgan JJ, Cohen DR, Hempstead JL, Curran T (1987) Mapping patterns of c-fos expression in the central nervous system after seizure. Science 237: 192–197

Morris BJ, Feasey KJ, Bruggencate GT, Herz A, Hoellt V (1988) Electrical stimulation in vivo increases the expression of proenkephalin mRNA and decreases the expression of prodynorphin mRNA in rat hippocampal granule cells. Proc Natl Acad Sci USA 85: 3226–3230

Moss SJ, Beeson DMW, Jackson JF, Darlison MG, Barnard EA (1987) Differential expression of nicotinic acetylcholine receptor genes in innervated and denervated chicken muscle. EMBO J 6: 3917–3921

Müller R, Bravo R, Burckhardt J, Curran T (1984) Induction of c-fos gene and protein by growth factors precedes activation of c-myc. Nature 312: 716–720

Mundiña-Weilenmann C, Chang CF, Gutierrez LM, Hosey MM (1991) Demonstration of the phosphorylation of dihydropyridine-sensitive calcium channels in chick skeletal muscle and the resultant activation of the channel after reconstitution. J Biol Chem 266: 4067–4073

Murphy TH, Worley PF, Baraban JM (1991) L-type voltage-sensitive calcium channels mediate synaptic activation of immediate-early genes. Neuron 7: 625–635

Neville CM, Schmidt MM, Schmidt J (1991) Kinetics of expression of ACh receptor α–subunit mRNA in denervated and stimulated muscle. NeuroReport 2: 655–657

Neville CM, Schmidt MM, Schmidt J (1992) Response of myogenic determination factors to cessation and resumption of electrical activity in skeletal muscle. Cell Mol Neurobiol 12: 511–527

Pette D, Vrbová G (1992) Adaptation of mammalian skeletal muscle fibers to chronic electrical stimulation. Rev Physiol Biochem Pharmacol 120: 115–202

Pezzementi L, Schmidt J (1981) Ryanodine alters the rate of acetylcholine receptor synthesis in chick skeletal muscle cell cultures. J Biol Chem 256: 12651–12654

Purves D (1976) Long-term regulation in the vertebrate peripheral nervous system. Int Rev Physiol 10: 125–177

Qian Z, Gilbert ME, Colicos MA, Kandel ER, Kuhl D (1993) Tissue-plasminogen activator is induced as an immediate-early gene during seizure, kindling, and long-term potentiation. Nature 361: 453–457

Robertson LM, Kerppola TK, Vendrell M, Luk D, Smeyne RJ, Bocchiaro C, Morgan JI, Curran T (1995) Regulation of c-fos expression in transgenic mice requires multiple interdependent transcription control elements. Neuron 14: 241–252

Robitaille R, Garcia ML, Kaczorowski GJ, Charlton MP (1993) Functional colocalization of calcium and calcium-gated potassium channels in control of transmitter release. Neuron 11: 645–655

Rosen LB, Ginty DD, Weber MJ, Greenberg ME (1994) Membrane depolarization and calcium influx stimulate MEK and MAP kinase via activation of ras. Neuron 12: 1207–1221

Rotzler S, Schramek H, Brenner HR (1991) Metabolic stabilization of endplate acetylcholine receptors regulated by calcium influx associated with muscle activity. Nature 349: 337–339

Saffen DW, Cole AJ, Worley PF, Christy BA, Ryder K, Baraban JM (1988) Convulsant-induced increase in transcription factor messenger RNAs in rat brain. Proc Natl Acad Sci USA 85: 7795–7799

Salmon A-M, Changeux J-P (1992) Regulation of an acetylcholine receptor lacZ transgene by muscle innervation. NeuroReport 3: 973–976

Salpeter MM, Loring RH (1985) Nicotinic acetylcholine receptors in vertebrate muscle: properties, distribution, and neural control. Progr Neurobiol 25: 297–325

Sanes JR, Johnson YR, Kotzbauer PT, Mudd J, Hanley T, Martinou J-C, Merlie JP (1991) Selective expression of an acetylcholine receptor-lacZ transgene in synaptic nuclei of adult muscle fibers. Development 113: 1181–1191

Sculptoreanu A, Scheuer T, Catterall WA (1993) Voltage-dependent potentiation of L-type Ca^{2+} channels due to phosphorylation by cAMP-dependent protein kinase. Nature 364: 240–243

Shainberg A, Cohen SA, Nelson PG (1976) Induction of acetylcholine receptors in muscle cultures. Pflugers Arch 361: 255–261

Sheng M, Greenberg ME (1990) The regulation and function of c-fos and other immediate early genes in the nervous system. Neuron 4: 477–485

Sheng M, Dougan ST, McFadden G, Greenberg ME (1988) Calcium and growth factor pathways of c-fos transcriptional activation require distinct upstream regulatory sequences. Mol Cell Biol 8: 2787–2796

Sheng M, McFadden G, Greenberg ME (1990) Membrane depolarization and calcium induce c-fos transcription via phosphorylation of transcription factor CREB. Neuron 4: 571–582

Sheng M, Thompson MA, Greenberg ME (1991) CREB: a Ca^{2+}-regulated transcription factor phosphorylated by calmodulin-dependent kinases. Science 252: 1427–1430

Shieh B-H, Ballivet M, Schmidt J (1987) Quantitation of an a-subunit splicing intermediate. Evidence for transcriptional activation in the control of receptor expression in denervated chick skeletal muscle. J Cell Biol 104: 1337–1341

Shieh B-H, Ballivet M, Schmidt J (1988) Acetylcholine receptor synthesis rate and levels of receptor subunit mRNAs in chick muscle. Neuroscience 24: 175–187

Simon AM, Hoppe P, Burden SJ (1992) Spatial restriction of acetylcholine receptor gene expression to subsynaptic nuclei. Development 114: 545–553

Smeyne RJ, Schilling K, Robertson L, Luk D, Oberdick J, Curran T, Morgan JI (1992) Fos-lacZ transgenic mice: mapping sites of gene induction in the central nervous system. Neuron 8: 13–23

Smeyne RJ, Vendrell M, Hayward M, Baker SJ, Miao GG, Schilling K, Robertson LM, Curran T, Morgan JI (1993) Continuous c-fos expression precedes programmed cell death in vivo. Nature 363: 166–169

Sonnenberg JL, Macgregor-Leon PF, Curran T, Morgan JI (1989a) Dynamic alterations occur in the levels and composition of transcription factor AP1 complexes after seizure. Neuron 3: 359–365

Sonnenberg JL, Mitchelmore C, Macgregor-Leon PF, Hepstead J, Morgan JI, Curran T (1989b) Glutamate receptor agonists increase the expression of fos, fra, and AP-1 DNA binding in the mammalian brain. J Neurosci Res 24: 72–80

Sonnenberg JL, Rauscher FJ, Morgan JI, Curran T (1989c) Regulation of proenkephalin by Fos and Jun. Science 246: 1622–1625

Strohman RC, Bandman E, Walker CR (1981) Regulation of myosin accumulation by muscle activity in cell culture. J Muscle Res Cell Motil 2: 269–282

Thomas SM, DeMarco M, D'Arcangelo G, Halegoua S, Brugge JS (1992) Ras is essential for nerve growth factor- and phorbol ester-induced tyrosine phosphorylation of MAP kinases. Cell 68: 1031–1040

Treisman R (1985) Transient accumulation of c-fos RNA following serum stimulation requires a conserved 5' element and c-fos 3' sequences. Cell 42: 889–902

Tsaur M-L, Sheng M, Lowenstein DH, Jan YN, Jan LY (1992) Differential expression of K+ channel mRNAs in the rat brain and down-regulation in the hippocampus following seizures. Neuron 8: 1055–1067

Tsay H-J, Schmidt J (1989) Skeletal muscle denervation activates acetylcholine receptor genes. J Cell Biol 108: 1523–1526

Tsay H-J, Neville CM, Schmidt J (1990) Protein synthesis is required for the denervation-triggered activation of acetylcholine receptor genes. FEBS Lett 274: 69–72

Tsukuda T, Fink JS, Mandel G, Goodman RH (1987) Identification of a region in the human vasoactive intestinal polypeptide gene responsible for regulation by cyclic AMP. J Biol Chem 262: 8743–8747

Walke W, Staple J, Adams L, Gnegy M, Chahine K, Goldman D (1994) Calcium-dependent regulation of rat and chick muscle nicotinic acetylcholine receptor gene expression. J Biol Chem 269: 19447–19456

Watson MA, Milbrandt J (1989) The NGFI-B gene, a transcriptionally inducible member of the steroid receptor gene superfamily: genomic structure and expression in rat brain after seizure. Mol Cell Biol 9: 4213–4219

Weintraub H (1993) The MyoD family and myogenesis: redundancy, networks, and thresholds. Cell 75: 1241–1244

Weintraub H, Davis R, Tapscott S, Thayer M, Krause M, Benezra R, Blackwell TK, Turner D, Rupp R, Hollenberg S, Zhuang Y, Lassar A (1991) The myoD gene family: nodal point during specification of the muscle cell lineage. Science 251: 761–766

Weis J (1994) Jun, Fos, MyoD1, and myogenin proteins are increased in skeletal muscle fiber nuclei after denervation. Acta Neuropathol 87: 63–70

Westenbroek RE, Ahlijanian MK, Catterall WA (1990) Clustering of L-type calcium channels at the base of major dendrites in hippocampal pyramidal neurons. Nature 366: 433–438

Wisden W, Errington ML, Williams S, Dunnett SB, Waters C, Hitchcock D, Evan G, Bliss TVP, Hunt SP (1990) Differential expression of immediate early genes in the hippocampus and spinal cord. Neuron 4: 603–614

Witzemann V, Sakmann B (1991) Differential regulation of MyoD and myogenin mRNA levels by nerve-induced muscle activity. FEBS Lett 282: 259–264

Witzemann V, Brenner H-R, Sakmann B (1991) Neural factors regulate AChR subunit mRNAs at rat neuromuscular synapses. J Cell Biol 114: 125–141

Wright WE, Binder M, Funk W (1991) Cyclic amplification and selection of targets (CASTing) for myogenin consensus binding site. Mol Cell Biol 11: 4104–4110

Editor-in-Charge: Professor D. Pette

Subject Index

Springer-Verlag and the Environment

We at Springer-Verlag firmly believe that an international science publisher has a special obligation to the environment, and our corporate policies consistently reflect this conviction.

We also expect our business partners – paper mills, printers, packaging manufacturers, etc. – to commit themselves to using environmentally friendly materials and production processes.

The paper in this book is made from low- or no-chlorine pulp and is acid free, in conformance with international standards for paper permanency.

GPSR Compliance

The European Union's (EU) General Product Safety Regulation (GPSR) is a set of rules that requires consumer products to be safe and our obligations to ensure this.

If you have any concerns about our products, you can contact us on ProductSafety@springernature.com

In case Publisher is established outside the EU, the EU authorized representative is:

Springer Nature Customer Service Center GmbH
Europaplatz 3
69115 Heidelberg, Germany

Batch number: 10620143

Printed by Printforce, the Netherlands